IoT
Security and Privacy Paradigm

Internet of Everything (IoE): Security and Privacy Paradigm

Series Editors: Vijender Kumar Solanki, Raghvendra Kumar,
and Le Hoang Son

IOT
Security and Privacy Paradigm
Edited by Souvik Pal, Vicente Garcia Diaz and Dac-Nhuong Le

Smart Innovation of Web of Things
Edited by Vijender Kumar Solanki, Raghvendra Kumar and Le Hoang Son

Big Data, IoT, and Machine Learning
Tools and Applications
Rashmi Agrawal, Marcin Paprzycki, and Neha Gupta

Internet of Everything and Big Data
Major Challenges in Smart Cities
Edited by Salah-ddine Krit, Mohamed Elhoseny, Valentina Emilia Balas, Rachid Benlamri, and Marius M. Balas

Bitcoin and Blockchain
History and Current Applications
Edited by Sandeep Kumar Panda, Ahmed A. Elngar, Valentina Emilia Balas, and Mohammed Kayed

Privacy Vulnerabilities and Data Security Challenges in the IoT
Edited by Shivani Agarwal, Sandhya Makkar, and Tran Duc Tan

Handbook of IoT and Blockchain
Methods, Solutions, and Recent Advancements
Edited by Brojo Kishore Mishra, Sanjay Kumar Kuanar, Sheng-Lung Peng, and Daniel D. Dasig, Jr.

Blockchain Technology
Fundamentals, Applications, and Case Studies
Edited by E Golden Julie, J. Jesu Vedha Nayahi, and Noor Zaman Jhanjhi

Data Security in Internet of Things Based RFID and WSN Systems Applications
Edited by Rohit Sharma, Rajendra Prasad Mahapatra, and Korhan Cengiz

Securing IoT and Big Data
Next Generation Intelligence
Edited by Vijayalakshmi Saravanan, Anpalagan Alagan, T. Poongodi, and Firoz Khan

Distributed Artificial Intelligence
A Modern Approach
Edited by Satya Prakash Yadav, Dharmendra Prasad Mahato, and Nguyen Thi Dieu Linh

Security and Trust Issues in Internet of Things
Blockchain to the Rescue
Edited by Sudhir Kumar Sharma, Bharat Bhushan, and Bhuvan Unhelkar

For more information about this series, please visit: https://www.crcpress.com/Internet-of-Everything-IoE-Security-and-Privacy-Paradigm/book-series/CRCIOESPP

IoT
Security and Privacy Paradigm

Edited by
Souvik Pal, Vicente García Díaz, and Dac-Nhuong Le

CRC Press
Taylor & Francis Group
Boca Raton London New York

CRC Press is an imprint of the
Taylor & Francis Group, an **informa** business

First edition published 2020

by CRC Press
6000 Broken Sound Parkway NW, Suite 300, Boca Raton, FL 33487-2742

and by CRC Press
2 Park Square, Milton Park, Abingdon, Oxon, OX14 4RN

First issued in paperback 2021

ISBN 13: 978-1-03-223964-4 (pbk)
ISBN 13: 978-0-367-25384-4 (hbk)

**Visit the Taylor & Francis Web site at
http://www.taylorandfrancis.com**

**and the CRC Press Web site at
http://www.crcpress.com**

Library of Congress Cataloging-in-Publication Data
Names: Pal, Souvik, editor. | García Díaz, Vicente, 1981- editor. | Le,
Dac-Nhuong, 1983- editor.
Title: IoT : security and privacy paradigm / edited by Souvik Pal,
Associate Professor, Department of Computer Science and Engineering,
Brainware University, Kolkata, India, Vicente García Díaz, Associate
Professor, Department of Computer Science and Engineering, University of
Oviedo, Spain, Dac-Nhuong Le, Deputy Head, Faculty of Information
Technology, Haiphong University, Haiphong, Vietnam.
Description: First edition. | Boca Raton, FL : CRC Press, 2020. | Series:
Internet of everything : Security and privacy paradigm | Includes
bibliographical references and index.
Identifiers: LCCN 2019059519 (print) | LCCN 2019059520 (ebook) | ISBN
9780367253844 (hardback) | ISBN 9780429289057 (ebook)
Subjects: LCSH: Internet of things--Security measures.
Classification: LCC TK5105.8857 .I7365 2020 (print) | LCC TK5105.8857
(ebook) | DDC 005.8--dc23
LC record available at https://lccn.loc.gov/2019059519
LC ebook record available at https://lccn.loc.gov/2019059520

Typeset in Times
by Deanta Global Publishing Services, Chennai, India

Contents

Preface

This book covers a wide range of security and privacy issues in IoT-enabled technologies. The diverse contents will differentiate this edited book from others. The chapters include security vulnerabilities, data-intensive security and privacy, privacy-preserving communication protocols, RFID-related technologies like signal interference, spoofing, eavesdropping, authentication token security in machine-to-machine (M2M) communication protocols, privacy in crowd-sensing, and self-adaptive cyber-physical systems. These topics are likely to be embedded with the security and privacy aspects of IoT-enabled technologies. This book brings together leading academic scientists, researchers, and research scholars to exchange and share their experiences and research results on all aspects of security and privacy issues in IoT. It also provides a premier interdisciplinary platform for researchers, practitioners, and educators to present and discuss the most recent innovations, trends, and concerns as well as practical challenges encountered and solutions adopted in the fields of security and privacy in IoT.

The book is organized into 16 chapters. Chapter 1 discusses the assorted dimensions of ubiquitous computing with the specific integration towards wireless technology-based cities using high-performance technologies. There is a huge need to develop mechanisms so that the overall scenarios and implementations of IoT in smart cities or similar wireless environments can be made secure. The key objective of this chapter is to present the assorted approaches with the secured mechanisms of password management for IP-based IoT sensors to enforce and implement security and overall performance.

Chapter 2 focuses on bringing together all the innovative technologies associated with a partially connected automated vehicle environment (PCAVE) to enhance advanced transportation as well as making the technologies clear and connecting them to intelligent transportation systems (ITS), providing an overview on cybersecurity issues, European Union regulations and economics, and a review of the known cyberattacks related to the transport sector. The aims of and scope of the research is based on broadly covering high-quality review and research articles on experiments, simulation, and modeling as well as analysis of traffic related to vehicles with the main focus being on actual time-data-driven methods in CAVE and PCAVE.

Chapter 3 presents the requirements of Fog platforms to serve the demands of big data generated through intelligent devices in an IoT environment in the future and also present the challenges in developing these platforms. The chapter looks closely at characteristics of existing Fog platforms and other crucial requirements for designing a platform, such as architectures and algorithms, that are available in the literature. In addition, research gaps and future directions are discussed.

Chapter 4 deals with traffic accidents and ways to ensure safety. Roads are unpredictable and there can be a fatal accident at every turn. One cannot rely on the capabilities of other drivers and pedestrians. One needs to be self-aware of the environment and other vehicles. Drivers should take all precautions and be mindful of the people on the road because every life is precious. Common reasons for accidents are drivers' lack of concentration due to distractions or lack of sleep. This research work incorporates an eye-blink sensor to ensure the driver's eyes are always open and that the driver doesn't fall asleep, and a mechanism to confirm the driver is wearing a seatbelt at all times. Through these various small efforts, driver safety can be ensured and the lives of others on the road can be saved.

Chapter 5 presents an approach to using proximity of sensors and the cloud for storing data. The chapter reveals Smart Attendance, a classroom attendance system that requires little installation cost, and saves paper and teacher work hours. If a student does not attend class for more than three days in a row or if the student has less than the required attendance, a message will be sent to parents.

Chapter 6 explores encryption of data in cloud-based industrial IoT devices. In the Internet of Things, devices of different capabilities are provided with a common platform. Using this platform they will be able to communicate with each other. As device communication is increasing, so is the data. To support abundant data storage and provide scalability, the cloud is introduced. The technology avails data to the user anytime, anywhere. The cloud increases reliability by 10.05% compared to regular encryption standard methods.

Chapter 7 analyzes the methodologies used in assessment and analysis of cyberattacks. The significance and contribution of this piece is manifold as this chapter details not only IoT-based devices but also introduces the basic concepts behind the development of the same and identifies common bottlenecks present. Subsequently, we cover open challenges in this domain and the possibility of future work for improving the existing technologies in cybersecurity.

Chapter 8 reviews the related literature to identify the approaches of cyber intrusion detection and prediction related to IoT devices as well as software. Various security mechanisms like cryptography will be explored with respect to their IoT-based hardware implementations, and some machine learning techniques will be presented for possible extension of security issues thereof. The key objective of this chapter is to explore research directions with the intention of providing the optimal result to the security issues that heterogeneous IoT devices and software face.

Chapter 9 discusses the need of authentication and authorization in IoT. Because the network is scalable, the authentication in a static and dynamic environment is presented. Various issues such as communication overhead, usage of computational resources, tolerance to several attacks, and trusting of devices are discussed. IoT comprising of heterogeneous devices has varied resources, hence centralized and decentralized schemes are discussed for authentication with reduced communication overhead and energy consumption.

Chapter 10 focuses on software-defined networking (SDN) methodology to formulate a network for an effective and heavy-duty IoT. IoT requirements which are provided by SDN, include flexibility and agility. Additionally, it allows using the applications layer for developers and network administrators to develop and manage advanced implements and software linking the IoT to be more efficient. The relation between these two technologies is directly proportional if we have better SDN leading to the best IoT performance. A survey is provided through this research discussing how the SDN has been helping IoT to have a more stable and secure framework to work through.

Chapter 11 discusses how a telecare medical information system (TMIS) helps patients access required medical services and access their medical information from a remote place. Due to the vulnerability of communication systems and the sensitivity of medical information, secure communication is necessary in TMIS. Mutual authentication of the user and the server and session key agreements are necessary for ensuring integrity, confidentiality, and security of the telecare medical architecture. Remote user authentication is an efficient technique to perform secure communication over an insecure network.

Chapter 12 explores the various security threats against existing IoT-based healthcare systems, such as illegitimate electronic patient record (EPR) modification attacks, and subsequently proposes remedies to prevent/detect them. In this respect, the chapter discusses the traditional approaches towards security and privacy in IoT healthcare systems, which are majorly based on pre-processing of data, vis-à-vis the much newer branch of security, viz. digital forensics, where the detection and prevention of EPR attacks are completely post-processing oriented. Finally, the chapter presents a recent state-of-the-art digital forensic technique for authentication of medical images in IoT-based healthcare infrastructure.

Chapter 13 exemplifies the understanding of the fundamentals and underlying architecture of IoT. This motivated us to impart the importance of IoT in terms of its beginning and basic concepts to readers and researchers. Also knowing the IoT architecture and possible applications in different domains of human life will encourage researchers to find solutions and design the IoT network efficiently. Thus the objective of this chapter is to cast light on various applications of IoT in the international scenario and its contribution in providing quality human life. This chapter summarizes the IoT fundamentals since its inception up to its application in current days. Another objective is to provide researchers a clear picture of the constraints faced for IoT design and its functioning.

Chapter 14 explains infrastructure technological developments; our world is connected through smart devices that are driven by the Internet of Things. These devices utilize the wireless medium to exchange information through broadcasting, and it makes the system more vulnerable to eavesdropping. Traditionally, different cryptographic techniques such as the asymmetric encryption algorithm (RSA) or symmetric encryption algorithm (AES) were employed to ensure privacy and security.

Chapter 15 elucidates how irrigation can be handled smartly using the Internet of Things by an agricultural justice collaborative model to detect the soil temperature, moisture content, and soil water potential. In this approach, the heat dissipation is counterbalanced by sprinkling water automatically onto the soil. Further, the dynamic variations in temperature will also render the water supply. Moreover, the collaborative monitoring and predictive maintenance of the soil's volumetric moisture content saves time and avoids the problem of constant vigilance. The chapter also generates impetus on conservation of water by supplying water to the plants and field as per requirements. Over and above, this chapter also incorporates the experimentally received information from sensors and various parameters. This information is fed as analog input to the equivalent electronic model which is implemented on an Arduino Uno microcontroller. The model also collaborates the information with the user's mobile phone and to a centralized data peak cloud.

Chapter 16 discusses privacy and security issues in different layers of the IoT architecture. Here, we are observing the data link layer, network layer, and application layer protocol. Most of the layers are affected based on transferring sensitive data; key distribution; pre-installed key in devices; authentication; network attacks; and application layer protocols such as RPL, PAIR, MQTT, and CoAP. For achieving security, different methods with authentication, encryption, PKI security, API security, hardware testing, and different case studies are provided.

We are sincerely thankful to the Almighty for supporting and standing with us at all times, whether it's good or tough times, and giving ways to console us. Starting from the call for chapters till the finalization of chapters, all the contributing authors have given their contributions amicably, which is a positive sign of significant teamwork. The editors are sincerely thankful to all members of CRC Press/Taylor & Francis Group for providing constructive inputs and allowing an opportunity to edit this important book. We are very thankful to the series editors Vijender Kumar Solanki, Raghvendra Kumar, and Le Hoang Son for their input for this book. We are equally thankful to reviewers who hail from around the globe who shared their support and stand firm towards quality chapter submission.

About the Book

IoT security is the technology area concerned with safeguarding connected devices and networks in the Internet of Things (IoT). IoT involves adding Internet connectivity to a system of interrelated computing devices, mechanical and digital machines, objects, animals, and/or people. Each 'thing' is provided a unique identifier and the ability to automatically transfer data over a network. Allowing devices to connect to the Internet opens them to a number of serious vulnerabilities if they are not properly protected. This edited book, *IoT: Security and Privacy Paradigm,* uses security engineering and privacy-by-design principles to design a secure IoT ecosystem and to implement cybersecurity solutions. This book will take readers on a journey that begins with understanding the security issues in IoT-enabled technologies and how they can be applied in various aspects. It walks readers through engaging with security challenges and builds a safe infrastructure for IoT devices. This book helps researchers and practitioners understand the security architecture through IoT and the state of the art in IoT countermeasures. It also differentiates security threats in IoT-enabled infrastructure from traditional ad hoc or infrastructural networks. It provides a comprehensive discussion on the security challenges and solutions in radio-frequency identification (RFID) and wireless sensor networks (WSNs) in IoT. This book brings together some of the top IoT-enabled security experts throughout the world who contribute their knowledge regarding different IoT security aspects. This edited book aims to provide the concepts of related technologies and novel findings of researchers. The primary audience for the book includes specialists, researchers, graduate and undergraduate students, designers, experts, and engineers who are occupied with research and security-related issues. The edited book is organized in independent chapters to provide readers great readability, adaptability, and flexibility.

Editors

Souvik Pal, PhD, is an associate professor in the Department of Computer Science and Engineering at Global Institute of Management and Technology, West Bengal, India. Dr Pal earned his BTech, MTech and PhD degrees in the field of Computer Science and Engineering. Prior to his current post, he was an Assistant Professor at the Nalanda Institute of Technology, Bhubaneswar, and JIS College of Engineering, Kolkata (NAAC "A" Accredited College). He has also worked as head of the Computer Science Department at Elitte College of Engineering, Kolkata. He has more than a decade of academic experience. He is editor/author of 12 books from publishers of repute such as Elsevier, Springer, CRC Press, and Wiley, and he was granted 3 patents. He is the recipient of a Lifetime Achievement Award in 2018. He is the series editor of "Advances in Learning Analytics for Intelligent Cloud-IoT Systems", Scrivener-Wiley Publishing, Beverly, Massachusetts. Dr Pal has published a number of research papers in Scopus and SCI-indexed international journals and conferences. His professional activities include roles as associate editor and editorial board member for more than 100 international journals and conferences of high repute and impact. Dr Pal has been invited as resource person/keynote plenary speaker in many reputed universities and colleges at national and international levels. His research area includes cloud computing, big data, Internet of Things, and data analytics.

Vicente García Díaz, PhD, is an associate professor in the Department of Computer Science at the University of Oviedo (Languages and Computer Systems area). He earned a PhD in computer science from the University of Oviedo and a diploma in advanced studies, as well as degrees in computer engineering and technical systems computer engineering. In addition, he earned a degree in occupational risk prevention. He has supervised 90-plus academic projects, including 4 doctoral theses, and published 80-plus research papers in journals (several journals included in the JCR index), conferences, and books from prestigious publishers.

Dr Díaz is also a member of the editorial and advisory boards of several journals, and the main editor of several special issues in prestigious journals such as *Scientific Programming* and *International Journal of Interactive Multimedia and Artificial Intelligence*. He also served as an editor for research books such as *Handbook of IoT and Big Data* (CRC Press), *Protocols and Applications for the Industrial Internet of Things* (IGI-Global), *Handbook of Research on Innovations in Systems and Software Engineering* (IGI-Global), *Progressions and Innovations in Model-Driven Software Engineering* (IGI-Global), and *Advances and Applications in Model-Driven Engineering* (IGI-Global). His research interests include e-learning, machine learning, and the use of domain specific languages in different areas. Finally, he carried out research stays at the University of Manchester (3 months), at the Francisco José de Caldas District University of Bogotá (2 months), at the University of Lisbon (2 months), and the Autonomous University of Santo Domingo (1 month).

Dac-Nhuong Le, PhD, is deputy-head of Faculty of Information Technology, and vice-director of Information Technology Apply and Foreign Language Training Center, Haiphong University, Vietnam. He earned an MSc, (2009) and PhD (2015) in computer science from Vietnam National University. He has a total academic teaching experience of 13 years with many publications in reputed international conferences, journals, and online book chapter contributions (indexed by SCI, SCIE, SSCI, Scopus, ACM, DBLP). His area of research includes evaluation computing and approximate algorithms, network communication, security and vulnerability, network performance analysis and simulation, cloud computing, IoT, and image processing in biomedicine. His core work is in network security, soft computing, IoT, and image processing in biomedical applications. Recently, he served on the technique program committee, the technique reviews, the track chair for international conferences: FICTA 2014, CSI 2014, IC4SD 2015, ICICT 2015, INDIA 2015, IC3T 2015, INDIA 2016, FICTA 2016, ICDECT 2016, IUKM 2016, INDIA 2017, CISC 2017, FICTA 2017, FICTA 2018 under the Springer-ASIC/LNAI Series. Presently, he is serving on the editorial board of international journals and he has authored nine computer science books published by Springer, Wiley, CRC Press, Lambert Publication, and Scholar Press.

Contributors

Ahmed Gaber Abu Abd-Allah
PhD Researcher
Helwan University
Helwan, Egypt

Aya Sedky Adly
Faculty of Computers and Artificial
    Intelligence
Helwan University
Helwan, Egypt

Jamimamul Bakas
National Institute of Technology
Rourkela, India

Rajib Bag
Supreme Knowledge Foundation Group
    of Institutions
West Bengal, India

M. K. Banga
Department of Computer Science and
    Engineering
Dayananda Sagar University
Bangalore, India

Siddhant Banyal
Division of Instrumentation and Control
Netaji Subhas University of Technology
    (formerly known as Netaji Subhas
    Institute of Technology)
New Delhi, India

Sudheer Kumar Battula
School of Technology, Environments
    and Design
University of Tasmania
Hobart, Australia

Kartik Krishna Bhardwaj
Division of Instrumentation and Control
Netaji Subhas University of Technology
    (formerly known as Netaji Subhas
    Institute of Technology)
New Delhi, India

Joy Chatterjee
Supreme Knowledge Foundation Group
    of Institutions
West Bengal, India

Suchismita Chinara
National Institute of Technology,
    Rourkela
Rourkela, India

Andrea Chiappetta
CEO ASPISEC (IT)
Rome, Italy

Atanu Das
Netaji Subhash Engineering College
Kolkata, India

Manab Kumar Das
Supreme Knowledge Foundation Group
    of Institutions
West Bengal, India

Saurabh Garg
School of Technology, Environments
    and Design
University of Tasmania
Hobart, Australia

Atef Zaki Ghalwash
Faculty of Computers and Artificial
    Intelligence
Helwan University
Helwan, Egypt

Sayon Ghosh
Supreme Knowledge Foundation Group
    of Institutions
West Bengal, India

M. Gowtham
Department of Computer Science and
    Engineering
Mysuru and Dayananda Sagar University
Bangalore, India

Daneshwari Hatti
Electronics and Communication
B.L.D.E.A's V.P. Dr. P.G. Halakatti
    College of Engineering and
    Technology
Vijayapur, India

Shyamalendu Kandar
Department of Information Technology
Indian Institute of Engineering Science
    and Technology
Shibpur, India

Byeong Kang
School of Technology, Environments
    and Design
University of Tasmania
Hobart, Australia

Ranjit Kumar
National Institute of Technology
    Rourkela
Odisha, India

Upendra Kumar
Department of Computer Science &
    Engineering
Birla Institute of Technology Mesra,
    Patna Campus
Bihar, India

Soumya Nandan Mishra
National Institute of Technology,
    Rourkela
Rourkela, India

Ambika N.
Department of Computer Applications
SSMRV College, Bangalore
Bangalore, India

Ruchira Naskar
Indian Institute of Engineering Science
    and Technology
Shibpur, India

Sumit Pal
Department of Information Technology
Indian Institute of Engineering Science
    and Technology
Shibpur, India

Smita Pallavi
Department of Computer Science &
    Engineering
Birla Institute of Technology Mesra,
    Patna Campus
Bihar, India

Pranjal Pandey
Department of Electronics and
    Communication Engineering
Indraprastha Institute of Information
    Technology
New Delhi, India

Mallanagouda Patil
Department of Computer Science and
    Engineering
Dayananda Sagar University
Bangalore, India

H. B. Pramod
Department of Computer Science and
    Engineering
Rajeev Institute of Technology
Hassan, India

James Montgomery
School of Technology, Environments
    and Design
University of Tasmania
Hobart, Australia

Meenakshi Rawat
Department of Electronics and
    Communication Engineering
Indian Institute of Technology, Roorkee
Roorkee, India

Deepak Kumar Sharma
Division of Information Technology
Netaji Subhas University of Technology
    (formerly known as Netaji Subhas
    Institute of Technology)
New Delhi, India

Umang Shukla
Ganpat University
Mehsana, India

Rupender Singh
Department of Electronics and
    Communication Engineering
Indian Institute of Technology
Roorkee, India

Gopi Sumanth
PSG College of Technology
Coimbatore, India

Suriya Sundaramoorthy
PSG College of Technology
Coimbatore, India

Ashok V. Sutagundar
Electronics and Communication
Basaveshwar Engineering College
Bagalkot, India

Asis Kumar Tripathy
School of IT Engineering
Vellore Institute of Technology
Vellore, India

1 Intrusion Detection and Avoidance for Home and Smart City Automation in Internet of Things

M. Gowtham, H. B. Pramod, M. K. Banga, and Mallanagouda Patil

CONTENTS

1.1   INTRODUCTION

With the huge escalation in the deployments and adoption of smart gadgets, devices in the wireless environment are vulnerable to huge security issues and lack mechanisms against hacking (Cui and Moran, 2016; Wortmann and Flüchter, 2015; Islam et al., 2015). Nowadays, almost every person has at least one smart gadget whether it is a smart phone, smart health gadget or any other (Dastjerdi and Buyya, 2016). These devices are quite open to the Internet of Things (IoT) search engines and need to be enforced with a higher degree of security (Al-Fuqaha et al., 2015; Li et al., 2015; Wortmann and Flüchter, 2015).

Figure 1.1 depicts the key perspectives for the deployment of an IoT-based city environment with automation and effectual outcomes for the assorted domains. The following includes the key implementation perspectives:

Smart city
- Smart roads
- Smart traffic lights

Industrial applications
- Ozone presence
- Smart grids
- Industrial disaster prediction

Smart home
- Wearables

Smart water
- Chemical leakage detection
- Portable water monitoring
- Pollution-level analysis
- River floods

Retail
- Industrial control systems

Environment protection with sustainable resources
- Sea-based disaster prediction
- Air pollutions
- Forest fire detection
- Avalanche and landslide prevention
- Earthquake detection
- Snow-level monitoring

Digital health and telemedicine
- Ultraviolet radiation
- Patient surveillance

Smart agriculture
- Soil quality measurement

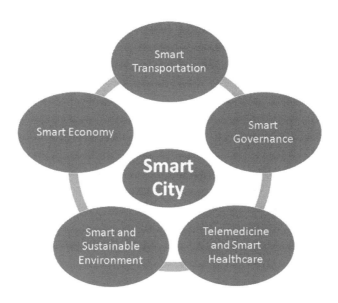

FIGURE 1.1 Key perspectives for deployment of IoT-based smart city.

1.2 KEY MODULES AND COMPONENTS OF AN IOT SCENARIO

The key components of an IoT scenario includes the following:

- Cloud
- Things or devices or gadgets
- User interface
- Gateway
- Interfacing modules
- Networks
- Storage panel
- Security mechanisms
- Communication platform

1.3 GLOBAL SCENARIO

Following are excerpts from assorted research datasets available on different analytics patterns. Statista is one of the prominent research portals in which the authenticated datasets and evaluations are presented.

From the extracts of Statista, Figure 1.2 presents the huge usage of IoT-based deployments and it is rapidly increasing because of the usage patterns in assorted domains.

Figure 1.3 depicts usage patterns of tablets in assorted locations from the year 2014 to the year 2019, and the figures are elevating. Because of these figures, it becomes necessary to enforce security mechanisms for IoT- and wireless-based environments.

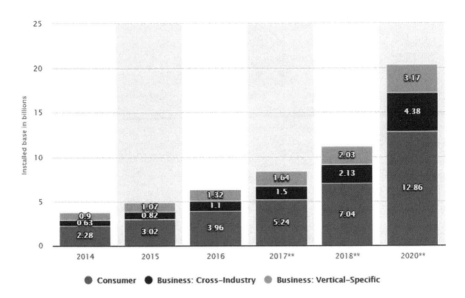

FIGURE 1.2 Installed Internet of Things (IoT) scenarios from year 2014 to 2020 in billions.

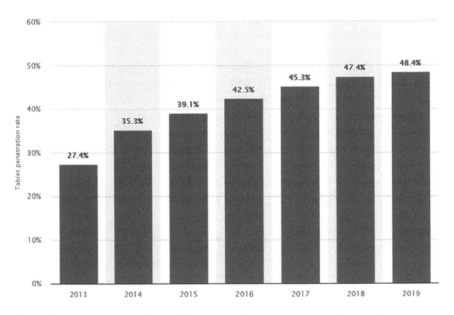

FIGURE 1.3 User penetration with the usage of tablets up to 2019 in Western Europe.

1.4   PROMINENT SEARCH ENGINES FOR INDEXING IOT DEVICES

Following are key Internet of Things Search Engines (IoTSE) used by enormous organizations and the hacker loopholes in web cams with other IP-based devices. These portals index smart gadgets, servers, web cams and many other devices.

- shodan.io/
- iotcentral.io
- censys.io/
- thingful.net/
- iotcrawler.eu/
- iotscanner.bullguard.com/
- um.es/iotcrawler/

1.5   SHODAN: AN IOT SEARCH ENGINE

Shodan (McMahon et al., 2017; Samtani et al., 2018; Arnaert et al., 2016; Wright, 2016) is one of the key IoT search engines that is widely used to identify and recognize open systems in IoT scenarios. Using Shodan, as in Figure 1.4, any web cam or server can be extracted using simple traditional search approaches.

1.5.1   KEY POINTS OF SHODAN

Shodan (Markowsky and Markowsky, 2015; Shemshadi, Sheng, and Qin, 2016) is cited in many research reports and news articles whereby enormous devices and gadgets are found vulnerable and need to be integrated with a higher degree of security algorithms without any probabilities of cracking. As per the research reports, on Shodan.io more than 52,000 servers were found accessible which were not enabled for secured authentication. In addition, more than 90,000 servers of databases were found vulnerable due to bad MongoDB configurations (Heller, 2017).

Figure 1.5 depicts the deep pattern analyses and fingerprinting of the devices from Shodan, and it is quite dangerous towards the security of IoT devices deployed for smart cities or smart home automations.

FIGURE 1.4   Shodan IoT indexing search engine.

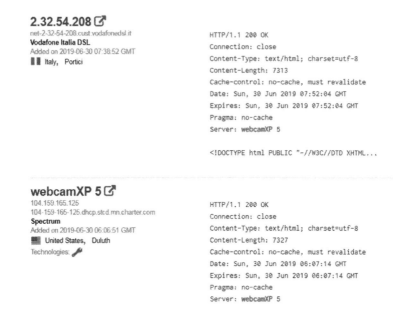

2.32.54.208 ☑
net-2-32-54-208.cust.vodafonedsl.it
Vodafone Italia DSL
Added on 2019-06-30 07:38:52 GMT
🇮🇹 Italy, Portici

```
HTTP/1.1 200 OK
Connection: close
Content-Type: text/html; charset=utf-8
Content-Length: 7313
Cache-control: no-cache, must revalidate
Date: Sun, 30 Jun 2019 07:52:04 GMT
Expires: Sun, 30 Jun 2019 07:52:04 GMT
Pragma: no-cache
Server: webcamXP 5

<!DOCTYPE html PUBLIC "-//W3C//DTD XHTML...
```

webcamXP 5 ☑
104.159.165.125
104-159-165-125.dhcp.stcd.mn.charter.com
Spectrum
Added on 2019-06-30 06:06:51 GMT
🇺🇸 United States, Duluth
Technologies: 🔧

```
HTTP/1.1 200 OK
Connection: close
Content-Type: text/html; charset=utf-8
Content-Length: 7327
Cache-control: no-cache, must revalidate
Date: Sun, 30 Jun 2019 06:07:14 GMT
Expires: Sun, 30 Jun 2019 06:07:14 GMT
Pragma: no-cache
Server: webcamXP 5
```

FIGURE 1.5   Extraction of fingerprints of devices.

1.5.2 DATA EXTRACTION USING SHODAN

A number of key points of the servers and devices are vulnerable and open including HTTP status, server, connection type, content type, and authentication aspects.

The extraction of cavernous data points and servers can be done using Shodan as shown in Figure 1.6 and these can be matched with the default password as presented in Table 1.1. The default passwords list is open and any person has access, as these

129.89.113.41
crtb80a-printer.ls.uwm.edu
University of Wisconsin - Milwaukee
Added on 2017-08-17 23:20:47 GMT
🇺🇸 United States, Milwaukee
Details

```
HTTP/1.1 401
Date: Sat, 21 Dec 1996 12:00:00 GMT
WWW-Authenticate: Basic realm="Default password:1234"
```

FIGURE 1.6   Extraction of internal data of devices.

TABLE 1.1

Open Access Authentications for IoT-Based IP Web Cams

Webcam	Default IP	Username/ID	Key/Password
Speco	DHCPIP	adminuser	1234
IndigoVision (BX/GX)	DHCPIP	Adminuser	1234
DynaColor	DHCPIP	Adminuser	1234
Samsung	192.168.1.200	adminuser	4321
Sentry360 (mini)	DHCPIP	adminuser	1234
Lorex	DHCPIP	adminuser	Adminpass
Oncam	DHCPIP	adminuser	Adminpass
IPX-DDK	192.168.1.168	root	adminpass
Digital Watchdog	DHCPIP	adminuser	adminpass
Dahua	192.168.1.108	666666	666666
Basler	DHCPIP	adminuser	adminpass
Sanyo	192.168.0.2	adminuser	adminpass
Honeywell	DHCPIP	adminuser	1234
Samsung	192.168.1.200	root	4321
Dahua	192.168.1.108	adminuser	adminpass
Uniview	DHCPIP	adminuser	123456
Messoa	192.168.1.30	adminuser	Model # of Camera
Sentry 360	192.168.0.250	Adminuser	1234
GVI	192.168.0.250	Adminuser	1234
Verint	DHCPIP	adminuser	adminpass
Toshiba	DHCPIP	root	ikwd
PiXORD	192.168.0.200	root	pass
Starvedia	DHCPIP	adminuser	
Axis	192.168.0.90	root	pass
CBC Ganz	192.168.100.x	adminuser	adminpass
VideoIQ	DHCPIP	supervisor	supervisor
American Dynamics	DHCPIP	adminuser	9999
Canon	DHCPIP	root	camera
CNB	192.168.123.100	root	adminpass
Avigilon	DHCPIP	adminuser	adminpass
Amcrest	DHCPIP	adminuser	adminpass
Longse	DHCPIP	adminuser	12345
Samsung	192.168.1.200	adminuser	1111111
Speco	192.168.1.7	adminuser	adminpass
Speco	192.168.1.7	root	root
3xLogic	192.0.0.64	adminuser	12345
Honeywell	DHCPIP	adminuseristrator	1234
Dahua	192.168.1.108	888888	888888
FLIR	DHCPIP	adminuser	fliradminpass
IQInvision	DHCPIP	root	system

(*Continued*)

TABLE 1.1 (CONTINUED)
Open Access Authentications for IoT-Based IP Web Cams

Webcam	Default IP	Username/ID	Key/Password
Costar	DHCPIP	root	root
VideoIQ	DHCPIP	supervisor	supervisor
Mobotix	DHCPIP	adminuser	meinsm
Grandstream	192.168.1.168	adminuser	adminpass
Intellio	DHCPIP	adminuser	adminpass
March Networks	DHCPIP	adminuser	
Scallop	DHCPIP	adminuser	password
JVC	DHCPIP	adminuser	jvc
Trendnet	DHCPIP	adminuser	adminpass
IPX-DDK	192.168.1.168	root	Adminpass
Samsung Techwin (old)	DHCPIP	adminuser	1111111
Northern	DHCPIP	adminuser	12345
QVIS	192.168.0.250	Adminuser	1234
Bosch	192.168.0.1	service	service
PiXORD	192.168.0.200	adminuser	adminpass
Ubiquiti	192.168.1.20	ubnt	ubnt
W-Box (Hikvision OEM)	DHCPIP	adminuser	wbox123
Sony	192.168.0.100	adminuser	adminpass
Toshiba	192.168.0.30	root	ikwb
Samsung Electronics	DHCPIP	adminuser	4321
Arecont Vision	DHCPIP	none	
Brickcom	192.168.1.1	adminuser	adminpass
FLIR (Dahua OEM)	DHCPIP	adminuser	adminpass
Sunell	DHCPIP	adminuser	adminpass
Interlogix	DHCPIP	adminuser	1234
DRS	DHCPIP	adminuser	1234
ACTi	192.168.0.100	adminuser	123456
American Dynamics	DHCPIP	adminuser	adminpass
HIKVision	192.0.0.64	adminuser	12345
Canon	192.168.100.1	root	Model # of camera
Bosch	192.168.0.1	Dinion	
Merit Lilin Recorder	DHCPIP	adminuser	1111
Stardot	DHCPIP	adminuser	adminpass
DVtel	192.168.0.250	Adminuser	1234
W-Box (Sunell OEM)	DHCPIP	adminuser	adminpass
IOImage	192.168.123.10	adminuser	adminpass
LTS	DHCPIP	adminuser	12345
FLIR (Quasar/Ariel)	DHCPIP	adminuser	adminpass
Panasonic	192.168.0.253	adminuser1	password
Samsung Electronics	DHCPIP	root	root

(Continued)

TABLE 1.1 (CONTINUED)
Open Access Authentications for IoT-Based IP Web Cams

Webcam	Default IP	Username/ID	Key/Password
JVC	DHCPIP	adminuser	Model # of Camera
Swann	DHCPIP	adminuser	12345
ACTi	192.168.0.100	Adminuser	123456
Q-See	DHCPIP	adminuser	adminpass
Pelco	DHCPIP	adminuser	adminpass
Samsung (new)	DHCPIP	adminuser	4321
Panasonic	192.168.0.253	adminuser	12345
Merit Lilin Camera	DHCPIP	adminuser	pass
Vivotek	DHCPIP	root	
GeoVision	192.168.0.10	adminuser	adminpass
Samsung	192.168.1.200	root	adminpass
AvertX	DHCPIP	adminuser	1234
Q-See	DHCPIP	adminuser	123456
Wodsee	DHCPIP	adminuser	

are provided by the manufacturer or installation engineers while deploying the smart web cams.

1.5.3 SEARCHING WEB CAMS AT TRAFFIC LIGHTS, AIRPORTS, HOMES, AND OFFICES

webcamxp city:sydney → Search by location
webcamxp geo: -37.81,144.96   → Search by longitude/latitude
webcamxp country:AU → Search by country

In the case of smart homes and smart cities automation, there is need to enforce the security, as servers or web cams can be covertly extracted using specific keywords or search perspectives (Figure 1.7).

1.6 ATTACKS ON IOT ENVIRONMENTS

A number of attacks on IoT-based infrastructure and sensor scenarios are quite prominent, but can be avoided using high-performance approaches and algorithms (Deogirikar and Vidhate, 2017; Pongle and Chavan, 2015; Nawir et al., 2016; Kolias et al., 2017; Apthorpe et al., 2017).

These assaults include denial of service (DoS) attacks, distributed denial of service (DDoS) attacks, Sybil attacks, node imitation attacks, and application-level attacks. With the use of blockchain-based technologies, the cumulative security can be elevated.

http://188.243.162.253:8080/

FIGURE 1.7   Scenario of extracted web cam.

1.6.1 OPEN SOURCE FRAMEWORKS FOR MONITORING AND PROGRAMMING OF IoT SCENARIOS

- DSA, www.iot-dsa.org
  - Inter-device communication
  - Effective logic
  - Apps on all layers
- Contiki, www.contiki-os.org
  - IPv6 compatible
  - IPv4 compatible
  - Low resource consumption
  - Protothreads compatible
  - Microcontrollers integration
  - Game console associations
- Node-RED, www.nodered.org
  - Flow-based programming
  - More than two Lac modules
- IoTivity, www.iotivity.org
  - On-board connectivity
  - Constrained Application Protocol (CoAP) compatible
- OpenIoT, www.openiot.eu
  - Sensing as a Service (S2aaS)
  - Dynamic graphs
  - Dynamic resource optimization

- CupCarbon, www.cupcarbon.com
  - Creation of dynamic networks
  - Smart city simulation
  - Generation of wireless vehicular network
  - SCI-WSN simulator
  - Effectual 2D and 3D mapping
  - OpenStreetMap
  - Effective visualization
- Zetta, www.zettajs.org
  - WebSocket
  - Low overhead connections
  - Dynamic and real-time communication on TCP
  - Reactive programming enabled
- KAA, www.kaaproject.org
  - Data analytics
  - Dynamic updates in real-time

1.7 IOT-INTEGRATED SECURED TELEMEDICINE DELIVERY IN SMART CITIES

With the use of IoT-based remote clinical services (IRCS), the medical health services can be made secure so that any attempt to crack the medical services can be pushed back and identified. The telemedicine services makes use of advanced wireless technologies which can be used by our proposed system for providing more security

The domain of IRCS is grouped fragments whereby each takes a shot at cutting-edge innovations with a higher level of precision and execution (Figure 1.8). Regardless of the multiple advantages of utilizing IRCS, there exists many difficulties including perspective of restorative specialists, patients' dread and newness, financial inaccessibility, lack of essential civilities, literacy rate and variety in dialects, technical limitations, quality viewpoint, government support, and comparative angles (Zanjal and Talmale, 2016; Farahani et al., 2018; Al-Majeed et al., 2015).

Taxonomy of IRCS

- Monitoring and observation of remote objects for healthcare
- Dynamic storage
- Real-time communication and logging

The other key points with IRCS include the following:

- IoT-integrated telepharmacy
- IoT-integrated telenursing
- IoT-integrated telerehabilitation
- IoT-integrated emergency services handling
- IoT-integrated teleneuropsychology
- IoT-integrated teletrauma care
- IoT-integrated specialist care delivery

- IoT-integrated teletransmission of ECG
- IoT-integrated telecardiology
- IoT-integrated teleradiology
- IoT-integrated telepathology
- IoT-integrated teledermatology
- IoT-integrated teleaudiology
- IoT-integrated telepsychiatry
- IoT-integrated teledentistry
- IoT-integrated teleophthalmology

IRCS technologies include the following:

- Data collection software
- Mobile applications
- Microcomputers
- Mobile IRCS
- Mobile operating system technology
- Patient monitoring devices
- Chatterbots

IRCS is one of the key zones of research in social assurance industry and remote seeing of patients with the genuine and accuracy cautious condition. The present conditions of helpful sciences and human affiliations are not secure and needs most silly idea while having the moment improvement of the enduring medicines and at least smart correspondence (Stradolini et al., 2018). There is the need to consolidate the unpreventable instruments and advances including IRCS and media transmission with the target that the down to earth correspondence can be feasible for the social security (Figure 1.9).

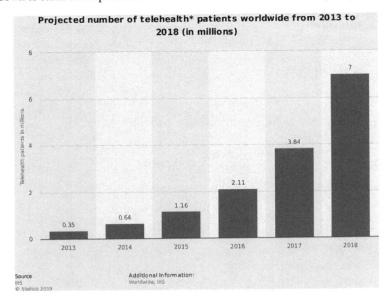

FIGURE 1.8   Global scenario of IoT-based remote clinical services (IRCS).

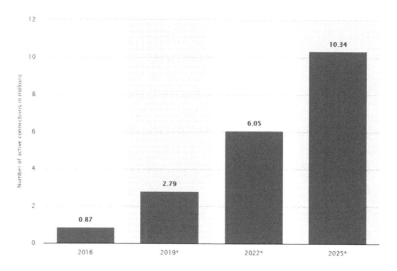

FIGURE 1.9   Active IoT healthcare connections in millions.

The present conditions of steady space and the unequivocally related medications are having the basics for the huge assessment of the work with imaginative viewpoints related with the social assurance and its relationship with bleeding edge satellites sorts of progression. In the same way, the relationship of satellite-based correspondence with IRCS for successful correspondence and least deferral is in a general sense required. In addition, the execution of the satellite-based correspondence and the relationship with the present models is moreover in the process with the huge viewpoints for the beast move of the execution for the helpful regions. The indicated affiliation examination has the woeful perspective on the related work in the driving a particularly drawn out stretch of time in satellite-based correspondence and IRCS so that a multidimensional examination ought to be possible.

The domain of IRCS is under research. IRCS is one of the key zones of research in the restorative organizations industry and remote seeing of the patients with point-by-point and exactness conditions. The present conditions of healing sciences and human organizations are flimsy and need development. World-class instruments and headways including IRCS and media transmission are encouraged with the target that the energetic correspondence can be feasible for the human organizations. The present conditions of supportive space and the definitely related medications are having the prerequisites for the colossal assessment of the work with imaginative points of view related with the restorative organizations and its relationship with front-line satellites advances. Also, the relationship of the satellite-based correspondence with IRCS for viable correspondence and least deferral is altogether required. What's more, the execution of the satellite-based correspondence and the relationship with the present models is in like way in the process with the colossal points of view for the monster rising of the execution for the supportive regions. The indicated arrangement examination is having the immense perspectives on the related work in the advancing quite a while in the satellite-based correspondence and IRCS so that multidimensional examination ought to be possible. The presented work is having the positive composition examination on the evaluation perspectives of IRCS and its utilization with the different strategies in the overall regions.

1.7.1 ADVANTAGES OF THE SECURED TELEMEDICINE FRAMEWORK

- Fast to compute with less overhead
- Collision resistant and fault tolerant
- Wide usage with security certificates
- Multiway hash with fewer complexities
- Resistant towards attacks on assorted types
- Resistant with hashing collisions with high performance
- Consistency check with higher proficiency
- Stronger protection against attacks with assorted types
- Longer hash as compared to traditional message digest

1.8 BLOCKCHAIN-BASED SECURED MECHANISMS FOR SECURITY WITH HOME AUTOMATIONS AND SMART CITIES

Wireless systems are presently very helpless towards the different attacks and in this way necessities emerge to verify the general situations (Biswas and Muthukkumarasamy, 2016). Blockchain technology is one of the unmistakable and elite methodologies that can be utilized for the incorporation of security into wireless systems to raise the level of security and, in general, performance. In blockchains an advanced ledger is kept up. The computerized ledger is very straightforward and there is no extent of any controls in the records by the intermediates or any director. The records of the considerable number of exchanges are signed in the blockchain ledger and the activities are submitted with various conventions and calculations which can't be hacked by outsiders (Sun et al., 2016).

In cutting-edge wireless situations, the general performance is significant and required with the goal that the general system condition can be verified. The work is exhibiting the utilization of blockchain-based usage with the wireless systems utilizing Python-based libraries which can be incorporated with Raspberry Pi or Arduino or some other open source board for the coordination and authorization of situations (Sharma et al., 2017) (Figure 1.10).

1.8.1 KEY ASPECTS OF BLOCKCHAIN

Blockchain is the front-line technology that is constantly connected with security and increasingly raised measure of confirmation in sorted out applications. Before long days, blockchain technology isn't kept to the mechanized kinds of money rather it is under execution for different social and corporate portions. These pieces combine e-association, individual to singular correspondence, online business, transportation, coordinated efforts, proficient correspondences and others (Puthal et al., 2018).

Blockchain proposes the pervasive and security cautious technology wherein an automated ledger is kept up. The impelled ledger is amazingly immediate and there is no level of any controls in the records by the intermediates or any official. The

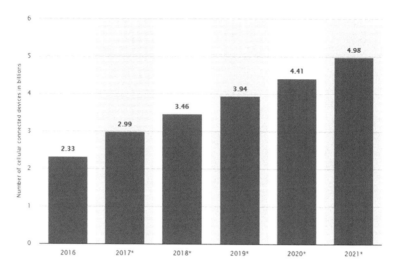

FIGURE 1.10   IoT and cellular devices worldwide in millions.

records of the noteworthy number of exchanges are set apart in the blockchain ledger and the errands are submitted last with various shows and calculations which can't be hacked by outsiders.

Following are some of the examples of blockchain implementations (Figure 1.11):

- **Entertainment**
  - KickCity
  - Guts
  - Spotify
  - B2Expand
  - Veredictum

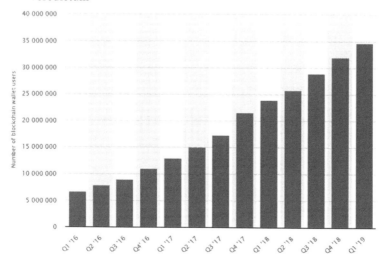

FIGURE 1.11   Number of blockchain wallet users worldwide (2016–2019).

- **Social networks**
  - Matchpool
  - Steepshot
  - MeWe
  - Minds
  - Mastodon
  - DTube
  - Sola
- **Cryptocurrency**
  - Namecoin
  - Bitcoin
  - Dogecoin
  - Primecoin
  - Ripple
  - Litecoin
  - Nxt
  - Ethereum
- **Retail**
  - Blockpoint
  - Warranteer
  - Loyyal
  - Shopin
  - Opskins
  - Spl.yt
  - Portion
  - Fluz Fluz
  - Ecoinmerce.io
  - Every.Shop
  - Buying.com

In a blockchain network, there exists blocks of different data elements and records. Each block participates in the blockchain network and it is immutable. The term 'immutable' here refers that it is secured and unbreakable.

Hence it forms the blockchain of secured chain of blocks without any probabilities of intentional or accidental tampering or leakage in the data. The first block in the chain of a network is known as the genesis block from where the blockchain initiates the transactions.

By inserting different blocks with encryption of every block with the previous block, it becomes secured and therefore difficult to crack the previous states because of so many encryptions.

1.8.2 Usage Aspects of Blockchain for Smart Cities and Home Automations

In smart-cities-based environments, cryptocurrencies and smart wallets are quite prominent. This approach can be effectively implemented using blockchains so that financial transactions in smart cities can be made securely. In addition, the smart wallets of users will give more security. As in blockchain technology, there are secured protocols and algorithms. The Python programme has enormous toolkits available on its official repository.

Additional libraries and toolkits can be installed in Python with Raspberry Pi or Arduino using pip installer. The blockchain technology is highly dependent on making use of the integrations with dynamic cryptography and encryption. For this, the hashlib library can be installed using the aforementioned instruction.

Following is the scenario of a secured blockchain that is generating the hash values so that overall transactions and records will be highly secured. In the following code, the dynamic hash value is generated that is the base of any blockchain with different transactions in a chain and that makes the overall blockchain.

```
    class Block:
def __init__(this, myidx, ts, mydata, backsecuredhash):
this.myidx = myidx
this.ts = ts
this.mydata = mydata
this.backsecuredhash = backsecuredhash
this.securedhash = this.securedhashop()
def securedhashop(this):
shasecuredhash = securedhasher.sha256()
shasecuredhash.update(mystr(this.myidx) + mystr(this.ts) +
mystr(this.mydata) + mystr(this.backsecuredhash))
return shasecuredhash.hexdigest()
    def keyblock():
return Block(0, date.datetime.now(), "Keyblock Block",
"0")
    def next_block(last_block):
this_myidx = last_block.myidx + 1
this_ts = date.datetime.now()
this_mydata = "Block" + mystr(this_myidx)
this_securedhash = last_block.securedhash
return Block(this_myidx, this_ts, this_mydata,
this_securedhash)
    blockchain = [keyblock()]
    IoT_Secured_Block = blockchain[0]
    maxblocks = 20
    for i in range(0, maxblocks):
block_to_add = next_block(IoT_Secured_Block)
blockchain.append(block_to_add)
IoT_Secured_Block = block_to_add
echo "Block #{} inserted in Blockchain".format(block_to_
add.myidx)
echo "Securedhash Value: {}\n".format(block_to_add.
securedhash)
```

With the execution of the code, the following outcome is obtained that has different hash values and provides a higher degree of security using cryptography functions. Using these hash values, hacking or sniffing the transaction will be almost impossible.

1.8.3 GENERATION OF HASH VALUES USING BLOCKCHAIN IMPLEMENTATION

Block #1 Gen in Blockchain

Node: 60afeff7f57bc04dc1c411763778f661d9088c77d7789b4881a67d465571 54ac

Block #2 Gen in Blockchain

Node: 8084d8e09b74f295082ea38cc3bbd892f27719b0fa9a732295fac34ed03d ebb9

Block #3 Gen in Blockchain

Node: 247086d83f38210185aed834e1efe218822c6e1223e465321a80d3d78f7e e803

Block #4 Gen in Blockchain

Node: e7af556dd479088f4fb8cfb11f8648e192c8bb66835cdc748b039e0fc5fac646

Block #5 Gen in Blockchain

Node: 439ed4470f39e44475852dee4d42e4474d535431b61807d02892af3 8e4a4395e

Block #6 Gen in Blockchain

Node: 452d166db9c397cac1870afab08b462b00e15d7f5450485b03f7243a1a68 4b36

Block #7 Gen in Blockchain

Node: 301fabc673d1d29b0c18e0e8cf3c59d85051cf2a1eb4bd11c92abe0c0e53 1441

Block #8 Gen in Blockchain

Node: bb876789e44fbd8d8d108d161cbde1b5d5d71cd5de388cdf2c0815c0e66a 8635

Block #9 Gen in Blockchain

Node: 0fa214cf91613631d59514eecab2e9bf74ce13fba4ba1582bcf2b40dae25 10db

Block #10 Gen in Blockchain

Node: 900139625df05d0dab34eab2e0921e4055cd2358dbbf64fd643c03d 364745ad8

Block #11 Gen in Blockchain

Node: 86d641c12979caaec401ba3a8966df562bb8439183be3e41361f1eb9a14f 8fbb

Block #12 Gen in Blockchain

Node: fcaee1e98c7719df6769323fa0aaa79dc068c1c000380b454dd0e1cbb45c 6e3a

Block #13 Gen in Blockchain

Node: 9429c9d0ecdfb284f189dc0c5359d565eeda405f5970fda0a6ce96294208 8d20

Block #14 Gen in Blockchain

Node: fc9d9357c58c46ea1ec067cbb5e305aac6ef54295da7792f2c6732a6c7c9 3ee5

Block #15 Gen in Blockchain
Node: aa7692f3f300e8178953328d99f9194c3c601fecc23b81319b739e6c801a
4b6c
Block #16 Gen in Blockchain
Node: 9d145ad7051eaf167c227277497a8eb5418ad17925d24c88276c412f9902
6cc1
Block #17 Gen in Blockchain
Node: 297e1510bc43890077d36a65ed13fab28f1b8dacaada1b31f4e247de77be
d878
Block #18 Gen in Blockchain
Node: 092eee32f96382135723d9ebe849b8e3b6c52c432f7930264d5565d
efb6e4f6c
Block #19 Gen in Blockchain
Node: 0ea694cea5e535107d840fd15d150f505269d6922ec348fdf8d212007cce
2f83
Block #20 Gen in Blockchain
Node: 52b567064c45680ee110a2d660f2c0ff061451510bf9b2848649940c0f03
2589

.

.

.

.

Block #19980 inserted in Blockchain
Hash Value: 58d86f4e14068cd1f56c5b27b7088a9894aceca01bd51bb304569b5
7c4e2a37e
Block #19981 inserted in Blockchain
Hash   Value: d8d86d541ed3abb30bddd98d35b6f34d57958cb266aa707742
51e5491e07d8ac
Block #19982 inserted in Blockchain
Hash Value: baf04029f374008683118504391e9a85f0af1593cad9b4d6e32088c
5dcf4814d
Block #19983 inserted in Blockchain
Hash Value: 43fa6ddebcc9b433711b3e613152223aea4ced529515a61b7b497a7
f82fc0a86
Block #19984 inserted in Blockchain
Hash Value: b124ea343e9608f48b98cdc20835d6af14d492cbc60a2aff2809a12
741f506a8
Block #19985 inserted in Blockchain
Hash Value: 149a0ebe5ba5cde548e24531f8097dc7ea0b5b2b88307597f6775bf
a55bd9686
Block #19986 inserted in Blockchain
Hash   Value: e7569db4f1353d1f462ef467cccf479cd45f143e3a711b319d3d0ba
59df7550d
Block #19987 inserted in Blockchain
Hash Value: a1ae0818dd75ac1edc82af2e6b6eacc549cbed4e6216cbe2aba5d44
467ed4639

Block #19988 inserted in Blockchain

Hash Value: 231ffb45f1bab279cd1e89d3fb44d47d900e1cc396b693bd2ea719b c8c83397b

Block #19989 inserted in Blockchain

Hash Value: 53816e89cb4b8b5c3725f27b2cc466dc7d1d32da72c6ce7549d65fa 357ac2789

Block #19990 inserted in Blockchain

Hash Value: 1f8a7ac887af2741c3efdbdef04dc785b2a415f607f11ca346cdf9e 4b7680777

Block #19991 inserted in Blockchain

Hash Value: 63cb49f0159d5389b35bdf7d38d1abc09132d561574913d30a46cae e5243b9d0

Block #19992 inserted in Blockchain

Hash Value: e25d51c6e2124653e73ae6ea69feb93f9fd28688afa824345bee481 a5dd97874

Block #19993 inserted in Blockchain

Hash Value: 7cadfde4d902eef011a88f7bc370d36165521e9bab96515254042fd 9b7f0f094

Block #19994 inserted in Blockchain

Hash Value: 9f8e6a12738a00fca346832860e2d2c4a5ac9f58a552ea642514689 297eb957d

Block #19995 inserted in Blockchain

Hash Value: ec85bd1133893a6178eedc3e7832c350c40f217418a9d363926caa4 4fe71108d

Block #19996 inserted in Blockchain

Hash Value: d5384c12997497c193c791971423c7ed619bad2ce3b202d14175c39 95182f0a1

Block #19997 inserted in Blockchain

Hash Value: e95d733459d11e25ea1de53d987e2acb6350e547315e124bbf50281 46f6c6528

Block #19998 inserted in Blockchain

Hash Value: b96522fce352e01351e25e40affc43f6708d4fb7844c7cfa6485b45 356d5274a

Block #19999 inserted in Blockchain

Hash Value: b88e5bfbc8c8853fc82676875f9b5f277ea8966a9ca60c0c597f162 3a654e675

Block #20000 inserted in Blockchain

Hash Value: fdd851eafca41027cd3ffb3d0a535f0f53ad6f70ff1f87ce091295e b9d53fa79

1.8.4 IoT-Enabled Blockchain for Secured Scenarios

As in the past model, the execution of hash work with the squares is done on a self-sufficient framework. If there should be an occasion of authentic blockchain, it is required to be passed on with the target so that various clients can start their exchanges and squares. For spread and electronic executions, there are various

```
E:\curl-7.65.0-win64-mingw\bin>curl "http://localhost:5000/myblockchain" -d "{\"
from\":\"Sender\",\"to\":\"Receiver\", \"amount\":5}" -H "Content-Type:applicati
on/json"
Transaction Successful

E:\curl-7.65.0-win64-mingw\bin>curl "http://localhost:5000/myblockchain" -d "{\"
from\":\"Person-1\",\"to\":\"Person-2\", \"amount\":5}" -H "Content-Type:applica
tion/json"
Transaction Successful

E:\curl-7.65.0-win64-mingw\bin>
```

```
{"timestamp": "2019-05-25 20:05:42.338000", "data": {"transactions": [{"to": "Re
ceiver", "amount": 5, "from": "Sender"}, {"to": "Person-2", "amount": 5, "from":
"Person-1"}, {"to": "*******************", "amount": 1, "from": "networ
k"}], "proof-of-work": 18}, "hash": "8ae9acf42e4c4b89384818b0bd1a11ce48640734fcc
5f2813bfb4368eed47d4d", "idx": 1}

E:\curl-7.65.0-win64-mingw\bin>
```

FIGURE 1.12   Executing the secured environment for IoT-based smart currencies in smart cities and smart home automation.

structures in Python. In blockchain programming, the proof-of-work (PoW) is one of the significant estimations. It is utilized to declare and support the exchanges with the target that the new squares are fused in the blockchain. It is suggested as the key understanding figuring for the checks and legitimacy of the exchanges. In blockchain making, various diggers eagerly underwrite and finish the exchanges. For their work, the diggers are reimbursed with the advanced propelled kinds of money as their compensation.

This process, in addition, keeps up a key partition from the twofold experiencing issue with the target that the moved money or exchange is executed in an affirmed way. For instance, if A transmits a record or wireless message to B, the particular report or money respects in the records of A certain need be killed and after that ought to be reflected in the records of B. By and large, it is finished by the wireless controller as broadly engaging. If there should be an occasion of blockchain plan, it is executed with no broadly engaging and it is embraced ordinarily utilizing express calculations. On the off chance that there are occasions of non-erasing the exchange from the sender, it will de-assess the wireless message in spite of the sort of wireless message.

As depicted in Figure 1.12, the execution of code and overall implementation with all the records and transactions can be analyzed so that the transparency of operations will be there without any attempt of hacking. Using PoW, the integrity of transactions are logged and committed. With this approach, the smart cities and smart homes or related aspects of IoT can be made secure using blockchain-based technologies.

1.9   CONCLUSION

IoT-based devices and smart gadgets are quite susceptible to assorted assaults and sniffing attempts because of assorted IoT search engines and indexing portals. Nowadays, there are many software applications and libraries which are associated with IoT and smart devices which face enormous performance issues including security, load balancing, turnaround time, delay, congestion, big data, and parallel computations. These key issues traditionally consume huge computational resources, and the low configuration computers are not able to work on high-performance tasks.

A number of cloud platforms are available on which the high-performance computing applications can be launched without having access to the actual supercomputer in the IoT environment. Using these IoT-integrated cloud services or Cloud of Things (CoT), the billing is done on a usage basis and it costs less as compared to purchasing the actual infrastructure required for working with high-performance computations. There is need to enforce blockchain-based implementations and smart dynamic algorithms which can strengthen the overall security. In this chapter, blockchain-based implementation is presented with the related scenarios whereby the effectual results and outcomes can be achieved with a higher degree of accuracy and performance.

REFERENCES

Al-Fuqaha, A., Guizani, M., Mohammadi, M., Aledhari, M., & Ayyash, M. (2015). Internet of things: A survey on enabling technologies, protocols, and applications. *IEEE Communications Surveys & Tutorials, 17*(4), 2347–2376.

Al-Majeed, S. S., Al-Mejibli, I. S., & Karam, J. (2015, May). Home telehealth by Internet of things (IoT). In *2015 IEEE 28th Canadian Conference on Electrical and Computer Engineering (CCECE)* (pp. 609–613). IEEE.

Apthorpe, N., Reisman, D., Sundaresan, S., Narayanan, A., & Feamster, N. (2017). Spying on the smart home: Privacy attacks and defenses on encrypted IoT traffic. arXiv preprint arXiv:1708.05044.

Arnaert, M., Bertrand, Y., & Boudaoud, K. (2016). Modeling vulnerable Internet of Things on SHODAN and CENSYS: An ontology for cyber security. In *SECURWARE 2016: The Tenth International Conference on Emerging Security Information, Systems and Technologies* (pp. 299–302).

Biswas, K., & Muthukkumarasamy, V. (2016, December). Securing smart cities using blockchain technology. In *2016 IEEE 18th International Conference on High Performance Computing and Communications; IEEE 14th International Conference on Smart City; IEEE 2nd International Conference on Data Science and Systems (HPCC/SmartCity/DSS)* (pp. 1392–1393). IEEE.

Dastjerdi, A. V., & Buyya, R. (2016). Fog computing: Helping the Internet of Things realize its potential. *Computer, 49*(8), 112–116.

Deogirikar, J., & Vidhate, A. (2017, February). Security attacks in IoT: A survey. In *2017 International Conference on I-SMAC (IoT in Social, Mobile, Analytics and Cloud) (I-SMAC)* (pp. 32–37). IEEE.

Farahani, B., Firouzi, F., Chang, V., Badaroglu, M., Constant, N., & Mankodiya, K. (2018). Towards fog-driven IoT eHealth: Promises and challenges of IoT in medicine and healthcare. *Future Generation Computer Systems, 78,* 659–676.

Heller, M. Insecure MongoDB configuration leads to boom in ransom attacks. https://searchsecurity.techtarget.com/news/450410798/Insecure-MongoDB-configuration-leads-to-boom-in-ransom-attacks (Accessed: June 20, 2019).

Islam, S. R., Kwak, D., Kabir, M. H., Hossain, M., & Kwak, K. S. (2015). The Internet of things for health care: A comprehensive survey. *IEEE Access, 3,* 678–708.

Kolias, C., Kambourakis, G., Stavrou, A., & Voas, J. (2017). DDoS in the IoT: Mirai and other botnets. *Computer, 50*(7), 80–84.

Li, S., Da Xu, L., & Zhao, S. (2015). The Internet of things: A survey. *Information Systems Frontiers, 17*(2), 243–259.

Markowsky, L., & Markowsky, G. (2015, September). Scanning for vulnerable devices in the Internet of Things. In *2015 IEEE 8th International Conference on Intelligent Data Acquisition and Advanced Computing Systems: Technology and Applications (IDAACS)* (Vol. 1, pp. 463–467). IEEE.

McMahon, E., Williams, R., El, M., Samtani, S., Patton, M., & Chen, H. (2017, July). Assessing medical device vulnerabilities on the Internet of Things. In *2017 IEEE International Conference on Intelligence and Security Informatics (ISI)* (pp. 176–178). IEEE.

Moran, S., with Cui, X. (2016). The Internet of things. In *Ethical Ripples of Creativity and Innovation* (pp. 61–68). Palgrave Macmillan, London.

Nawir, M., Amir, A., Yaakob, N., & Lynn, O. B. (2016, August). Internet of Things (IoT): Taxonomy of security attacks. In *2016 3rd International Conference on Electronic Design (ICED)* (pp. 321–326). IEEE.

Pongle, P., & Chavan, G. (2015, January). A survey: Attacks on RPL and 6LoWPAN in IoT. In *2015 International Conference on Pervasive Computing (ICPC)* (pp. 1–6). IEEE.

Pop, C., Cioara, T., Antal, M., Anghel, I., Salomie, I., & Bertoncini, M. (2018). Blockchain based decentralized management of demand response programs in smart energy grids. *Sensors, 18*(1), 162.

Puthal, D., Malik, N., Mohanty, S. P., Kougianos, E., & Yang, C. (2018). The blockchain as a decentralized security framework [future directions]. *IEEE Consumer Electronics Magazine, 7*(2), 18–21.

Samtani, S., Yu, S., Zhu, H., Patton, M., Matherly, J., & Chen, H. (2018). Identifying supervisory control and data acquisition (SCADA) devices and their vulnerabilities on the Internet of Things (IoT): A text mining approach. *IEEE Intelligent Systems.*

Sharma, P. K., Moon, S. Y., & Park, J. H. (2017). Block-VN: A distributed blockchain based vehicular network architecture in smart City. *JIPS, 13*(1), 184–195.

Shemshadi, A., Sheng, Q. Z., & Qin, Y. (2016, July). Thingseek: A crawler and search engine for the Internet of things. In *Proceedings of the 39th International ACM SIGIR Conference on Research and Development in Information Retrieval* (pp. 1149–1152). ACM.

Stradolini, F., Tamburrano, N., Modoux, T., Tuoheti, A., Demarchi, D., & Carrara, S. (2018, May). IoT for telemedicine practices enabled by an Android™ application with cloud system integration. In *2018 IEEE International Symposium on Circuits and Systems (ISCAS)* (pp. 1–5). IEEE.

Sun, J., Yan, J., & Zhang, K. Z. (2016). Blockchain-based sharing services: What blockchain technology can contribute to smart cities. *Financial Innovation, 2*(1), 26.

Wortmann, F., & Flüchter, K. (2015). Internet of things. *Business & Information Systems Engineering, 57*(3), 221–224.

Wright, A. (2016). Mapping the Internet of things. *Communications of the ACM, 60*(1), 16–18.

Zanjal, S. V., & Talmale, G. R. (2016). Medicine reminder and monitoring system for secure health using IOT. *Procedia Computer Science, 78*, 471–476.

2 Heterogeneous Intelligent Transportation Systems

Review of Cybersecurity Issues, EU Regulations, and Economics

Andrea Chiappetta

CONTENTS

2.1   INTRODUCTION

The transport sector has undergone quite a number of changes as technology has progressed over time. The road system is directed towards an automated and connected paradigm. At the moment, there are numerous experiments being made with automated cars and this is likely to change the whole landscape. With such improvements, there has been the introduction of wireless connectivity in different types like vehicle-to-vehicle (V2V) infrastructure which drastically boosts the perception of the traffic environment. With automobile manufacturers aiming to automate and connect all vehicles, the resultant system of transportation will operate in a way that will be fully automated, attaining great performances in traffic (Aloso et al., 2018).

Vehicle automation and connectivity helps improve mobility and enhances the safety of the passengers, but it also creates possible loopholes for attackers to come in and compromise the security through hacking. This has made the manufacturers of connected automobiles devise different ways and solutions that will help them to keep the cybersecurity of autonomous and connected vehicles within the provided framework and standards of different regions and countries (Ring, 2015).

In Europe, the main goal set by the European Union (EU) is cutting down on greenhouse gases by a massive 20% come the year 2020. This has in turn translated to a massive improvement in the transport sector. However, the major focus has only been on fuel efficiency and saving the environment through creating fuel-efficient vehicles. However, transport systems need to be connected in real time to their surroundings and also to central systems to maximize the overall performance in the transport sector (Li et al., 2017). For instance, intersections in roads are bottlenecks prone to accidents in the traffic systems. Through the development of intelligent algorithms whereby vehicles are able to exchange information and then make a decision on a safer schedule for each car coming through the intersection, which will improve the flow of traffic and reduce the possibly fatal accidents.

Changing from the present unconnected to completely automated or connected transportation is expected to be gradual, with unclear impacts on traffic operations as well as environmental pollution being mooted. As a result, it has been projected that a partially connected automated vehicle environment (PCAVE) will be around for quite some time (possibly decades) (Li et al., 2017).

2.2   CYBERATTACKS IN THE TRANSPORT SECTOR

The transport sector is not undamaged from cyber threats, due to the strategic role it plays in the society; in the economy is the transport sector covers all kinds of modalities (land, sea, air). Clearly everything that is connected means that dialogue with at least two different devices/sensors, and for the same reason they could be attacked if not correctly hardened.

Another aspect is related to the costs of cyberattacks in such settings that is estimated to reach over 2 trillion USD in the next coming years, and today IoT is just beginning to emerge with vulnerabilities and exploits reported at a steady pace and showing that information security and operational security are already the most important challenges to be faced. Since the IoT ecosystem can often have critical components, it will unavoidably be a target for attacks and espionage, and denial-of-service along with many other types of cyberattacks (Chiappetta, 2017). IoT Industry 4.0 and interconnected devices along with the infrastructures are expected to be a standard in the near future, introducing disruptive changes as we move from the era of personal devices to an era where there is a promotion of large-scale interconnected (and highly integrated) devices and platforms (supporting real-time monitoring, autonomous adaption, instrumentation, actuation, control logic, etc.).

Anderson, in his work published in 1980 about network security monitoring, showed that a large number of research works has been conducted around intrusion detection. In Table 2.1 are the main cybersecurity attacks on transports that show how they are continually growing along with a description on typologies, methodologies, and damages that were reported.

2.3 THE FLOW OF TRAFFIC FOR AUTOMATED AND CONNECTED VEHICLES

There has been great progress seen over the past few years in the development of connected and automated vehicles (CAVs). The development witnessed has been made public through the modern applications being rolled out by the industry. Nevertheless, there is insufficient knowledge on the effect of CAV technologies on the performance of surface transportation network (Li et al., 2017). More so, the specifications of technology related to CAVs as well as the way drivers respond to such new technologies have not been fully integrated into models of traffic flow. Such models are required to evaluate and assess the mobility and safety effect on the conditions of the current roadways (Blokpoel, Mintsis, & Schinder, 2018).

It is fundamental to have an understanding of the possible implications that will come with CAV technologies on the dynamics of traffic flow at both the network level and local link level. Implications like that will not be comprehended without looking into two dimensions: the human dimension and the technology dimension. With regard to the technology dimension, the dynamics of vehicles and communication as well as sensing of various specifications of CAVs needs to be pointed out and then translated into models of traffic flow. With regard to the human dimension, the driver responsiveness to the CAV technologies ought to be tested and measured through extensive experiments and demonstrations, considering the fact that CAVs have various types of connectivity as well as different automation levels (Raposo et al., 2018).

TABLE 2.1

Historical Cyberattacks in the Transport Sector up to 2018

Location/ Victim of Attack	Typology	Methodology	Damage Done
Worcester Airport (1997)	Phishing	The hacker managed to disable a telephone company computer servicing at the Worcester Airport. In doing this, he sent a series of commands from his personal computer and disabled key services at the FAA control tower, spanning six hours long.	He disabled the key services at the FAA control tower and crippled the airport for a total of six hours. In the course of the attack, services at the airport stood still and did not move, leading to massive losses and confusion.
Port of Houston (2001)	Denial-of-service attack	A teenager from Britain is said to have brought all Internet systems and services of a major port in the US to their knees in an attempt to seek revenge on a fellow user of IRC. In doing that, he directed an attack to a fellow user in the chat room, with the attack managing to slow down the systems at the port through a DoS. He took out the network connection of the fellow chat room user through a device he had created, only to disable the entire system at the port.	The system was running alongside other server systems, and the PING flood attack affected all the systems, but the most affected was the port system that could not work because of slowed operations. The attack made it impossible to access data (on weather, tides. and water depths) at the port. Even though no physical injuries or damages were created, the actions still led to electronic sabotage.
CSX, US railway (2003)	DoS	The hackers gained access into the system and disrupted the operations for some time. The system was accessed through three IP addresses, probably from another country. The country of the attackers was not named.	System operations were derailed for quite some time before they were normalized. The attack disrupted traffic in 23 states in the eastern side of the US, generating delays during the day from 15 minutes to 6 hours.
LA traffic engineers' strike (2007)	Hacking	The two engineers went on strike and were locked out of accessing the traffic lights control systems. However, they hacked themselves in and changed the settings back to what they were before and could easily access them. They said that their motive was protecting the system from any form of attack.	Only system settings were changed and it took four days to have them back to normal and operating well. No accidents were reported at the time.

(Continued)

TABLE 2.1 (CONTINUED)
Historical Cyberattacks in the Transport Sector up to 2018

Location/ Victim of Attack	Typology	Methodology	Damage Done
Trams in Lodz, Poland (2008)	Hacking	A polish teenager is said to have derailed a tram after he attacked a train network. In doing this, he turned the tram system in the city of Lodz into his personal train set, which brought about chaos and derailed a total of four vehicles in the process. He modified a TV remote control in that it could be used in changing track points. He managed to trespass the depots of the tram and collect information required to create the device. He said that he had done it as a prank.	Four vehicles were derailed and a total of twelve people were injured in the process.
US Pacific Northwest (2011)	DoS	An unidentified railway company was hacked into, disrupting all its railway signals for a period of two days in December 2011. The railway located in the Northwest of Pacific was slowed and could not perform its operations normally.	System and operations were shut down at the railway company for two days.
Port of Antwerp (2011 and 2013)	Hacking (use of Trojan horses)	In this case, a group of drug traffickers hired hackers to breach the IT security systems that controlled the location and movement of containers. The hackers began by emailing malicious software to the port's staff. Through that, they were able to gain access to the data through remote access, which they applied in identifying and intercepting the containers carrying drugs and cleared them. After being discovered, the attackers physically broke into the port's offices and made away with the computing materials used by staff, including computers and keyboards.	There was physical damage to the port and the port's computing equipment was stolen. At the same time, the systems were compromised and it took time to normalize operations by neutralizing the Trojan horse used in the attack.

(Continued)

TABLE 2.1 (CONTINUED)
Historical Cyberattacks in the Transport Sector up to 2018

Location/ Victim of Attack	Typology	Methodology	Damage Done
Tesla hijacking competition (2014)	Hacking	A group of Chinese researchers managed to interfere with a Tesla car by taking remote control of the Model S from a distance of 12 miles, hacking the car's door locks, brakes, and other electronic features, showing an attack that could possibly lead to hijacking and compromise of Tesla cars.	The car's systems were totally interfered with. However, there were no major damages, as this was for testing purposes.
Sweden airports (2015)	DoS	A DoS attack was carried out on Swedish airports in the year 2015, raising alarm to NATO and other stakeholders. The attack seems to be linked to a group of Russian intelligence individuals and the system's services were totally crippled.	The systems of the airports were crippled for some time before they could be normalized.
Port of LA (2015)	Ransomware attack	Maersk confirmed that a ransomware attack blocked their services around the world. The attack meant that the LA port could not work and it was shut down for a whole day.	Operations were stopped at the APM terminal leading to imminent closure of the port.
Uber (2016)	Ransomware	Hackers gained access into Uber systems and obtained data of 57 million users worldwide, among those being customers and drivers. However, the attack was concealed by Uber when they paid $100,000 to the hackers and told them to delete the data and not make the breach public.	User data was illegally obtained and Uber lost $100,000 as ransom to the users.
Port of Rotterdam (2016)	Ransomware	A ransomware attack was initiated on the system and the virus crippled several businesses around the world. The businesses included Maersk and APM.	Many businesses were affected by the attack, bringing down their operations and services that depended on the affected system.

(Continued)

TABLE 2.1 (CONTINUED)
Historical Cyberattacks in the Transport Sector up to 2018

Location/ Victim of Attack	Typology	Methodology	Damage Done
San Francisco (2016)	Ransomware	Attackers locked the San Francisco Municipal Transportation Agency computers and demanded to be compensated with 100 Bitcoins as payment to have the services back to normal. The municipal transport agency was forced to offer free rides to passengers because they could not access their systems to book the passengers and keep data. A malware called HDDCryptor was used in infecting a total of 2,112 computers and encrypted all the data.	A total of 2,112 computers were crippled and could not work. Customers were given free rides, making the agency lose a lot of revenue in the process.
Maersk (2017) and Jawaharlal Nehru Port Trust (JNPT) (2017)	Ransomware	The port was attacked trough a ransomware that blocked the operations at the port of JNPT. The system terminal was shut down by the attack after the ransomware attack was carried out.	The port's operations were totally crippled and could not be done normally. The system's terminal was also shut down.
Deutsche Bahn (2017)	DDoS	The attackers spammed users with emails that they were tricked to open to give the attackers access to the system. The attackers used a ransomware called WannaCry, which later encrypted the computers and their data demanding fees of $300 and $600 to have the services reinstated back to normal.	A total of 57,000 computers were affected and could not be accessed, with the hackers only promising to reinstate them if they were paid.
Danish state rail operator (2018)	DoS	A DoS attack hit the Danish State Railways, paralyzing several operations including the communication infrastructure and ticketing system. The attackers also took offline control of telephone infrastructure and mail system. The attack was meant to destroy the entire system and bring it down to its knees, but managed to slow its operations for some time before everything was normalized.	The ticketing systems were totally affected and could not work. The communication infrastructure was also damaged and no communication could be done.

2.3.1 Modelling a Platoon of Connected and Automated Vehicles (CAVs) Using the Four-Component Framework

The consideration in this case is placed on a platoon of connected and automated vehicles with an aim to have them moving at the same speed while keeping the required spacing between the vehicles. The platoon in this case has a vehicle leading while others follow from behind as demonstrated in Figure 2.1. The platoon can be seen as a combination of four major components:

1. The node dynamics (ND) that describes the character of every car involved in the CAV platoon.
2. The information flow network (IFN) that defines the way the nodes make information exchange with one another, including the quality and topology of information flow.
3. The distributed controller (DC) that implements the feedback control with the obtained neighbouring information.
4. The formation geometry (FG) that informs on the required vehicle distance while platooning of the CAVs.

Every component in Figure 2.2 has substantial influence on the collective character of the given platoon. As per the framework of four components, a grouping of the

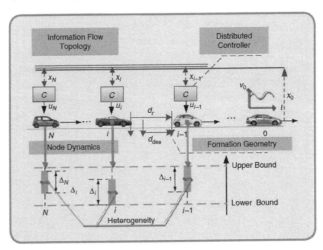

FIGURE 2.1   The four main components of a platoon. (Source: Calvert et al., 2010.)

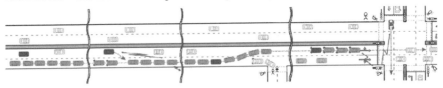

FIGURE 2.2   A demonstration of a vehicle leaving/entering a platoon and traffic management with other users on the road. (Source: Calvert et al., 2010.)

current literature can be seen in Blokpoel, Mintsis, and Schinder, (2018) and Calvert et al. (2010).

2.3.2   Research Opportunities and Challenges on Automated and Connected Traffic Flow

This part will summarize the research and views provided on the opportunities and challenges as well as current efforts placed to translate CAV features into models of traffic flow. The summary will entail the contributions as well as motivation connected with the current research, the major conclusions as well as future directions for research to be done in the area (Ring, 2015).

2.3.2.1   Challenges Faced in Modelling of Automated Vehicles for Traffic Flow

The growth and development of automated vehicles is something that has been done for quite a long time now. The latest developments in the area with regard to the maturity levels in automation of vehicles has come to a stage whereby vehicles that have lower automation levels are on roads being tested if they can operate and even handle high automation levels. Consequently, there are still several aspects that are unclear with regard to the automated vehicle's physical performance in how they interact with other vehicles as well as operate within traffic. In order to safely transition from the current state to having automated vehicles operating on roads, there is still need to do a lot of research. This should also be done with regard to deploying the vehicles and thus considering all the impacts of the vehicles that are deployed (Raposo et al., 2018).

One of the greatest challenges is that the automated vehicles needs models that are very accurate. To start with, the movement of conventional vehicles should be very accurate in regular traffic since interacting with automated vehicles is subtler. At the same time, the various stages of automated vehicles should be accurately captured and considered from vehicles with assistance in driving, then move to vehicles that are fully automated. Consequently, the interaction between various automated vehicle levels and conventional vehicles should be correct. All the aforementioned requirements are very hard to attain at the moment, more so considering the fact that there has not even been consideration on the influence of cooperation of vehicles as well as connectivity and also the fact that every automated system will operate as well as perform in a different way, even for the same automation levels (Raposo et al., 2018).

At the moment, the simulation of traffic flows for longitudinal driving works well, but it does not always have the lateral modelling of old traffic. This is a great issue with regard to simulation. However, there is already empirical research being carried out with regard to SAE level 1 automated vehicles and also some systems of level 2 that are expected to provide insights into their performance and dynamics in traffic. Nevertheless, the ground truths for higher automation levels and more so for the interaction between old traffic and automated vehicles are rare to find. There are some challenges that manufacturers of vehicles should take into consideration as well as for the simulation of automated vehicles (Ring, 2015).

A lot needs to be done to understand the dynamics of automated vehicles and also managing to model them, but the outlook is not clear and strong enough. There is need to put more emphasis and focus on attaining greater truths on the ground for the performance of automated vehicles in real traffic, which goes past what can be attained from just theory. At the same time, a strong reference with the right old driving models is authoritative. Thus, there should be awareness created that automated vehicles will also form new dynamics in the flow of traffic, starting from the interaction that they have with other vehicles and because of differences in system and vehicle system design as well as their capabilities. Such aspects will be required to be continuously addressed on automated vehicles and should later create more capabilities to forecast with the next generations of models for traffic simulation (Raposo et al., 2018).

2.3.2.2   CACC-V2X Solutions to the Challenges of ACC

CAMP V21 is carrying out a small-scale test project whose main objective is understand the basic technical steps as well as possible challenges likely to be faced while implementing connected automated cruise control (CACC) in vehicles (see Figure 2.3).

The project looked at the behaviour exhibited by automated cruise control (ACC) systems at the time they are operated in strings of vehicles, one after the other. The tests were done through the implementation of a prototype system of ACC into four vehicles from various manufacturers and then characterized them on a test track. The results obtained from the test indicated that at the time of decelerating, the time of reaction from one vehicle to another was just 1.5 seconds. About 0.8 seconds of that time can be credited to the detection of the previous vehicle's manoeuvre and the other 0.7 seconds can be credited to the reaction of the host vehicle to the attained reaction (Mattas et al., 2018). Because of such latencies, the vehicles would work and operate in a way that is not desired, increasing the decelerations from one vehicle to another, which would later lead to more perturbations in traffic or in other cases, phantom jams in traffic.

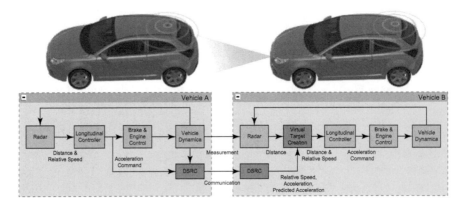

FIGURE 2.3   CACC control model. (Source: Calvert et al., 2010.)

2.4 CONNECTED VEHICLES CAN BOOST THROUGHPUT AND REDUCE DELAYS ON URBAN ROADS

In urban road systems, interceptions are considered to be bottlenecks since the capacity of an intersection is just a fraction of which the roads leading to the intersection can carry. Taking for instance an intersection that has four approaches and every approach with one through and one on the left-turn lane, so that the given approaches can be able accommodate a total of eight movements. However, the intersection can just allow two movements at any moment. Thus, the capacity of the intersection is just 25% of all the approaches (Raposo et al., 2018).

Through connected cars, the throughput of road systems in urban areas can be improved only when the vehicles can manage to cross the intersections in platoons instead of one by one as is the case at the moment. Platoon information is enhanced by the technology of connected vehicles. Thus, this research looks into the possible benefits of mobility in platoons. It states that rates of saturation flow and the capacity of an intersection can be enhanced by a factor C within a range of 1.7 to 2.0 (Raposo et al., 2018).

The analysis of queuing as well as simulations shows that having a signalized network with constant control of time will help provide support to increased demand by a factor C, if all flows of saturation are enhanced by the same factor, with no changes in the given control. At the same time, besides the increased demand vehicles will go through the same travel time as well as delay (Schoitsch et al., 2016). The same improvement in scaling is attained whenever a fixed control of time is exchanged with the maximum pressure adaptive control. Nevertheless, the lengths of the queues will increase by the same factor C, which might in turn lead to saturation. Part of the increase in capacity can be alternatively applied to cut down the lengths of queues and the connected delay in queuing through increasing the time of each cycle. Impairments to the control of connected vehicles to achieving platooning at intersections tend to be quite small (Petit et al., 2015).

2.5 THE REQUIRED ICT INFRASTRUCTURE FOR CONNECTED, COOPERATIVE, AND AUTOMATED TRANSPORT IN AREAS OF TRANSITION

Managing the existence of highly automated vehicles and conventional vehicles has proven to be a tough challenge while trying to ensure that there is uninterrupted efficiency and safety levels. The vehicles operating at higher levels of automation might have to alter to lower automation levels in some areas considering the existing circumstance like types or road, the weather conditions and much more. This research aims to looking into the phases of transition between various automation levels (Raposo et al., 2018).

The major aims of intelligent transport systems (ITS) target enhancing safety, comfort, effectiveness, and efficiency of individual mobility, freight transport, and public transport. The major technologies in the domain of ITS are the technologies of sensors, control technology, information processing, and communication systems. The applications of such technologies range from automobile manufacturers depending on the functions to large networks of traffic management (Petit et al., 2015).

The intended step-by-step introduction of automated vehicles is likely to face a period of transition for highly automated and conventional vehicles and will need to be managed to ensure that there is uninterrupted safety levels as well as efficiency. Infrastructure of ICTS will play a huge part in the period of transition through vehicle infrastructure communication (VIC) (Mattas et al., 2018).

2.5.1 Levels of Automation and Transition of Automation Levels

Vehicle transition levels in the ITS domain have been grouped in various ways like BASt definition, National Highway Traffic Safety Administration (NHTSA) definition, and SAE definition. SAE is a world organization for membership of engineers in different industries, which is involved in the development of many standards. The definition of SAE is adopted by TransAID as it offers a system of classification based on the intervention of drivers instead of capabilities of vehicles. Within TransAID, the emphasis of automation (see Table 2.2) is placed on the driver and the way control is transitioned if need be and also modelling it. The definition of SAE has been used widely on the ITS domain and also by the NHTSA. Table 2.2 provides the SAE definition (Ring, 2015).

2.5.2 The Scope and Concept of TransAID

High automation levels, more so in the conditions of urban traffic, will require to be supported by appropriate road infrastructures to make sure that there are uninterrupted safety levels as well as more efficiency. However, as indicated in Figure 2.4, there are many other situations and sectors on the roads whereby there can be guarantee of high automation levels (situation A and C), but others will also be there whereby highly automated driving will not be permitted or is not possible (situation B) because of the criticality in safety and lack of sensor inputs among many other factors (Petit et al., 2015).

An increasing number of automated vehicles will be required to perform a control in switching from automatic to human drivers whenever situations and sectors of high automation and low automation are reached (situation A). Consequently, several automated vehicles might switch to a higher automation level whenever high automation becomes available under the given circumstances (situation C). The sectors are referred to the number of vehicles that do automation level switches as automation level switch areas (Petit et al., 2015).

In the areas of transition, several vehicles that are highly automated are altering their automation level for different reasons. TransAID will do an investigation of the effects of different vehicle automation levels on current traffic systems, for different rates of penetration for every vehicle per type of automation, according to the anticipated near-future market shares and many other aspects. Many new concepts for hierarchical traffic management systems (TMS) are being created to provide certain benefits and advantages in such situations (Zhou, Li, & Ma, 2017).

TABLE 2.2
Description of Automation Levels

SAE Level	Name	Definition of Narrative	Execution of Acceleration or Deceleration and Steering	Monitoring of Driving Environment	Fallback Performance of Dynamic Driving Task	System Capability (Modes of Driving)
		Human driver monitoring the environment of driving.				
0	No automation	The complete performance by the human driver in all facets of the task of dynamic driving, even at a time when it has been enhanced by intervention or warning systems.	Human driver	Human driver	Human driver	n/a
1	Driver assistance	The specific execution based on driving mode by driver assistance system of either acceleration/deceleration or steering through the use of information on the environment of driving and with the hopes that the human driver can do the other features of the dynamic driving task.	Human driver	Human driver	Human driver	Some modes of driving
2	Partial automation	The execution that is driving-mode specific by at least one driver assistance system of both acceleration/deceleration and steering utilizing information about the environment of driving and with the hope that the human driver can do all the remaining features of the dynamic driving task.	System	Human driver	Human driver	Some modes of driving
		Automated driving system monitoring the environment of driving.				
3	Conditional automation	The performance of driving mode-specific driving system of all features of the dynamic task of driving with the hope that the human driver will make response in an appropriate way to ask for permission to make an intervention.	System	**System**	Human driver	Some modes of driving
4	High automation	The performance of the driving mode-specific driving system of all features of the dynamic driving task, even if the human driver never responds in the right way to a request to make an intervention.	System	System	**System**	Some modes of driving
5	Full automation	The performance at full time by an automated system of driving of all features of the dynamic driving task under all environmental and roadway conditions which a human driver can manage.	System	System	System	**All modes of driving**

Source: Ring, 2015.

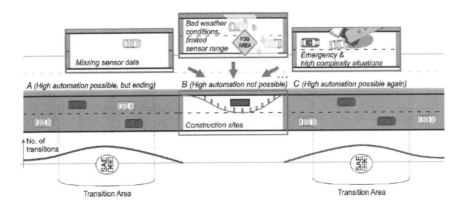

FIGURE 2.4   A demonstration of transition areas. (Source: Blokpoel, Mintsis, & Schinder, 2018.)

2.6   ATTACKS ON CONNECTED CARS

The advancement in the technology of connected cars has led to a relative increase in cases of attacks as hackers devise new ways of hacking into connected cars and controlling them. This has created relative tension among manufacturers, who are trying on a daily basis to counter the attacks and create new security mechanisms that will be hard for hackers to crack. However, there have been many reports on hacking, as some companies and individuals have fallen prey to the hackers (Mattas et al., 2018).

2.6.1   History of Hacking Connected Cars

Hacking into vehicles has been around for quite some time now, with the first cases reported in the year 2002. There presently are over 36 million vehicles operating on roads and connected to the Internet, but automobile manufacturers seem not to be learning more from the greatest crises of security facing them in this age of the Internet. In many cases, cybersecurity is just an afterthought instead of an integral part of the manufacturing and engineering processes (Mattas et al., 2018).

In the year 2002 hackers began targeting technologies of engine management, which control fuel injectors and performance supercharges. In the year 2005, Trifinite used Bluetooth to irrationally transmit or intercept in-car signals of a radio. Also Inverse Path (a UK company) demonstrated how an attacker could potentially jeopardize the integrity of in-car navigation systems through sending fake updates on traffic over FM, which made cars to change their routes and consider other routes with no traffic (Petit et al., 2015).

In the year 2010, there was experimentation done which created dramatic interventions that could affect the car's mobility. For instance, there is a case in Texas where a discontented former employee in a car dealership company used stolen credentials to access a mobilization console based on the Internet and started to systematically 'Brick' cars that been sold by the former employer. There can be sentiments that the remote attack depended on an aftermarket immobilizer and thus cannot be considered to be a weakness of connected vehicles. However, that is still a weakness that could expose connected cars to attackers in different circumstances (Schoitsch et al., 2016).

2.6.2 The Reality in Remote Car Hacks

Hackers have developed strong mechanisms to help them attack cars remotely, thanks to advancements in technology (see Figure 2.5). In the year 2015, Chris Valasek and Charlie Miller managed to remotely hack and commandeer a Jeep Cherokee driving on a public highway. The two managed to do this by taking advantage of zero-day vulnerability in the entertainment software being used by the vehicle. They managed to take over the functions of the dashboard, transmission, brakes, and steering.

Two years after that, Valasek and Miller had fine-tuned their research to a level where they were able to effectively and fully control the vehicle, managing to bypass the limited digital security on board the car and mechanisms of error connection, and turned the wheels at any speed, accelerated, and slammed the brakes (Mattas et al., 2018).

Whereas the two hackers never carried out such second attacks remotely – they emphasized their attacks using laptops directly connected to the control area network (CAN) bus in the vehicle – their previous research shows that this is still possible to achieve for any determined hacker (Maa et al., 2018).

Still in the year 2015, Valasek and Miller managed to remotely determine 471,000 Jeep Cherokees operating on the road, which they could possibly commandeer remotely considering their first successful attempt. This shows the potential danger that the owners and manufacturers of connected vehicles face in the current digital era (Lu & Blokpoel, 2016).

2.6.3 Methods of Hacking Cars

There are different methods that attackers can use to gain access into car networks and manipulate them to do as they want. In this case, these are the major and common methods that hackers can use:

1. By leveraging the vulnerabilities in the production line
2. By fooling the car's internal network

FIGURE 2.5   A car hacking scenario. (Source Ring, 2015.)

2.6.3.1   Leveraging the Vulnerabilities in the Production Line

Attackers can take advantage of the vulnerabilities in the components of the connected car before leaving the factory production line. In this way, they will have the advantage to control the cars when they are finally released into the roads and manipulate them to their advantage (Lu & Blokpoel, 2016). This is a vital stage whereby security needs to be tight, otherwise hackers will identify weak points and use them to penetrate (Petit et al., 2015).

The complexity of self-driving and connected vehicle software is something that can be easily beaten by hackers and a great challenge for manufacturers to create. Considering that it is not easy for anyone to guarantee security in the current world scenario, a sensitive technology offers hope for the attackers to come in and compromise the security (Mattas et al., 2018).

To deal with hackers at the vulnerable point of entry at the manufacturing line, Blackberry has decided to venture into security and offer Jarvis software aimed at ensuring security for connected and autonomous vehicles. The organization has clearly pointed out the suitability of its technology to offer robust defence from its weakest points going forward (Maa et al., 2018).

2.6.3.2   Fooling the Vehicle's Internal Network

A vehicle's internal network is the backbone of all connected vehicles, even in self-driving cars. This makes it a section that all automotive manufacturers should look into as they move into connected cars, and all autonomous vehicles should ensure that they have heavy and sufficient security against any supple hackers. Hackers can also manage to attack connected vehicles through means like hardcoded credentials, coding logic errors, buffer overflows, information disclosures, and backdoors (Maa et al., 2018).

This method of hacking is not theoretical since it has already been tested in practice in 2016 at the Black Hat conference. Valasek and Miller managed to intercept important correct messages that were being transmitted to a Jeep Cherokee's internal network through sending false signals and messages. This enabled the car to violently turn, brake, and speed up (Williams, Wu, & Closas, 2018).

2.6.4   Present Technologies Not Sufficient for the Future

The modern-day world and the advances in technology have eliminated the concept of companies being physical premises. In the past, conventional manufacturers never considered digital security while designing their products, more so the automobile industry. Conventional automotive technology might have been subtly internally interconnected, but there was substantial external data exchange that went through on-board diagnostics ports (Mattas et al., 2018).

Unfortunately, the current security levels based on CAN bus technology are very weak and easy to overcome. It has been confirmed by research that the weaknesses in the CAN bus architecture are very basic and that they can completely be addressed through updating the standard of CAN architecture (Ferguson, 2018).

The future of automated vehicles is reliant on rapidly increasing connectivity. The use of V2V connectivity will enable the creation of ad-hoc networks that are wireless on the road, allowing the vehicles to exchange road traffic and condition data. In the same way that Internet of Things (IoT) quickly ushered in Internet of Everything (IoE), there are developments with vehicle-to-everything (V2X) replacing V2V. This entails vehicle-to-infrastructure (V2I), V2V, vehicle-to-device (V2D), vehicle-to-pedestrian (V2P), and vehicle to-grid (V2G) (Ferguson, 2018).

With more connectivity, there will come more lines of code, and with more coding lines, there will be more weaknesses that can be exploited by attackers. Expansion of the ecosystem will mean that there will be more joints to secure and also more possible avenues to maliciously access the systems (Schoitsch et al., 2016).

2.6.5 FUTURE ATTACKS ON CONNECTED CARS AND POSSIBLE DEFENCE

It would be wrong to think that cybercriminals are only motivated by money and will never have any interest in attacking vehicles. However, the motives of these people can be different, hence their interest in attacking cars is very likely and will only advance over time to become more sophisticated (Williams, Wu, & Closas, 2018).

The first scenario of possible attacks on cars is ransomware for cars, which is very common and familiar. Consider unlocking a car in the morning and then giving instructions to the digital assistant to drive to the workplace. Instead of being welcomed by the usual procedure, a ransomware message pops up in the car demanding money before the car is released for use. This is a possible scenario and attackers will still benefit and demand for money, just like in other attacks (Ferguson, 2018).

Further still, take the automated cars as possible weapons used by terrorists to attack places remotely. This will have more far-reaching negative effects than anyone could imagine. For instance, the latest attacks seen in Barcelona, London, Nice, and New York can be taken as case scenarios for precautionary measures as connected cars are being created (Maa et al., 2018).

It is vital that researchers and organizations dealing with cybersecurity work closely with automotive manufacturers to help bridge the gap in skills lacking in both sets of professionals. Thus, ensuring security in vehicles should not be afterthought of the manufacturing process but should be a factor considered when manufacturing the cars in the industry (Petit et al., 2015). There are many lessons that can be taken by the way the Internet is revolutionizing things through rapid innovation and adoption of innovative ideas. However, such rapid adoption and innovation of ideas without considering security makes it worse as it provides a possible avenue that can be used by criminals for their own personal gain (Ferguson, 2018).

2.7 THE ROLE AND REGULATIONS OF THE EU IN DEPLOYMENT OF CONNECTED AND AUTOMATED VEHICLES

The EU fully supports connected and automated vehicles being deployed at different levels.

1. Policy initiatives: Creating policies, roadmaps, initiatives, and strategies in close coordination with all stakeholders. The role of DG CONNECT is bringing together countries and stakeholders to enhance exchange of proposals, ideas, and experiences.

2. Creating standards at the European level.
3. Funding innovation and research projects (H2020) supporting infrastructure and actions at initial stages.
4. Ensuring legislation at the European level when required.

Cooperative Intelligent Transport Systems (C-ITS) is a system that allows for information exchange between vehicles and the infrastructure on the road. Operators and authorities are working in unison on the C-Roads platform that enables them to harmonize C-ITS's activities deployment in Europe. The main aim is achieving the deployment of interoperable C-ITS services for all users of roads (Tatjana, 2018).

The commission initiated a High Level Group GEAR 2030 on January 2016 to make sure that there are intelligible vehicle policies in the EU. The group brought together member states, commissioners, and other stakeholders representing insurance, telecoms, and IT industries. Recommendations were made by the group to ensure that relevant public, legal, and policy frameworks exist before connected and automated vehicles are rolled out in 2030.

While the rules are created to enhance the security and safety of people, they should not be so rigid to a level that hinders innovation. The industry requires freedom to bring in a wide variety of technologies which will help the EU attain its goals in transportation. Thus, despite the emphasis of the European Commission to take a hybrid approach in communication in its C-ITS strategy, it is quite worrying that two years after that the technological neutrality principle tends to no longer be respected by the commission itself (see Table 2.3).

2.8 ECONOMIC IMPLICATIONS OF CONNECTED AND AUTOMATED VEHICLES

Traffic management techniques based on supply and demand are taken to be the major approaches of control that are used in dealing with congestion, because of lack of sufficient space to install new infrastructure and increase the needs of mobility. From the more supply oriented to the more demand oriented, the methods include: dynamic urban space usage, smart control of traffic signals at the network level, route guidance, preferential public transport treatment, congestion pricing, and sharing of cars (Tatjana, 2018).

Considering the high number of private cars with a limited capacity in network, techniques that are supply oriented do not have the capacity to totally eradicate congestion as the demand for the travel peak hours in major cities is more than expected. Ensuring proper integration of demand- and supply-oriented policies is a great challenge which would need consolidating realistic congestion models, user acceptability, and economic principles.

With the introduction of CAVs, the value of in-vehicle time may be affected, as users can do other economically benefitting activities instead of driving (Schoitsch et al., 2016). If conventional and autonomous vehicles have to share one road, it is likely to push away the users of traditional vehicles from the common anticipated arrival time and cut down the cost of autonomous vehicle users. Nevertheless, it is common

TABLE 2.3

European Commission Laws Created to Regulate the Use and Liability of Automated Vehicles

Year	Document/ Regulation	Key Deliberations
2017	GEAR 2030 HLG Final Report 18/10/2017	• About compensation of victims, GEAR 2030 HLG states that product liability and motor insurance directives are enough at this level, at least for systems anticipated by the year 2020. • There are many views on whether it is important or desirable to match the different regimes in national liability. • The European Commission will keep an eye on the need to change the Product Liability Directive (PLD) and Motor Insurance Directive (MID) and also the need to have extra EU legal instruments with technologies of future development.
2016	GEAR S2030 HLG Roadmap	• For the present level of development of CAVs, the current legal framework in risk appropriation and liability is not enough. • With the growing automation and connectivity of vehicles, it might be important to revise or change the liability rules between the involved parties.
2015	Business Innovation Observatory/ Study by PwC commissioned by DG GROW	• With autonomous and assisted driving technology growing fast, liability uncertainty is an increasing concern. With no clarity, insurance organizations lack the ability to assess their liability. • A harmonized European legal framework was proposed to deal with the concerns on the liabilities that come with self-driving cars.
2015	C-ITS Final Report	• In as much as the driver is in control of the vehicles, there is no need to change liability clauses and rules. • Nevertheless, taking into consideration the tendencies towards attaining higher connectivity levels as well as higher automation levels where information is offered through C-ITS might trigger other actions from the vehicle. • The last C-ITS report recommends that the commission re-evaluates the concept of liability for the aspects in the second phase of the C-ITS platform.

Source: Evas et al., 2018.

knowledge that if all the users are the same, the individual costs are not dependent on the value of the in-vehicle time. Here, traditional and autonomous vehicles are separated physically such that the users are quite homogeneous and the value of in-vehicle time does not affect the individual cost of congestion. Whereas the cost of having a free-flow part of the trip might still be dependent on the type of vehicle used, the benefit depends on the conditions of traffic and thus can be captured by the constant individual-specific coordination cost (Schoitsch et al., 2016). Table 2.4 describes several scenarios and their cost–benefit analysis.

TABLE 2.4
Cost–Benefit Analysis of CAV Scenarios for the EU

Customer Effects	Insurance/Liability Scenarios					Sensitivity Tests		
	Scenario 1: Early Deployment	Scenario 2: Slower Deployment	Scenario 3: No Cost of Insurance	Scenario 4: Fully Internalized Costs	Scenario 5: Lower Productivity	Scenario 6: Higher Accident Rate	Scenario 7: Enhanced AV Safety	Scenario 8: 50% Shared AVs
Transport user effects	116.53	−35.58	35.22	−23.95	−1188.14	−879.04	17.18	215.29
Health effects	−1.99	0.00	−0.59	0.19	2.09	0.03	−0.36	−4.21
Effects on external accident costs	2.34	−0.81	−22.12	6.92	0.05	−49.24	1.27	−0.10
Effects on costs of external environment	8.60	−3.01	−0.20	0.06	0.71	−0.03	−0.12	−1.44
Revenue from tax	6.57	0.82	−4.96	1.55	−2.67	130.85	−2.97	−26.81
Wider economic effects	16.11	−5.55	0.75	−0.24	−226.30	−15.41	0.45	5.43
Totals	**148.15**	**−44.13**	**8.10**	**−15.47**	**−414.27**	**−812.85**	**15.46**	**288.17**

Source: Evas et al., 2018.

2.9   CONCLUSION

Connected automated vehicles are a phenomenon that every automotive manu-facturer is targeting in the modern-day world. However, a challenge stands on how the companies can stand up to the attacks being made by hackers. Various attempts have been made since the year 2002 and it has been proven that hacking into vehicles is quite simple and possible for hackers who use different sophisti-cated methods.

To ensure more security, automotive manufacturers and software producers need to come together and create long-lasting solutions. This way, they can bridge the gaps that exist between the skills that the two sets of technicians have. It will also help the automotive manufacturers to consider security while manufacturing cars rather than taking it as a secondary activity that is only done after the car has been manufactured.

There are several methods that hackers use to infiltrate the systems of connected vehicles. The first and most common one is hacking them while they are in the manufacturing line before release. This way, they will have the ability to control the vehicles when they have been released to the users. Another one is fooling the internal network of the vehicle by sending wrong messages, which will be followed by the vehicle as demonstrated by Valasek and Miller.

Achieving perfect networks and functional systems in connected cars is not possi-ble, but automotive manufacturers should put more resources on research and devel-opment to stay ahead of hackers in terms of data and network security. The future can be good for connected and automated vehicles only if they are well protected and used for the right purposes as intended while manufacturing them where cybersecu-rity is a topic that is considered.

REFERENCES

Aloso Raposo, M., Ciuffo, B., Makridis, M., & Thiel, C. (2018). The r-evolution of driving: From connected vehicles to coordinated automated road transport (C-ART): Part I, framework for a safe & efficient coordinated automated road transport (C-ART) sys-tem: Study.

Anderson, James P. (1980). *Computer Security Threat Monitoring and Surveillance.* Fort Washington.

Blokpoel, R, Mintsis, E, & Schinder, J. (2018). ICT infrastructure for cooperative, connected and automated transport in transition areas. *TRA2018*, Vienna, Austria.

Calvert, S., Mahmassani, H., Meier, J. N., Varaiya, P., Hamdar, S., Chen, D., & Mattingly, S. P. (2018). Traffic flow of connected and automated vehicles: Challenges and opportuni-ties. In *Road Vehicle Automation 4* (pp. 235–245). Springer, Cham.

Chiappetta, A. (2017). Hybrid ports: The role of IoT and Cyber Security in the next decade. *Journal of Sustainable Development of Transport and Logistics*, 2(2), 47–56. doi:10.14254/jsdtl.2017.2-2.4.

Cognitive Heterogeneous Architecture for Industrial IoT - D1.4 CHARIOT Design Method and Support Tools – H2020 project - Grant agreement ID: 780075.

Evas, T., Rohr, C., Dunkerley, F., Howarth, D. (2018). A common EU approach to liabil-ity rules and insurance for connected and autonomous vehicles. *EPRS European Parliamentary Research Service*. P.E 615.635.

Ferguson, R. (2018). A brief history of hacking Internet-connected cars. *New World Crime*. Retrieved from https://medium.com/s/new-world-crime/a-brief-history-of-hacking-Internet-connected-cars-and-where-we-go-from-here-5c00f3c8825a.

Li, S. E., Zheng, Y., Li, K., Wu, Y., Hedrick, J. K., Gao, F., & Zhang, H. (2017). Dynamical modeling and distributed control of connected and automated vehicles: Challenges and opportunities. *IEEE Intelligent Transportation Systems Magazine*, 9(3), 46–58.

Lu, M., & Blokpoel, R. J. (2016). A sophisticated intelligent urban road-transport network and cooperative systems infrastructure for highly automated vehicles. In *Proceedings: World Congress on Intelligent Transport Systems*, Montreal.

Maa, J., Lib, X., Zhoua, F., Huc, J., & Parkd, B. B. (2018). Parsimonious shooting heuristic for trajectory control of connected automated tra c part II: Computational issues and optimization.

Mattas, K., Makridis, M., Hallac, P., Raposo, M. A., Thiel, C., Toledo, T., & Ciuffo, B. (2018). Simulating deployment of connectivity and automation on the Antwerp ring road. *IET Intelligent Transport Systems*, 12(9), 1036–1044.

Petit, J., Stottelaar, B., Feiri, M., & Kargl, F. (2015). Remote attacks on automated vehicles sensors: Experiments on camera and LiDAR. *Black Hat Europe*,

Raposo, M. A., Ciuffo, B., Makridis, M., & Thiel, C. (2017, June). From connected vehicles to a connected, coordinated and automated road transport (C 2 ART) system. In *Models and Technologies for Intelligent Transportation Systems (MT-ITS), 2017 5th IEEE International Conference on* (pp. 7–12). IEEE.

Ring, T. (2015). Connected cars–the next target +for hackers. *Network Security*, 2015(11), 11–16.

Schoitsch, E., Schmittner, C., Ma, Z., & Gruber, T. (2016). The need for safety and cyber-security co-engineering and standardization for highly automated automotive vehicles. In *Advanced Microsystems for Automotive Applications 2015* (pp. 251–261). Springer, Cham.

Williams, N., Wu, G., & Closas, P. (2018, April). Impact of positioning uncertainty on eco-approach and departure of connected and automated vehicles. In *Position, Location and Navigation Symposium (PLANS), 2018 IEEE/ION* (pp. 1081–1087). IEEE.

Zhou, F., Li, X., & Ma, J. (2017). Parsimonious shooting heuristic for trajectory design of connected automated traffic part I: Theoretical analysis with generalized time geography. *Transportation Research Part B: Methodological*, 95, 394–420.

3 Fog Platforms for IoT Applications

Requirements, Survey, and Future Directions

*Sudheer Kumar Battula, Saurabh Garg,
James Montgomery, and Byeong Kang*

CONTENTS

3.1   INTRODUCTION

Recent technological advancements in the IoT environment have led to a rapid increase in the number of intelligent devices in a variety of domains. Consequently, the data generated from these devices has grown in both volume and variety. According to Cisco (Evans 2011), there will be over 50 billion IoT devices in use in 2020. The data generated from these devices and their communications will reach 0.5 yottabytes, and about 45% of that generated data will be IoT-created data (Hong et al. 2013). IDC 2019) predicted that 41.6 billion connected devices would be in existence and generate 79.4 zettabytes (ZB) of data in 2025. Some examples: an aircraft engine produces 1 terabyte (TB) of data per trip, while big manufacturing industries create 1 TB of raw data per day, and a surveillance camera generates 1 TB of raw data per hour (Perera et al. 2017). Currently, Cloud computing is used to process this data. Cloud computing is used as the primary solution for IoT applications due to the Cloud's characteristics, such as being almost infinitely scalable and having rapid elasticity, and abundant processing and storage power. However, Cloud-enabled IoT applications encounter numerous performance challenges because of the high latency between the IoT devices and Cloud.

Several applications in the IoT environment are time-critical (Bonomi et al. 2012; Yi 2015). For example, smart traffic light signals should make swift decisions to allow fast movement of emergency vehicles (Sarkar and Misra 2016). Similarly, in a smart parking system, the applications should be able to detect the parking areas continuously, in order to suggest parking spots to the users, thus avoiding clashes. Especially while responding to disaster situations and emergencies, any delays in the exchange and processing of critical information can result in significant damages to both lives and physical infrastructure (Ujjwal et al. 2019). For all the aforementioned examples, almost real-time response is required; if it is delayed, it can cause damage to the individual and affect society. If we process these applications in the Cloud, the opportunity to react according to the event may be lost because of the high latency. Therefore, high latency in Wide Area Network (WAN) makes the Cloud solutions inefficient for the aforementioned real-time applications (Evans 2011). As such, Fog computing has evolved to process time-sensitive applications in an effective way. Fog computing is a distributed computing paradigm in which data is stored and processed nearer to the edge devices (Bonomi et al. 2012; Rayamajhi et al. 2017), so that the overheads of data transmission can be significantly reduced. As a result, the performance of time-sensitive applications can be improved. This paradigm provides a promising solution for deploying the applications and offering different services at various scales, in order to meet user needs (Chiang and Zhang 2016). Some researchers have contributed to establishing the principles, underlying architectures and algorithms in the Fog. Some research has contributed to definition, characteristics and platform challenges in Fog computing environments. Some other work (Chiang et al. 2017; Yi, Li, and Li 2015) have compared the Fog with other related technologies, and in another work (Hu et al. 2017a) infrastructure requirements to process time-sensitive applications have been discussed. Fog computing architecture that suits big data applications have also been proposed (Naha, Garg, Georgekopolous, et al. 2018). A summary of the survey papers and their contributions are presented in Table 3.1. Based on these surveys, full-fledged implementation of Fog computing platforms still needs to be developed for the efficient performance of time-sensitive applications.

TABLE 3.1

Existing Surveys on Fog and Their Contributions

Author	Contributions
Bonomi et al. 2012	Definition and characteristics of Fog computing
	Need for Fog in building a platform for critical applications
Yi, Li, and Li 2015	To contrast Fog computing architecture with other related technologies
	Explained Fog computing with different use cases and scenarios
	Discussed Fog computing research challenges and open issues
Chiang et al. 2017	Presented Fog characteristics and compared Fog with Cloud
	Presented a novel Fog computing architecture for IoT applications
Hu et al. 2017b	Discussed the suitable infrastructure for Fog computing to support the deployment of IoT applications
	Discussed different application cases and open issues
Naha, Garg, Georgakopoulos, et al. 2018	Proposed architecture that is suitable for data-intensive applications in the Fog computing environment

To enable the design of such a platform, this chapter discuss the goals, requirements, characteristics, and challenges in building an efficient, fully-fledged Fog platform to serve the needs of present and future applications. This chapter also reveals research gaps and future directions by looking closely at the characteristics of existing Fog platforms and other crucial requirements for designing platforms.

3.2 WHAT IS FOG COMPUTING?

In this section, we discuss Fog definitions in general and in other domains, such as the IoT and vehicular networks, given in the literature by different researchers. Based on the application and characteristics of the Fog platform, we present our definition. This section also presents the differences between the Fog and other similar distributed computing paradigms.

3.2.1 FOG COMPUTING

According to a definition provided by Bonomi et al. (2014):

> Fog computing is considered as an extension of the Cloud computing paradigm from the core of the network to the edge of the network. It is a highly virtualised platform that provides computation, storage, and networking services between end devices and traditional Cloud servers.

According to a definition provided by NIST (Iorga et al. 2018):

> Fog computing is a layered model for enabling ubiquitous access to a shared continuum of scalable computing resources. The model facilitates the deployment of distributed,

latency-aware applications and services, and consists of fog nodes (physical or virtual) residing between smart end-devices and centralised (Cloud) services.

3.2.2   FOG COMPUTING IN IoT: FOG OF THINGS

Fog of Things (FoT) is a new paradigm introduced by Prazeres and Serrano (2016) who proposed a self-organizing platform for IoT in the Fog computing environment called SOFTIoT (Self-Organizing Fog of Things). This platform supports interoperability by allowing more complex operations to run in both virtual entities of Fog devices and Cloud virtual machines. FoT service-oriented middleware is for interaction between the applications and the IoT services and message-oriented middleware in order to communicate among devices and gateways.

3.2.3   FOG VEHICULAR COMPUTING

Sookhak et al. (2017) proposed a paradigm similar to Fog computing called vehicular Fog computing (VFC); this uses the infrastructure of vehicles for communication and processing the data. Vehicular Cloud computing (VCC) differs from VFC in that it only uses the nearest vehicle cluster, instead of the remote servers. This paradigm consists of three entities: smart vehicles, roadside nodes (RS), and serves as a data, Fog, and Cloud layer, respectively. Similarly, Chen et al. (2016) used unmanned aerial vehicles (UAVs) as the Fog nodes to cover a large number of IoT devices. Hou et al. (2016) proposed vehicular micro clouds and vehicular grids which group the nearest vehicles to form the cluster in order to process the data very quickly. Comparisons between VFC and VCC are listed in Table 3.2 (Ning, Huang, and Wang 2019).

Bonomi et al. (2014) explained the Fog with respect to the Cloud and network, whereas Vaquero and Rodero-Merino (2014) explained the Fog with respect to the environment and capabilities. NIST (Iorga et al. 2018) explains the Fog in terms of its architecture and facilities. Based on the definitions and features of the Fog, we describe Fog computing as a model that consists of a set of services that are offered to the user in a large-scale distributed manner, by exploiting the processing, storage, and network capabilities of users, and communicating with other devices to process the applications on time. Fog computing alone is unable to satisfy user requirements,

TABLE 3.2

Comparison between VFC and VCC

Feature	VFC	VCC
Resource capacity	Low	High
Control	Decentralized	Centralized
Latency	Low	High
Mobility management	Hard	Easy
Reliability	Low	High
Resources available	Locally	Globally

specifically in big data applications. So, Fog computing acts as an extension to the Cloud.

This chapter will not focus on the VFC and Fog of Things especially, because these paradigms have similar characteristics and also face similar challenges and limitations as Fog computing.

3.3 FOG COMPUTING VERSUS OTHER SIMILAR DISTRIBUTED COMPUTING PLATFORMS

Fog computing is similar to other distributed computing platforms which use local end-user computational devices to process the request. In Fog computing, the generated data from the IoT are processed in the intermediate layer, which consists of Fog devices or a cluster of Fog devices and data. The obtained results and data are stored in the Cloud for future purposes. The main difference between the Fog and other similar distributed paradigms in terms of processing edge devices are mobile devices for mobile cloud computing (MCC), specialized data centres in Cloudlet computing, Edge server in mobile edge computing (MEC), and base station server in MCC. However, Fog computing uses any end-devices which have a basic configuration. Fog computing has a few other differences with respect to the capacity, latency, distance, mobility, availability, power consumption, and architecture from the other distributed computing platforms, as shown in Table 3.3.

3.4 FOG COMPUTING ENVIRONMENT AND LIMITATIONS

This section discusses the requirements of IoT applications, which are handled by the Fog computing environment. This section presents the architecture of Fog computing and explains the limitations of Fog computing.

Fog computing works in the way that it extends Cloud computing to process IoT data anywhere from the Cloud to Things. The main objective of this paradigm is to bring computation nearer to end-users to solve high latency and scalability problems.

3.4.1 FOG COMPUTING ENVIRONMENT

Three categories of the Fog computing environment exist in the literature. Simple non-IoT applications, such as content delivery systems, will only consist of two layers with the absence of the IoT stratum. The three-layered Fog environment is best suited to IoT applications (Battula et al. 2019), as shown in Figure 3.1.

The connection between the Cloud and the intermediate layer (Fog) is through the Wide Area Network (WAN), and connection between the intermediate layer and the IoT layer is through Local Area Network (LAN). The IoT layer mainly senses the surrounding information from the sensors; this is also known as the data generator layer. The Fog layer helps to offer the services nearer to the end-users and away from the Cloud. The generated IoT data will send to the Fog devices in an intermediate layer, instead of the Cloud, to process the data. For future purposes, the data stores in the Cloud. Hence, Fog computing achieves low latency, resulting in higher performance of the applications.

TABLE 3.3

Attributes of Fog and Related Computing Paradigms

Platforms/Attributes	Fog Computing	Mobile Computing	Mobile Edge Computing	Mobile Cloud Computing	Cloudlet Computing
Capacity	Limited	Limited	Moderate	High	Moderate
Coupled	Very loosely	Moderately	Moderately	Tightly	Tightly
Latency	Low	Moderate	Low	Relatively high	Low
Distance	Very close	Very close	Close	Far	Close
Mobility	High	High	Medium	High	Medium
Availability	High	High	Low	Very high	Low
Power consumption	Very low	High	Very high	Low	Low
Cloud	Yes	No	Yes	Yes	Yes
Architecture	Decentralized hierarchy	Distributed	Localised/hierarchical	Centralized Cloud with distributed mobiles	Centralized data centre
Server element	Any end device (SCN)	Mobile devices	Edge server	Server at base station	Data centre in a box

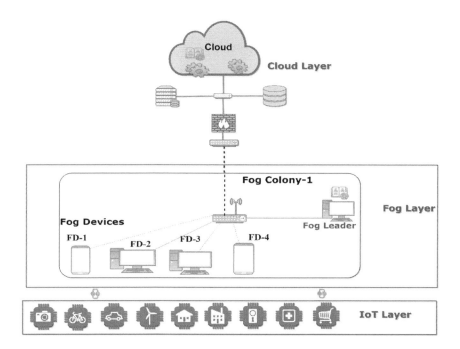

FIGURE 3.1 Fog computing environment.

3.4.2 Fog Computing Elements: Fog Nodes

The Fog computing environment consists of Fog nodes or devices. The Fog node is a physical infrastructure element that helps to provide services nearer to the users, such as network devices; Raspberry Pis; smartphones; PCs; access points; or any device with computation, storage and network capabilities. These massively distributed Fog devices form an intermediate layer between the Cloud layer and the IoT layer in order to increase the speed of execution for time-sensitive IoT applications. Fog nodes are located at gateways, network elements such as (routers and switches), and intermediate computer nodes.

3.4.3 Fog Computing Limitations

Fog computing has evolved as a feasible solution to process time-sensitive applications with minimal delays. However, there are some inherent limitations in the paradigm, which are explained in detail as follows.

1. *Limited capacity*: The Fog nodes in Fog computing have limited bandwidth and storage capacity. The Fog node alone is not suitable to process the IoT data. It is essential to find the best techniques to organize or distribute the data on the fly because data locality plays a vital part in Fog computing to decrease the latency of the system. Researchers address these issues in other distributed computing paradigms such as sensor networks (Sheng, Li, and Mao 2006) and the Cloud (Agarwal et al. 2010). However, those techniques

do not fully solve the problem because, unlike Cloud computing, Fog computing needs to solve new challenges related to whether the data should store in Fog devices, nodes, or the Cloud. Hence, it is essential to redesign the data placement and management techniques by leveraging the user service, requests, and mobility patterns to increase the performance of the system.

2. *Connectivity:* Fog computing has a heterogeneous network in each layer of the Fog strata. For optimization of cost and performance regarding connectivity, researchers' use clustering, partitioning, and network relaying techniques in other distributed computing paradigms. This technique can be used in Fog computing by dynamically selecting the Fog nodes to increase the availability of the system.

3. *Power management:* Most of the Fog devices are lower energy devices which run mostly on batteries, such as smartphones, laptops, and Raspberry Pis. Due to limited power, the devices may shut down at any time, resulting in the failure of the application processing the device.

4. *Reliability:* The reliability of the environment will affect the performance of the system. Hence, reliability in Fog computing must be considered in various cases, such as individual sensors and Fog node failures; connectivity failures in Fog nodes; failures in entire zones and regions; and the failure of users' connections, services, and platforms. Madsen et al. (2013) discuss the various requirements and challenges imposed when providing a reliable Fog computing platform by comparing present computing paradigms.

5. *Security and privacy:* In Fog computing, the security of resources and privacy of the data is the primary limitation in this environment, due to heterogeneity of devices, multiple ownership, and no full control over the devices.

3.5 FOG COMPUTING PLATFORM DESIGN GOALS, REQUIREMENTS, AND CHALLENGES

An efficient Fog computing platform should overcome all the inherent limitations of Fog. This section discusses the goals and requirements of such an efficient platform and presents the associated challenges.

3.5.1 Fog Computing Design Goals

The major design goals for building an efficient platform for managing the resources in the Fog computing environment are as follow.

1. *Efficiency:* The Fog computing platform should be more efficient in terms of resources and energy utilisation because the resources in a Fog environment, such as storage, network, computing, and energy, are limited.

2. *Latency:* This is an important factor in support of time-sensitive applications. Fog computing has evolved mainly to process time-critical applications. Thus the primary goal of the Fog platform is to offer applications and

services with low latency. To achieve this goal, the decision-making time, scheduling time, execution time of the task, and task offloading time should be minimal.

3. *Application Programming Interfaces (APIs)*: The platform's goal is to provide a common APIs to support all kinds of services and applications.
4. *Abstraction*: Fog computing consists of heterogeneous Fog nodes which may not support similar protocols in all the Fog nodes. So, the platform's goal is to provide an abstraction layer which abstracts all the Fog devices to support interoperability.
5. *Mobility*: Most of the Fog devices and sensors, such as smartphones, handheld devices, and wireless body sensors, are mobility support devices. The vital goal of the Fog platform is to handle fault tolerance and migration efficiently to offer better services to the clients.

3.5.2 Fog Computing Platform Requirements

By default, the applications should scale the resources with distributed capabilities automatically, quickly, and easily without the application developers writing special code. They should provide a platform to support the deployment and operations of the custom applications. So, to achieve these features, every Fog platform should have the following requirements (Cloud Standards Customer Council 2015; Kepes 2011).

1. *On-demand self-service*: Platform should deliver the computing and storage resources based on the demands of the users without manual interaction of service providers.
2. *Elasticity*: Platform should provision and de-provision the resources based on user or application requirements.
3. *Scalability*: Platform should handle the sudden spikes in the traffic and provide efficient results to the users.
4. *Auto-deployment*: Platform should follow a single-click deployment model where the developers can deploy the applications with a single click.
5. *Bullet services*: Platform should provide quick services to the resources for testing and managing the applications in a single environment.
6. *Web user interface*: Platform should provide a user interface to manage, deploy, and test the applications in different scenarios and use cases.
7. *Multi-tenancy*: Multi-tenancy refers to the sharing of common infrastructure by different applications hosted by the different tenant at a time. It is very crucial in a federated environment to improve the utilization of resources and to enable business opportunities. Multi-tenancy will also ensure isolations, security, and privacy among the tenants. The platform should provide isolation for different simultaneous clients to use the same application environment.
8. *Security:* Platform should secure the environment for deploying applications.
9. *Measured service*: The platform should automatically monitor, control, and manage the resources, and provide a transparent metering service based on the usage of the customers.

3.5.3 Challenges in Building an Effective Fog Computing Platform

In the practical development and deployment of time-sensitive applications, providing services to users has become a challenging problem due to their geo-distributed nature and unmanageable nodes. Many researchers have tried to build the platforms, but those are very specific to the use cases or the domain. Fog computing differs from other distributed computing paradigms. Traditional Cloud platforms are not suitable for time-sensitive applications due to high latency. Building an efficient Fog computing platform is more complex than the Cloud. The complexity involved in building features of Cloud and Fog platforms are presented in Table 3.4.

Therefore, there is a need to build an effective platform for developing and deploying applications in the Fog computing environment. It is a challenging task due to the different characteristics of Fog devices and the environment. The Fog computing operational layers are shown in Figure 3.2. The layer-wise challenges are presented as follows:

Virtualization technology: Virtualization technology helps in the efficient utilization of hardware resources. Therefore, the correct choice of virtualization is a significant research challenge because the decision will depend on the hardware configuration of the devices. The advantages of the container over hypervisor technology

TABLE 3.4

Comparison between Cloud and Fog Computing Characteristics

Feature	Cloud	Fog
Data management	Low	Very high
Migration	Low	High
Resource monitoring	Low	High
Self-organizing	Low	High
Virtual machine placement and provisioning	Low	High
Virtualization	Low	High
Energy management	Hard	Very high
Location dependency	Low	High
Federation complexity	Moderate	High
Geo-distribution and ubiquitous network access	Moderate	Very high
Load balancing	Moderate	High
Privacy and security risk	Moderate	Very high
Resource demand profiling	Moderate	High
Scaling	Moderate	Very high
SDN-based federation	Moderate	Moderate
Mobility	NA	High
Multi-tenancy	Low	Moderate
Service-oriented	Low	High
Heterogeneity	Low	High

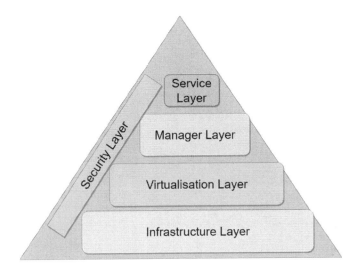

FIGURE 3.2 Fog computing platform operational layers.

are listed in Table 3.5. However, the Fog is a heterogeneous environment where some devices are not supporting containerization at present. So, the choice of an appropriate 'container or hypervisor' is an important factor which decides the performance of the system.

Monolithic or microservice architecture: Architecture plays a key role in building a complex, scalable, and robust system. In a monolithic architecture, it is difficult to scale the individual service, but in a microservice architecture, it is easy to scale the individual services. Table 3.6 compares the different features of the monolithic and microservice architectures.

TABLE 3.5
Comparison between Container and Hypervisor

Feature	Container	Virtualization
Requirement	Container engine	Hypervisor
Software	Docker	KVM
Weight	Lightweight	Heavyweight
Virtualization	Operating system level	Hardware-level
Size	Small (MBs)	Large (GBs)
Provisioning	Real-time	Slow
Boot time	Fast	Slow
Performance	Native	Limited
Secure	Less	More
Isolation	Process	Full

TABLE 3.6

Comparison between Monolithic and Microservice Architecture

Feature	Monolithic Architecture	Microservice Architecture
Integration	Easy	Difficult
Development	Single language support	Multi-language support
Upgrades	Difficult	Easy
Maintenance	Difficult	Easy
Communication	Memory	Between microservices
Scalability	Scale	Scale individual microservices
Changeability	Redeploy the entire system	Redeploy single microservice
Service type	Fine	
Management	Centralized	Distributed

Network management: Due to the billions of IoT devices, managing the network is a complex task. However, many methods are proposed in the literature to efficiently manage the network using a Network Function Virtualization (NFV) and Software-Defined Network (SDN). The main challenges come when these technologies are integrated with the Fog environment to achieve low latency.

Resource monitoring and management: The monitoring tools play an important role in making key decisions in many resource management components Battula et al. (2019). For example, these resource-monitoring tools are helpful in scheduling the tasks based on the workloads and availability of the resource, so that energy and cost optimization can be achieved. Resource discovery and appropriate resource selection, allocation, and de-allocation of tasks are responsible for resource management. Improper resource management can cause poor QoS.

VM or container scheduling: Traditional scheduling algorithms are not suitable for the Fog environment due to geographical distribution and mobility. There are many different container orchestration and scheduler technologies such as Swarm, Kubernetes, and Mesos which are available in the market for scheduling in micro-data centres and the many Fog devices. However, these tools consume more resources to run in a limited resource environment. Therefore, efficient algorithms or strategies are to be implemented to meet the requirements of IoT application users.

Latency: The Fog computing paradigm resolves one of the limitations of Cloud, that is, latency. However, providing low latency in Fog is the main challenge due to the fully distributed and highly mobile devices.

Task migration: Efficient task migration and offloading are important factors due to the mobility and unmanageability of devices. For example, the task is scheduled to one of the Fog devices and, before completion of the task, the Fog device is turned off due to low battery. The task cannot be completed until it is rescheduled to available devices. So, for an efficient platform, we need to migrate or offload the entire task, or a portion of the task, to meet the requirements of the application.

TABLE 3.7

Fog Computing Privacy and Security Requirements

Feature	Requirements
Data storage	Minimum overhead, light encryption, integrity verification
Data query	Secure query, dynamic support, optimized result
Data sharing	Access control, access efficiency, authorization efficiency
Compute	Confidentiality of applications and application data
Network	Authenticity, threats, and attacks defence

Fog security and privacy: The data will be stored and processed in autonomous Fog devices which update data frequently. Whenever there is an update in the data, access control for the data will also change. This type of environment demands new security and privacy requirements, as shown in Table 3.7. Hence, security and privacy in the Fog environment are the biggest challenges due to the open participation of devices (both trusted and untrusted). To solve these issues, we must provide security to each layer in the Fog architecture.

3.6   STATE-OF-THE- ART FOG COMPUTING ARCHITECTURES AND PLATFORMS

This section discusses the works that have been carried out so far to build efficient Fog computing platforms and architectures.

3.6.1   Fog Computing Architectures for Specific Domains or Applications

Bonomi et al. (2012) proposed a Fog software architecture based on the requirements of multiple use cases. This consists of two main layers, such as the Fog layer, to manage the device heterogeneity by providing a common platform for device management and monitoring resources, while hiding resource information. The service orchestration layer for the Fog consists of Foglets which are responsible for managing the containers, the status monitoring of applications, and the policy manager in order to meet the various business policies specified by clients in the distributed database. However, the system is scalable when receiving multiple messages but is not scalable in processing the data.

The healthcare, environmental calamities, and smart cities applications are time-sensitive. Most of the researchers in the literature did not consider the scalability of their architectures and frameworks. Stantchev et al. (2015) proposed a service-oriented architecture (SOA) in the Fog computing environment for any smart sensor-based healthcare application. This model was validated by considering an example use case of monitoring patients' glucose levels. The IoT layers consist of blood glucose sensors. The applications are deployed at the Fog layer to collect and send blood glucose levels to the doctor to evaluate whether emergency medical attention is required. However, the authors did not consider the scalability of the proposed system in terms of the number of devices that can be handled by the Fog nodes. The mobility of smart wearable devices was not considered. Fratu et al. (2015) and Masip-Bruin et al. (2016) proposed Fog computing–based architectures for monitoring COPD patients in order to handle patient emergencies, such as low and high pulse rates. However, the authors did not discuss the scalability and fault tolerance of the proposed system. Similar to the aforementioned platforms, there exists an Industrial IoT (IIoT) platform specially designed for smart industries. Nebbiolo Technologies created a Fog computing platform for Industrial IoT (Steiner and Poledna 2016) for data analysis, communication, and application deployment. This platform consists of three components: FogSM, which is responsible for end-to-end system management; FogOS, which is a stack of modules responsible for secure, flexible deployment and data management on Fog nodes; and fogNodes, a proprietary hardware architecture that enables the virtualization built by Nebbiolo. However, heterogeneity of devices are not supported.

3.6.2 PLATFORMS AND FRAMEWORKS OF FOG

In the literature, some Fog computing platforms exist and these address some of the challenges discussed earlier in the chapter. This section presents a critical review of some of the currently available open source and commercial Fog platforms in the market.

3.6.2.1 Commercial Fog Platforms

We have a few commercial Fog computing platforms which support time-critical applications. This section describes the features and limitations of them. The comparative analysis is presented in Table 3.8.

3.6.2.1.1 Cisco IOx

Cisco IOx (Cisco 2016a) is a commercial product which runs in the router and provides network storage and computing facilities to the operating system running on a virtual machine, together with an IOS virtual machine.

3.6.2.1.2 Data in Motion

Data in Motion is a Cisco product to control the flow of data by deploying it in a distributed fashion, performing the basic analysis and findings with the support of the simple rule-based RESTful API (Cisco 2017).

3.6.2.1.3 Local Grid

Local Grid (Cisco 2015) runs on network devices with the help of an embedded software installed in the devices. This provides a secure connection between all types of sensors. This platform resides between the Cloud and end devices, can perform real-time analysis with minimum latency, as well as supporting integration with the Cloud. It can solve complex problems in communication between both the Fog and the Cloud.

3.6.2.1.4 ParStream

ParStream (Cisco 2016b) is a Cisco tool that offers big data analytics near the users with simple querying of data to perform data analytics in a fully distributed manner. It builds on top of the patented database named ParaStream DB.

TABLE 3.8
Fog Platform Characteristics

Product Name	Installed Devices	Purpose	Analytics	Big Data	Scalable	Cloud Integration	Fault Tolerance
Cisco IOx	Ruggedized routers	Real-time event and data management		✗	✗	✓	✗
Data in Motion	IoT devices	Data management and analysis	✓	✗	✓	✗	✗
Local Grids	Network devices and sensors	Real-time analysis and decisions	✓	✓	✓	✓	✗
ParStream	Embedded devices	Real-time IoT analytics	✓	✓	✗	✗	✗
PrismTech Vortex	Prism tech Fog devices	Data-sharing platform	✗	✗	✓	✓	✗

PrismTech Vortex (Adlink 2017) is a ubiquitous data-sharing platform for Fog computing which provides fine-grained access control from a single point and authentication for symmetric and asymmetric environments.

3.6.2.2   Fog Platforms and Frameworks

This section discusses the features and limitations of Fog platforms and frameworks that support particular applications and presents a comparative analysis in Table 3.9.

Fog Torch (Brogi and Forti 2017) is a tool for probabilistic QoS-assurance, resource consumption, and cost estimation of eligible deployments of Fog applications. However, Fog Torch does not support scalability and mobility of the nodes.

EmuFog (Mayer et al. 2017) is a Fog emulator which enables the design of Fog computing scenarios from a low level. This emulator runs on top of the Docker-based containerization technology. The Fog nodes in the EmuFog connect through different network topologies based on the use-case scenario. The authors evaluated the performance of the system through real-world and synthetic network topologies. However, this framework does not support hybrid use-case scenarios.

A new architecture is proposed by extending the traditional architecture of PAAS and the REST Paradigm to interact between PAAS and Fog, with the key idea of reusing the existing architecture and standards (Yangui et al. 2016). It also explains the four key layers in the architecture: the application development layer, the application deployment layer, the application hosting and execution layer and the application management layer. The application development layer is containing facilities like IDE (tools), API, and SDKs for developers to develop their applications in Hybrid PAAS/Fog with ease. The application deployment layer is an extension of deployer and controller modules of the existing architecture. Deployer module deploys the applications and controller module interacts with the deployer to control the hosted resources. To maintain the available resources in the Fog computing environment, a novel module Fog resource repository is used. The application hosting and execution layer is a novel orchestration layer for the flow of execution between the containers and is responsible for exchanging and coordinating the messages between them. The application management layer is the extension layer of a PAAS management layer, which interacts with all three layers and the elastic engine to provide the scalability and availability based on the SLA and supports VM migration.

Pahl et al. (2016) argues that the traditional technologies that are used in the Cloud will not be suitable due to the heterogeneity of devices. They proposed a new container-based virtualization technique to solve the limitations of the Cloud in Edge Clouds. The authors showed that container-based technology helps in managing and orchestrating the services. The use of Raspberry Pis as Fog devices with containers is a better option in terms of energy consumption, robustness and cost. However, the authors agreed that it is a basic prototype and many improvements should still be made to manage the other aspects of the technologies, such as network and storage with proper resource scheduling and migration for better performance of the Fog platform.

Liyanage, Chang, and Srirama 2016) proposed mePaaS as a platform service for IoT applications to develop, deploy and execute the applications written in their programme models at the Edge devices. This platform uses a resource-aware component to manage the availability of Fog devices with the help of the MIST node. However, in their work, the authors did not consider the inter-operability between tasks.

3.7 IOT APPLICATIONS IN FOG COMPUTING

Fog computing provides a solution for various time-sensitive IoT applications in various domains, such as healthcare, smart cities, smart transportation, smart agriculture, smart energy, and gaming. Some of these are discussed next.

3.7.1 Healthcare

Yassine et al. (2019) proposed a platform which supports the big data IoT analytics, using Fog computing for smart homes to benefit the house owners by providing smart recommendations. This platform supports data collection from the sensors in smart homes and performs analytics to identify health conditions based on house owners' physical activities. It can also prepare an energy-saving plan by identifying the patterns of energy consumption. The authors suggested these platforms can be extended to various services and applications based on business requirements, such as the preferable timing to place advertisements based on their interests. They validated their platform with a real-time data set of smart homes.

Verma and Sood (2019) proposed a novel five-layer Fog Cloud-based stress monitoring framework to classify and predict students' stress rates. To classify students' stress rates the Bayesian Belief Network (BBN) algorithm was used and to find the stress index, the authors used the two-stage Temporal Dynamic Bayesian Network (TDBN) model. From their correlation results, they identified the key parameters that affect stress levels.

Rani, Ahmed, and Shah 2019) proposed a model in Smart Health (S-Health) to monitor and analyze the health of the patients in order to suggest the precautions and preventive measures to be undertaken for a healthier life. For example, based on the information about the cause of mosquitoes' growth, the model will suggest precautionary measures to prevent chikungunya disease. The summary of the recent platforms and frameworks for specific applications are listed in Table 3.10.

The Fog layer is not only used to process the data in healthcare. In some work, the Fog layer is also used as a transmission layer and to maintain medical records (Venckauskas et al. 2019; Silva et al. 2019).

3.7.2 Smart City

Wu et al. (2019) proposed a novel traffic control architecture using Fog computing. They have used reinforcement learning to generate the flow of communication and traffic light control at each intersection point in order to avoid traffic jams. These flows are generated by four workflow components of their system: (1) Fog node

TABLE 3.9
Fog Platforms, Frameworks, and Architectures

	Monitoring	Fault Tolerance	Resource Selection and Allocation	Deployment	Dynamic Resource Scheduling	Heterogeneity	Mobility	Scalability	Security and Privacy
Fog Torch	✓	✗	✓	✓	✓	✗	✗	✗	✗
MePaaS		✗	✓	✓	✗	✗	✗	✗	✗
Hybrid Fog computing	✗	✗	✓	✓	✓	✗	✗	✗	✗
Emu Fog	✗	✗	✓	✓	✓	✗	✗	✓	✗
Container-based	✓	✗	✓	✓	✗	✗	✗	✗	✗
Raspberry Pi clusters									

TABLE 3.10
Fog Platform and Frameworks for Smart City Applications

Platform/ Frameworks	Purpose	Algorithms	Applications/ Case Studies
IoT big data analytics platform	Provides recommendations to the house owners by analyzing their smart home data	• Frequent pattern mining • Cluster mining	• Smart energy • Health care • Commercial advertisement
IoT-aware student-centric stress monitoring	Predicts the stress level of the students in a particular context	Temporal Dynamic Bayesian Network (TDBN) and Bayesian Belief Network (BBN)	m-health
Smart health system framework	Collects and predicts information about the cause of the spread of epidemic diseases and suggests preventive measures	Fuzzy K-Nearest Neighbour (FKNN) method	S-health

control, (2) vehicle information, (3) traffic Cloud control, and (4) traffic condition information.

The vehicle information component collects and sends the vehicle information, such as destination and speed, to the Fog node control flow. The traffic condition video flow component collects data from intersection cameras to the Fog node control flow. This collects all the data that have been sent by these components and sends to the Cloud. The traffic Cloud control flow component collects the information from all the Fog nodes and applies a reinforcement algorithm, generating the traffic light control flow and sending it back to its respective Fog node control flow component.

3.8 FUTURE RESEARCH DIRECTIONS IN FOG COMPUTING PLATFORM

Due to the limited resources of the Fog environment, building an effective platform for resource management and distribution of Fog devices, such as smartphones and Raspberry Pis are required to improve the performance of IoT applications. Hence, managing the heterogeneity of devices and logical resources, such as virtual machines, micro-elements, and containers with security, is a complicated task. The various open challenges and research directions in this area are listed next, based on the examination of existing systems' requirements for effective Fog platforms which were discussed earlier.

Resource provisioning: The platform should estimate and provision the Fog device resources based on user requirements. The Fog controller can do this provisioning which is the crucial component of the Fog computing platform. However, the challenges are that resource provisioning should be done with more accuracy and minimal overheads because the service that the Fog controller is running has limited storage, compute, and network resources. Moreover, due to the mobility of Fog nodes in Fog computing, the system should also consider and handle the mobility of devices when it is estimating and provisioning; this is a complicated task.

Resource allocation: The platform should allocate the resources fairly by considering the application's QoS requirements based on the resource estimation information. The requirements of the application or end-user may change dynamically Naha et al. (2020). Handling those dynamic requirements by satisfying the QoS requirements is a challenging task.

Storage management: Fog devices have limited storage capabilities. Hence, once the data is processed in the Fog device, the processed data and input is stored in the Cloud for future purposes, and the data in the Fog device is removed. Managing the offloading and uploading of data to and from the Cloud is also a complex task. Moreover, to process the large data in the Fog computing, the data must be portioned and stored in multiple peer Fog devices. Abstracting the storage of multiple Fog nodes as a single system storage image, by handling fault tolerance, is another complex task.

Power/energy management: In Fog computing, Fog devices such as smartphones and laptops are battery-powered and do not have rich energy. As a result, the devices may turn off while the applications are running. Hence, scheduling algorithms should consider and predict the energy of the device before allocating the resources to the application, a complex task.

Network management: Fog devices must support wired or wireless network connectivity to communicate with the other sensors, Fog devices, and the Cloud. Ensuring network connectivity and distribution of the network to serve the demands of the end-user, considering the failure of the applications, is a complex task, due to the lack of network coverage in a few locations. In the literature, researchers used SDN to distribute and manage the network resources of Fog devices.

Security: In the Fog computing environment, autonomous end devices participate in processing the IoT applications to serve the requests of the user. Due to distributed ownership and autonomy, these devices do not guarantee the security and privacy of the data. Moreover, due to the openness of the environment, this network is exposed to more attacks and vulnerabilities. With the limitations of Fog devices, with respect to the processing, storage, and network, attackers can initiate attacks such as session hijacking, denial of service, and man-in-the-middle. Moreover, identifying malicious devices and malware running in those devices is a critical task. Traditionally, PKI-based authentication schemes are used to authenticate the Fog resources. However, as the Fog devices are very large and distributed, these techniques are not suitable. Moreover, 5G technology plays a crucial role in a Fog computing environment to achieve low latency. The traditional (3G and 4G) authentication and security protocols are not suitable for the 5G environment because of the additional authentication required between users and services. So, to address these issues, efficient authentication and authorization techniques are needed.

Privacy: In the Fog environment, the data is transmitted from collecting or sensor nodes to a single Fog device or multiple Fog devices in order to process the data and act according to it. Hence, the data that is stored should not be shared with unauthorized devices in the network. However, we cannot control this due to the limited access to the devices. Therefore, achieving privacy is a very complex task. The main challenge in security and privacy is ensuring the trust between the unauthorized and autonomous Fog devices in the network.

Fog-as-a-Service (FaaS): From the aforementioned definition and characteristics, we define Fog-as-a-Service (FaaS) as a new way of delivering services to the customers. Unlike the Cloud, the services in FaaS will be provided with both private and public existing computing infrastructure idle resources at end devices of all scale industries and individuals to meet the requirements of the customers. FaaS should be able to deploy and launch various applications of different vertical marketplaces by efficiently managing the resources to satisfy the requirements of the IoT applications. Efficient resource management can be achieved by abstracting the compute, network, and storage services with virtualization techniques and by providing common API for the developers. Bonomi et al. (2014) specified that multi-tenancy features, such as data isolation and resource isolation, should be supported along with the virtualization of resources. There are other features of the platform which are required to satisfy the requirements of IoT applications, such as user management, auditing and billing, monitoring and logging, disaster recovery, and high availability. FaaS should be implemented in a microservice architecture, rather than in traditional architecture, due to its advantages of scalability and manageability. Building FaaS with all these requirements is an open challenge.

3.9 CONCLUSION

An efficient Fog platform that can meet the present and future needs of IoT applications should provide features such as low latency, efficient resource provisioning, scalability, elasticity, easy deployment methods, and mobility support. Existing Cloud-based platforms are not suitable in the Fog environment because of the massively distributed heterogeneity of devices, high latency, and mobility support. However, there are very few Fog computing platforms existing in the literature to meet the requirements of specific applications. There is no generic platform which supports all types of applications.

REFERENCES

Adlink. 2017. "Prismtech vortex." Accessed 2019-06-12. http://www.prismtech.com/vortex.

Agarwal, Sharad, John Dunagan, Navendu Jain, Stefan Saroiu, Alec Wolman, and Habinder Bhogan. 2010. "Volley: Automated data placement for geo-distributed cloud services." In Proceedings of the 7th USENIX conference on Networked systems design and implementation (NSDI'10). USENIX Association, USA, 2.

Battula, S. K., Garg, S., Montgomery, J., & Kang, B. H. (2019). "An Efficient Resource Monitoring Service for Fog Computing Environments". IEEE Transactions on Services Computing.

Battula, Sudheer Kumar, Saurabh Garg, Ranesh Kumar Naha, Parimala Thulasiraman, and Ruppa Thulasiram. 2019. "A micro-level compensation-based cost model for resource allocation in a fog environment." *Sensors* 19(13):2954.

Bonomi, Flavio, Rodolfo Milito, Preethi Natarajan, and Jiang Zhu. 2014. "Fog computing: A platform for Internet of things and analytics." In *Big Data and Internet of Things: A Roadmap for Smart Environments*, Studies in Computational Intelligence, 546: 169–186. Springer, Cham.

Bonomi, Flavio, Rodolfo Milito, Jiang Zhu, and Sateesh Addepalli. 2012. "Fog computing and its role in the Internet of things." *Proceedings of the First Edition of the MCC Workshop on Mobile Cloud Computing*, 13–16.

Brogi, Antonio, and Stefano Forti. 2017. "QoS-aware deployment of IoT applications through the fog." *IEEE Internet of Things Journal* 4(5):1185–1192.

Chen, Ning, Yu Chen, Yang You, Haibin Ling, Pengpeng Liang, and Roger Zimmermann. 2016. "Dynamic urban surveillance video stream processing using fog computing." *2016 IEEE Second International Conference on Multimedia Big Data (BigMM), 2016.*

Chiang, M., S. Ha, I. Chih-Lin, F. Risso, and T. Zhang. 2017. "Clarifying fog computing and networking: 10 questions and answers." *IEEE Communications Magazine* 55(4):18–20.

Chiang, Mung, and Tao Zhang. 2016. "Fog and IoT: An overview of research opportunities, in *IEEE Internet of Things Journal*, 3(6): 854–864, Dec. 2016.

Cisco. 2015. "LocalGrid fog computing." Accessed 2019-08-02. http://www.localgridtech. com/wp-content/uploads/2015/02/LocalGrid-Fog-Computing-Platform-Datasheet.pdf.

Cisco. 2016a. "Cisco IOx local manager pages and options." Accessed 2019-07-09. http:// www.cisco.com/c/en/us/td/docs/routers/access/800/software/guides/iox/lm/refere nce-guide/1-0/iox_local_manager_ref_guide/ui_reference.html.

Cisco. 2016b. "Parstream." Accessed 2019-07-09. https://www.cisco.com/c/en_intl/obsolete/ analytics-automation-software/cisco-parstream.html.

Cisco. 2017. "Data in motion." Accessed 2019-07-09. https://www.cisco.com/c/m/en_us/sol utions/data-center-virtualization/data-motion.html.

Cloud Standards Customer Council 2015. "Practical guide to platform-as-a-service." Accessed 2019-07-10. http://www.cloud-council.org/CSCC-Practical-Guide-to-PaaS.pdf.

Evans, Dave. 2011. "The Internet of things: How the next evolution of the Internet is changing everything." *CISCO White Paper* 1(2011):1–11.

Fratu, O., C. Pena, R. Craciunescu, and S. Halunga. 2015. "Fog computing system for monitoring mild dementia and COPD patients - romanian case study –." *2015 12th International Conference on Telecommunications in Modern Satellite, Cable and Broadcasting Services (Telsiks)*, 123–128.

Hong, Kirak, David Lillethun, Umakishore Ramachandran, Beate Ottenwälder, and Boris Koldehofe. 2013. "Mobile fog: A programming model for large-scale applications on the Internet of things." *Proceedings of the Second ACM SIGCOMM Workshop on Mobile Cloud Computing, 2013.*

Hou, Xueshi, Yong Li, Min Chen, Di Wu, Depeng Jin, and Sheng Chen. 2016. "Vehicular fog computing: A viewpoint of vehicles as the infrastructures." *IEEE Transactions on Vehicular Technology* 65(6):3860–3873.

Hu, P. F., S. Dhelim, H. S. Ning, and T. Qiu. 2017a. "Survey on fog computing: Architecture, key technologies, applications and open issues." *Journal of Network and Computer Applications* 98:27–42. doi:10.1016/j.jnca.2017.09.002.

Hu, Pengfei, Sahraoui Dhelim, Huansheng Ning, and Tie Qiu. 2017b. Survey on fog comput- ing: Architecture, key technologies, applications and open issues, *Journal of network and computer applications*, 98, 27–42.

IDC. 2019. "The growth in connected IoT devices is expected to generate 79.4ZB of data in 2025, According to a new IDC forecast." Accessed 2019-10-05. https://www.idc.com/ getdoc.jsp?containerId=prUS45213219.

Iorga, Michaela, Larry Feldman, Robert Barton, Michael J. Martin, Nedim S. Goren, and Charif Mahmoudi. 2018. *Fog Computing Conceptual Model*, (No. Special Publication (NIST SP)-500-325).

Kepes, Ben. 2011. "Understanding the cloud computing stack: Saas, paas, iaas" Diversity Limited, 1-17. Accessed 2020-03-26. http://www.etherworks.com.au/index.php?option =com_k2&id=74_6ec28b0c12234c2140cdf0f1c19cf0cb&lang=en&task=download&v iew=item

Liyanage, M., C. Chang, and S. N. Srirama. 2016. "mePaaS: Mobile-embedded platform as a service for distributing fog computing to edge nodes." *2016 17th International Conference on Parallel and Distributed Computing, Applications and Technologies (Pdcat)*:73–80. doi:10.1109/Pdcat.2016.29.

Madsen, Henrik, Bernard Burtschy, G. Albeanu, and F. L. Popentiu-Vladicescu. 2013. "Reliability in the utility computing era: Towards reliable fog computing." *Systems, Signals and Image Processing (IWSSIP), 2013 20th International Conference on, 2013*.

Masip-Bruin, X., E. Marin-Tordera, A. Alonso, and J. Garcia. 2016. "Fog-to-cloud Computing (F2C): The key technology enabler for dependable e-health services deployment." *2016 15th Ifip Mediterranean Ad Hoc Networking Workshop (Med-Hoc-Net 2016)*.

Mayer, Ruben, Leon Graser, Harshit Gupta, Enrique Saurez, and Umakishore Ramachandran. 2017. "EmuFog: Extensible and scalable emulation of large-scale fog computing infrastructures." *arXiv preprint arXiv:1709.07563*.

Naha, R. K., S. Garg, D. Georgakopoulos, P. P. Jayaraman, L. X. Gao, Y. Xiang, and R. Ranjan. 2018. "Fog computing: Survey of trends, architectures, requirements, and research directions." *IEEE Access* 6:47980–48009. doi:10.1109/Access.2018.2866491.

Naha, R. K., Garg, S., Chan, A., and Battula, S. K. (2020). "Deadline-based dynamic resource allocation and provisioning algorithms in Fog-Cloud environment". *Future Generation Computer Systems*, *104*, 131–141.

Ning, Z. L., J. Huang, and X. J. Wang. 2019. "Vehicular fog computing: Enabling real-time traffic management for smart cities." *IEEE Wireless Communications* 26(1):87–93. doi:10.1109/Mwc.2019.1700441.

Pahl, Claus, Sven Helmer, Lorenzo Miori, Julian Sanin, and Brian Lee. 2016. "A container-based edge cloud PaaS architecture based on Raspberry Pi clusters." *Future Internet of Things and Cloud Workshops (FiCloudW), IEEE International Conference on, 2016*.

Perera, C., Y. R. Qin, J. C. Estrella, S. Reiff-Marganiec, and A. V. Vasilakos. 2017. "Fog computing for sustainable smart cities: A survey." *Acm Computing Surveys* 50(3). doi:Artn 3210.1145/3057266.

Prazeres, Cássio, and Martin Serrano. 2016. "Soft-iot: Self-organizing fog of things." *Advanced Information Networking and Applications Workshops (WAINA), 2016 30th International Conference on, 2016*.

Rani, S., S. H. Ahmed, and S. C. Shah. 2019. "Smart health: A novel paradigm to control the Chickungunya virus." *IEEE Internet of Things Journal* 6(2):1306–1311. doi:10.1109/Jiot.2018.2802898.

Rayamajhi, Anjan, Mizanur Rahman, Manveen Kaur, Jianwei Liu, Mashrur Chowdhury, Hongxin Hu, Jerome McClendon, Kuang-Ching Wang, Abhimanyu Gosain, and Jim Martin. 2017. "ThinGs in a fog: System illustration with connected vehicles." *Vehicular Technology Conference (VTC Spring), 2017 IEEE 85th, 2017*.

Sarkar, S., and S. Misra. 2016. "Theoretical modelling of fog computing: A green computing paradigm to support IoT applications." *Iet Networks* 5(2):23–29. doi:10.1049/iet-net.2015.0034.

Sheng, Bo, Qun Li, and Weizhen Mao. 2006. "Data storage placement in sensor networks." *Proceedings of the 7th ACM International Symposium on Mobile Ad Hoc Networking and Computing, 2006*.

Silva, C. A., G. S. Aquino, S. R. M. Melo, and D. J. B. Egidio. 2019. "A fog computing-based architecture for medical records management." *Wireless Communications & Mobile Computing*. doi:Artn 1968960 10.1155/2019/1968960.

Sookhak, Mehdi, F. Richard Yu, Ying He, Hamid Talebian, Nader Sohrabi Safa, Nan Zhao, Muhammad Khurram Khan, and Neeraj Kumar. 2017. "Fog vehicular computing: Augmentation of fog computing using vehicular cloud computing." *IEEE Vehicular Technology Magazine* 12(3):55–64.

Stantchev, Vladimir, Ahmed Barnawi, Sarfaraz Ghulam, Johannes Schubert, and Gerrit Tamm. 2015. "Smart items, fog and cloud computing as enablers of servitization in healthcare." *Sensors & Transducers* 185(2):121–121.

Steiner, W., and S. Poledna. 2016. "Fog computing as enabler for the industrial Internet of things." *Elektrotechnik Und Informationstechnik* 133(7):310–314. doi:10.1007/s00502-016-0438-2.

Ujjwal, K. C., Saurabh Garg, James Hilton, Jagannath Aryal, and Nicholas Forbes-Smith. 2019. "Cloud Computing in natural hazard modeling systems: Current research trends and future directions." *International Journal of Disaster Risk Reduction*: 38:101188.

Vaquero, L. M., and L. Rodero-Merino. 2014. "Finding your way in the definition fog: Towards a comprehensive of fog computing." *Acm Sigcomm Computer Communication Review* 44(5):27–32. doi:10.1145/2677046.2677052.

Venckauskas, A., N. Morkevicius, V. Jukavicius, R. Damasevicius, J. Toldinas, and S. Grigaliunas. 2019. "An edge-fog secure self-authenticable data transfer protocol." *Sensors* 19(16). doi:ARTN 3612 10.3390/s19163612.

Verma, P., and S. K. Sood. 2019. "A comprehensive framework for student stress monitoring in fog-cloud IoT environment: m-health perspective." *Medical & Biological Engineering & Computing* 57(1):231–244. doi:10.1007/s11517-018-1877-1.

Wu, Q., J. Shen, B. B. Yong, J. Q. Wu, F. C. Li, J. Q. Wang, and Q. G. Zhou. 2019. "Smart fog based workflow for traffic control networks." *Future Generation Computer Systems-the International Journal of Escience* 97:825–835. doi:10.1016/j.future.2019.02.058.

Yangui, Sami, Pradeep Ravindran, Ons Bibani, Roch H. Glitho, Nejib Ben Hadj-Alouane, Monique J. Morrow, and Paul A. Polakos. 2016. "A platform as-a-service for hybrid cloud/fog environments." *Local and Metropolitan Area Networks (LANMAN), 2016 IEEE International Symposium on, 2016*.

Yassine, A., S. Singh, M. S. Hossain, and G. Muhammad. 2019. "IoT big data analytics for smart homes with fog and cloud computing." *Future Generation Computer Systems-the International Journal of Escience* 91:563–573. doi:10.1016/j.future.2018.08.040.

Yi, Shanhe, Zijiang Hao, Zhengrui Qin, and Qun Li. 2015. "Fog computing: Platform and applications." *2015 Third IEEE Workshop on Hot Topics in Web Systems and Technologies (HotWeb)*:73–78.

Yi, Shanhe, Cheng Li, and Qun Li. 2015. "A survey of fog computing: Concepts, applications and issues." *Proceedings of the 2015 Workshop on Mobile Big Data - Mobidata* 15:37–42.

4 IoT-Based Smart Vehicle Security and Safety System

Asis Kumar Tripathy

CONTENTS

4.1   INTRODUCTION

Before the discovery of the wheel, primitive man would remain secluded from other groups and communities. They could commute only within walking distance. The discovery of the wheel entirely evolved the early man life. His social boundary also grew with time. With passing time, primitive man evolved to a mannered, civilized individual and refined the design of the wheel. With the advent of technology, transportation has become an indispensable part of our lives. Though it has countless advantages and uses, we have to deal with the major problem it brings with it that costs human life. Statistically, according to Ministry of Statistics and Programme Implementation, there were 114 million motor vehicles registered in India in the year 2009 and 159 million in the year 2012. The data provided by *Delhi Statistical Hand Book* clearly indicates the rise in the number of registered motor vehicles from 534,000 to 877,000 in the year 2014–2016 thus increasing the number of accidents and in turn the causalities associated with the surge. Data collected by the National Crime Bureau and Ministry of Road Transport and Highway revealed that in the year 2013 more than 100,000 people lost their lives in road rage. Despite the efforts of awareness campaigns, road signs, and traffic rules, motor accidents accounted for 83% of total traffic-related bereavements in the year 2015 as published by IndiaSpend.

4.1.1   MOTIVATION

The Internet of Things (IoT) is making human life easy in all aspects. The applications it offers are beyond comprehension. IoT is an abstract idea, a notion which interconnects all devices, tools, and gadgets over the Internet to enable these devices to communicate with one another. It utilizes information technology, network technology, and embedded technology. Various sensors and tracking devices are coupled to deliver the desired outcome thus making lives easier. IoT finds application in various areas, such as intelligent cars and their safety, security, navigation, and efficient fuel consumption. This project puts forth a solution to achieve the desired outcome of saving precious human lives that are lost to road crashes. In the proposed system, we are designing and deploying a system that not only avoids accidents but also to take action accordingly.

4.1.2   AIM OF THE WORK

This research aims at dealing with the issues that cause fatal crashes and also integrates measures to ensure safety. Life without transportation is impossible to imagine; it makes far off places easy to reach and greatly reduces the travel time. But the problems which surface due to the ever-increasing number of vehicles on the road cannot be ignored. The project aims to eradicate a few of the major reasons of car crashes and also aims to integrate post-crash measures. The reasons for automotive accidents focused here in this project are

- Nonchalant attitude towards the use of seat belts.
- Driving under the influence of alcohol.
- Distracted driving due to drowsiness.

The post-accident measure incorporated in the project is

- Intimation to the near and dear ones of the occurrence.

4.1.3 OBJECTIVES

The proposed project aims to achieve the following:

- Switch on the ignition only if the seat belts are locked in.
- Deploy a gas senor to make sure that driver is not drunk. If the driver is not drunk, only then will the engine ignite.
- To ensure the driver is not drowsy, eye-blink sensors are deployed in the automobile.
- To circumvent a crash, a proximity sensor is deployed to discover the interruption in front of the automobile on the path.
- To ensure post-crash safety an alert system is deployed which makes use of a GPS system to attain the geographical location of the crashed vehicle and it is sent to a responsible and authorized individual. The accident is detected with the use of a vibration sensor.

4.1.4 CHAPTER ORGANIZATION

In the first section, an introduction has been provided to the whole project. All the fundamentals have been presented in which key modules of the project have been explained like the aims and objectives of the implemented system along with the motivation for choosing this project title. The second section of this project report reviews the literature surveys that have been performed to provide the basis for the implementation that is being performed. Equivalent and competing approaches that exist and have been worked upon are examined, recorded, and contrasted with the techniques and methods being implemented in this chapter. These methods are further verified for any dichotomy that may be prevalent in their system. Further, methods are integrated to overcome those gaps.

Starting from the third section, the technical aspect of the project is addressed. The basic framework and architecture of the methods are incorporated that will be realized in the building of smart vehicle safety and security systems. This is explained with the help of text and diagrams and flow charts. This helps in the step-by-step visualization and organization of the project.

The methodology of the project implementation is studied in greater depths in Section 4.4. The technique is further listed and demonstrated by finding and classifying the functional and non-functional necessities that are to be met in the formation of this chapter. Point-by-point software and hardware constraints and requirements to be met to accomplish the obvious building guidelines are further enrolled and comprehended in detail.

The final section comprises of the conclusions. The same has been used to supply the basis for the brief of what has been done and further work scope of the project is discussed to provide a summary.

4.2 LITERATURE SURVEY

4.2.1 SURVEY OF THE EXISTING MODELS/WORK

For Pannu et al. [1], the emphasis is on making a monocular vision, self-sufficient auto model utilizing Raspberry Pi as a handling chip. A high-definition camera alongside an ultrasonic sensor was utilized to give fundamental information from this present reality to the automobile. The automobile is fit for achieving the given goal securely and insightfully in this manner avoiding the danger of human mistakes. Numerous current calculations like path identification and impediment location are consolidated to give vital control to the auto. The chapter undertakes the implementation of the system using Raspberry Pi, by the ethicalness of its processor.

Kumar et al. [2] proposed the design and development of an accelerometer-based system for driver safety. This framework is structured by using Raspberry Pi (ARM11) for quickly accessing the control and accelerometer for event discovery. If any event occurs the message is sent to the authorized personnel so they can take quick and immediate response to save the lives and abate the harms. The system only incorporates one module ignoring the other fatal causes thus making the proposed model incompetent and incomplete.

Sumit et al. [3] proposed a compelling strategy for the crash evasion arrangement of a vehicle to identify the hindrances present in the front and blind spot of the vehicle. The driver is alarmed with the help of a buzzer and an LED sign, as the distance between vehicle and obstacle reduces and is reflected on a display board. The ultrasonic sensor identifies the state of the object if it is moving or is stationary with respect to the vehicle. This system is valuable for discovering vehicles, bicycles, motorcycles, and pedestrians that cross by the lateral side of the automobile. The paper executes the proposed system using Raspberry Pi as the microcomputer but it limits out-of-the-box performance.

Mohamad et al. [4] proposed a proficient vehicle collision aversion framework inserted with an alcohol detector. This system has the capability of making the driver alert regarding the amount of alcohol consumed and depicting the same on an LCD screen. In addition it generates a warning using a buzzer to make the driver mindful of his or her own particular situation and to flag others in the encompassing zone. The security segment proposed by this framework is the driver in an unusually abnormal state of tipsiness isn't allowed to drive an automobile as the start framework will be shut down. This method works in a way to intimidate the driver about his own condition, which is ironic because the person won't be mindful to take any action against it. The idea is novel but practically it is not workable.

4.2.2 SUMMARY/GAPS IDENTIFIED IN THE SURVEY

The current system showcases a mechanism for receiving the geographical coordinates of the automobile during a crash. This existent framework additionally provides a means of discovery of pre-crash with an object. But it does not target on the intensions that cause these fatal accidents. It does not focus on the crashes that are caused by drunk driving with the help of an alcohol/gas sensor and neither the negligence of use of seat belts.

Also these framework don't guarantee if the driver is wide awake or feeling drowsy. There is no use of eye-blink sensor for the same reason. Additionally, the current framework requires manual involvement. However, the proposed framework works on the shortcomings of the current work and is completely mechanized.

4.3 OVERVIEW OF THE PROPOSED SYSTEM

4.3.1 INTRODUCTION AND RELATED CONCEPTS

The proposed system utilizes an embedded system based on the Internet of Things and the Global System for Mobile Communication (GSM). To avoid an accident, when the system is initiated, the seat belt is checked using a pressure sensor. If the driver is not wearing a seat belt, the engine is turned off. Then the alcohol sensor comes into play and checks for alcohol consumption, and if positive the engine is turned off. After these two main tasks, three things – tiredness, collision, and obstacles – are checked using the eye-blink sensor, vibration sensor, and the infrared (IR) sensor, respectively. If there is a collision and the vibration sensor is active, then there is a message sent to the contact mentioned. If there is any obstacle present, then the buzzer beeps to tell the driver. If the driver is feeling sleepy or drowsy, the eye-blink sensor detects it and switches off the engine.

4.3.2 FRAMEWORK AND ARCHITECTURE/MODULE FOR THE PROPOSED SYSTEM

1. The system utilizes GSM technology for the communication of code pattern to transmit location coordinates.
2. The system is Arduino Uno based.
3. The system should be able to communicate even from physically far off distances.
4. The system uses an IR sensor, vibration sensor, alcohol sensor, eye-blink sensor, and pressure sensor.
5. To practically put together all the components and execute them, the composition of various sensing devices in our system is as shown in Figure 4.1.

4.3.3 PROPOSED SYSTEM MODEL

The software development system model that best suits this project and aligns itself with the needs of the given project is the Agile development model. Figure 4.2 is a block diagram, depicting the steps concerned in implementing the Agile development model.

In the Agile development model the entire requirement set is broken into numerous builds (Figure 4.3 and Figure 4.4). Various development stages take place here, making the development cycle a "multi-step waterfall" cycle. Cycles are split into tinier portions, making the modules easier to manage and implement. Every module goes through the planning, requirements analysis, design, implementation or building, and testing stages. A running version of the system is delivered at the end of the first iteration, so we get a working model early on during the product development

FIGURE 4.1   System design implementation.

cycle. Each iteration releases a model with added modules integrating more functions to the last release. This process goes on until the complete system is developed. These iterations are repeated in a loop till an end version of the system is refined and is the expected outcome is obtained.

- As the project deploys a real-time checking and a monitoring system, the outputs produced are further used to take the necessary actions and are thus fed back to the code to give an appropriate action for the further events.
- This process is redundant and cyclic in nature which is implemented whenever a driver enters the automobile.

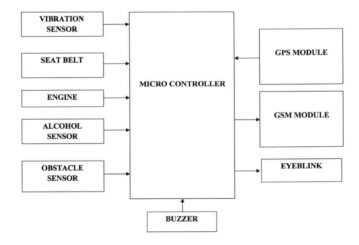

FIGURE 4.2   Block diagram.

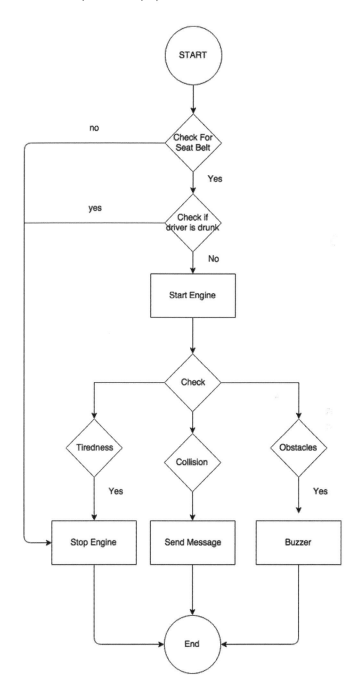

FIGURE 4.3 UML diagram (activity diagram).

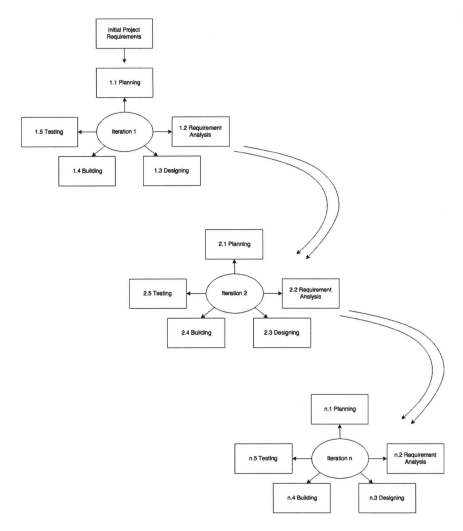

FIGURE 4.4 Agile model.

4.4 PROPOSED SYSTEM ANALYSIS AND DESIGN

4.4.1 REQUIREMENT ANALYSIS

4.4.1.1 Functional Requirements

The arrangement of the idea to implement the functional requirements is elucidated under the heading system design. The arrangement of the idea to implement the non-functional requirements is elucidated in the system architecture section of this project report.

The imperative functional requirements of this project's objective accomplishment are:

1. The automotive system should have the capability to determine whether the seat belt is put on or not by the driver.

2. The implemented system should have the capability to determine whether alcohol has been consumed by the driver or not.
3. The automotive system should have the capability to determine the mental awareness of the driver in terms of if he is feeling sleepy.
4. The automotive system should have the capability to check whether the vehicle is not coming too close to the vehicle in front.
5. The automotive system should have the capability to determine whether an accident has already taken place and thus should have the capability of sending the location coordinates of accident to a responsible person with the help of GSM technology.

4.4.1.1.1   Product Perspective

This project deals with problems which cause accidents and attempts to ensure safety. This project addresses various reasons that lead to fatal accidents. Roads are unpredictable and at every turn of the road can be fatal accidents present and one cannot rely on the driving sense of other drivers and the pedestrians. One needs to be self-aware of the environment and the vehicles around. The driver should take all the precautions and be mindful of the people on the road as well because every life has value. Common reasons for accidents are the lack of concentration of the driver on the road because of some distraction or because of lack of sleep of the driver.

4.4.1.1.2   Product Features

The product aims to provide the following functions:

1. Accident Detection And Driver Safety: There is a vibration sensor inside the vehicle, and if there is any impact on the vehicle it immediately initiates the GPRS (General Packet Radio Service) and the GSM module.
2. Pedestrian and vehicle safety: There is an IR sensor placed on the front of the vehicle which is used to check for any obstacle present.
3. Alcohol detection: The alcohol sensor comes into play and checks for alcohol consumption and if alcohol is detected the engine is turned off.
4. Sleep detection: The eye-blink sensor is placed in a transparent glass which is used to check if the driver is not feeling drowsy or sleepy.
5. Seat belt mechanism: The pressure sensor checks for the seat belt lock and once confirmed starts the engine.

4.4.1.1.3   User Characteristics

The following user characteristics are put into action in the development of the system:

• External forces should not result in the damaging of the system.
• The framework must be able to explicitly identify and discover problems related to the components.
• The issue detected should be reported back to the system.

4.4.1.1.4   Assumptions and Dependencies

The assumptions and dependencies are established in the beginning itself to give us a lucid understanding of the implementation of the product:

- Need of an appropriate GPS module to deliver exact geographical location coordinates.
- The driver should be wearing the spectacles eye gear integrated with the eye-blink sensor.
- The system should always be connected to Internet.
- Proper placement of numerous proximity sensors can be added.

4.4.1.1.5   Domain Requirements

- The system must be reachable for and should accept the data.
- The system must renew the database each and every time a crash occurs or whenever the driver is drowsy.
- The system must be equipped to reach the authoritative person in the case of a crash from anywhere and in any condition.
- The system must renew the database each and every time a driver sits in the car and the seat belt status should thus be updated.

4.4.1.1.6   User Requirements

- Basic operational information is expected from the user of the system.
- Users must have knowledge of the basic working of a phone to read and send texts and make calls.
- Users must be able to have a network connectivity in order to receive text messages in case of a crash.
- The sensors used in the system must be functional and should be able to receive and thus record their corresponding values.
- The system should be able to send processed values via the Internet.
- The system as well as the users both should be aware of the technology and the functionalities offered by it so that the system can be completely exploited.

4.4.1.2   Non-Functional Requirements

In frameworks designing and prerequisites building, a non-useful necessity is a necessity that determines criteria that can be utilized to judge the task of a framework instead of particular practices. They stand out from practical necessities that characterize a particular conduct or capacity. The arrangement for executing useful necessities is point by point in the framework plan.

The course of action of the plan to execute the non-practical prerequisites is clarified under framework design, as they are for the most part critical in term of architectural requirements. The arrangement of the idea to implement the functional requirements is elucidated in the design section of this project report.

Some imperative functional requirements of this project's objective accomplishments are accounted as follows:

1. The sensor value status should be updated in real time.
2. The hardware that will be used for this project is Arduino Uno and various sensors.
3. The software that will be used for this project is Arduino IDE.
4. The project will be developed on a Mac system.

4.4.2 PRODUCT REQUIREMENTS

Product requirements apply every one of the prerequisites to a specific item. They are composed to enable individuals to comprehend what an item ought to do. The Smart Vehicle Security and Safety System should, in any case, for the most part abstain from envisioning or characterizing how the item will perform with a specific end goal to later permit interface fashioners and designers to utilize their aptitude to give the ideal answer for the necessities.

4.4.2.1 Efficiency

The framework is productive, i.e. both hardware and the product give fast outcomes. The inserted C code composed for the hardware has complexity of O(n) and a space complexity of O(n) also. This gives a speedy outcomes and the SIM GPRS 800A modem just takes 30 seconds to send information to the website page.

4.4.2.2 Reliability

The framework is vigorous and is dependable with regards to creating exact outcomes and yield. Every one of the sensors and transfers should work fine in vigorous conditions, i.e. at the point when the sources of info are quickly changing and when the information sources are gradually changing also. In particular the framework is giving correct esteems in every one of the conditions.

4.4.2.3 Portability

The structure is compact and it is achievable for a customer to present it in a desired place without a great deal of extra effort and a considerable amount of issues. The thing which is to be considered is the affiliations shouldn't be free or else the yield will be broken.

4.4.3 OPERATIONAL REQUIREMENTS

Economic

For components such as Arduino Uno, sensors are frugal and cheap to buy. From an economic perspective the price of buying the hardware components is low and the software available is free and open source. Ultimately, the implementation of this project will have little impact on the budget of the user.

- The trade-off experienced here is accuracy is compromised over cost.
- The building and deployment of the Smart Vehicle Security and Safety System is not expensive.

Environmental
- The device poses no threat to the environment.
- The device does not take in nor release any harmful substance to the environment.

Social
- The device has a big social impact, as it will help in preventing road accidents. Moreover, it will also save many lives.
- People will be able to circumvent road fines.

Ethical
- The device is installed by the user.
- It is an open source system that is available to everyone, charging a nominal license fee.
- There is no private companies or third-party companies involved.

Health and safety
- The Smart Vehicle Security and Safety System plays an important role in preventing road accidents and saving lives of people.
- The Smart Vehicle Security and Safety System is completely safe to use without causing any harm to the user.
- This project will ensure no health issues are caused.

Sustainability
This framework could streamline the sustainable design process and lead to more integrated infrastructure delivery. It will dependably have extent of change which makes it considerably more maintainable. The module is exceptionally easy to understand interface and does not require additional preparation for utilization.

Legality
- The Smart Vehicle Security and Safety System is engineered to comply to all legal safeguards.
- All programming will be a standard permit issue and not a pilfered duplicate.

Inspectability
- There is no need for inspection, as there will be no change in the Smart Vehicle Security and Safety System as long as there is no change in the initial idea.
- The is a one-time installation effort after which software can be used effortlessly.
- If by any chance the device gives wrong data, it will be inspected as soon as possible to not add a wrong error into the database.

4.4.4 System Requirements

4.4.4.1 Hardware Requirements

The output of the Smart Vehicle Security and Safety System is heavily dependent on Android application. The hardware components utilized for the project titled Smart Vehicle Security and Safety System are as follows.

Arduino Uno board: The project utilizes Arduino Uno as the microcontroller. All the sensor components are attached and soldered to this microcontroller board and the microcontroller then takes the input and computes to give an appropriate output (Figure 4.5).

Global vibration sensor: This project utilizes the vibration sensor to sense the accident and the crash of the automobile. This input received by the sensor is given to the microcontroller Arduino Uno board that further utilizes the input to give a specific output (Figure 4.6).

Alcohol/gas sensor: This project utilizes the alcohol sensor to sense the alcohol content in breath. This input received by the sensor is given to the microcontroller Arduino Uno board that further utilizes the input to give a specific output (Figure 4.7).

Eye-blink sensor: This project utilizes the eye-blink sensor to sense the tiredness of the driver. This input received by the sensor is given to the microcontroller Arduino Uno board that further utilizes the input to give a specific output (Figure 4.8).

Buzzer: This project utilizes the buzzer that signals and alerts the driver and the surroundings. The output is sent to the buzzer by the Arduino Uno board according to the computation (Figure 4.9).

GPS module: This project utilizes the GPS module to track the coordinates of the location of the where the project is present. This input received by the sensor is given

FIGURE 4.5 Arduino Uno.

FIGURE 4.6   Vibration sensor.

to the microcontroller Arduino Uno board that further utilizes the input to give a specific output (Figure 4.10).

GSM module: This project utilizes the GSM module to communicate the coordinates of the location as detected by the GPS module. This input received by the sensor is given to the microcontroller Arduino Uno board that further utilizes the input to give a specific output (Figure 4.11).

This instructional exercise will disclose how to interface a GSM modem with Toradex modules (Figure 4.12).

FIGURE 4.7   Gas sensor.

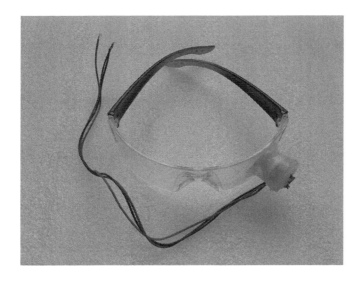

FIGURE 4.8   Eye-blink sensor.

FIGURE 4.9   Buzzer.

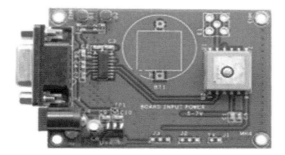

FIGURE 4.10   GPS module.

FIGURE 4.11   GSM module.

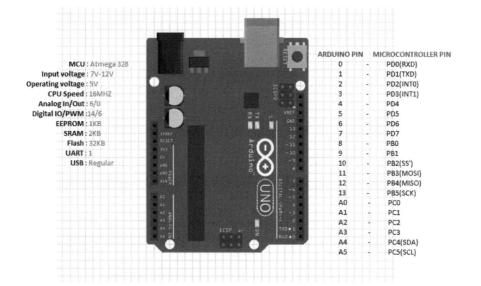

		ARDUINO PIN		MICROCONTROLLER PIN
MCU : Atmega 328		0	-	PD0(RXD)
Input voltage : 7V–12V		1	-	PD1(TXD)
Operating voltage : 5V		2	-	PD2(INT0)
CPU Speed : 16MHZ		3	-	PD3(INT1)
Analog In/Out : 6/0		4	-	PD4
Digital IO/PWM : 14/6		5	-	PD5
EEPROM : 1KB		6	-	PD6
SRAM : 2KB		7	-	PD7
Flash : 32KB		8	-	PB0
UART : 1		9	-	PB1
USB : Regular		10	-	PB2(SS')
		11	-	PB3(MOSI)
		12	-	PB4(MISO)
		13	-	PB5(SCK)
		A0	-	PC0
		A1	-	PC1
		A2	-	PC2
		A3	-	PC3
		A4	-	PC4(SDA)
		A5	-	PC5(SCL)

FIGURE 4.12   Arduino mapping.

4.5   RESULTS AND DISCUSSION

4.5.1   Experimental Results

The experimental results show that the proposed model gives us a better result as compared with other available devices. The output of the force sensitivity sensor is shown in Figure 4.13. This figure shows that the output provides more values as per the increase in time. Figure 4.14 shows the serial monitor of the vibration sensor.

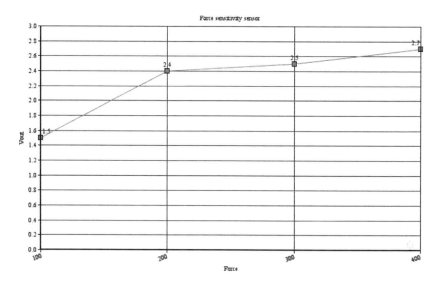

FIGURE 4.13   Force sensitivity sensor.

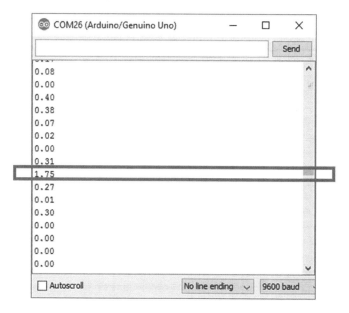

FIGURE 4.14   Serial monitor for vibration sensor.

Whereas, Figure 4.15 and Figure 4.16 display the percentage of crashes due to fatigue and causes of crashes, respectively.

4.5.2   FINAL OUTPUT OF THE RESEARCH AND CONCLUSION

A competent Smart Vehicle Security and Safety System integrated with a pressure sensor, eye-blink sensor, alcohol sensor, proximity sensor, and vibration sensor

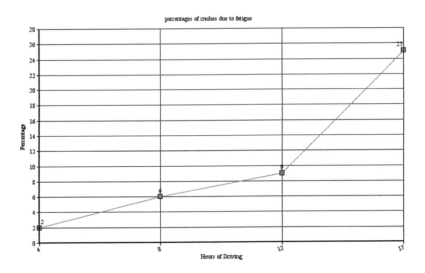

FIGURE 4.15   Percentage of crashes due to fatigue.

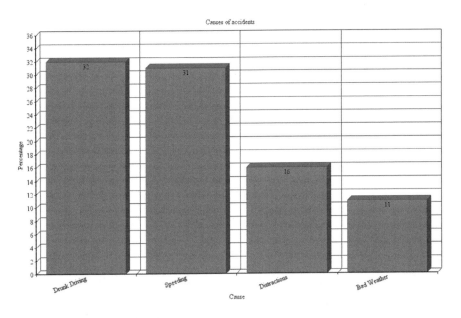

FIGURE 4.16   Causes of crashes.

(Figure 4.17) using the concept of GPS and GSM has been implemented. The sensors are integrated with the Arduino board. Areas with outreach problems that experience bad network connectivity or in remote areas with no network connectivity available can be an issue. This can in turn lead to the accident intimidation text not being sent to the specified number. The proposed and thus implemented system can be enhanced and modified by adding concepts of technology such as big data and

FIGURE 4.17 Hardware prototype.

GPS to study the thus collected data to understand and read the patterns associated with the crashes. The same system can be modified accordingly and implemented for two wheelers. Further, the location of the crash can be sent to an ambulance as well for quick medical response and attention.

REFERENCES

1. Pannu, Gurjashan Singh, Mohammad Dawud Ansari, and Pritha Gupta. "Design and implementation of autonomous car using Raspberry Pi." *International Journal of Computer Applications* 113, no. 9 (2015): 22–29.
2. Kumar, V. Naveen, V. Sagar Reddy, and L. Padma Sree. "Design and development of accelerometer based system for driver safety." International Journal of Science, Engineering and Technology Research (IJSETR), 3(12), (2014).
3. Garethiya, Sumit, Lohit Ujjainiya, and Vaidehi Dudhwadkar. "Predictive vehicle collision avoidance system using raspberry-pi." *ARPN Journal of Engineering and Applied Sciences* 10, no. 8 (2015): 3656–3659.

4. Mohamad, Mas Haslinda, Mohd Amin Bin Hasanuddin, and Mohd Hafizzie Bin Ramli. "Vehicle accident prevention system embedded with alcohol detector." *International Journal of Review in Electronics & Communication Engineering (IJRECE)* 1, no. 4 (2013).

5. Miesenberger, K. et al. (eds.) "Computers 'helping people with special needs', LNCS, vol. 4061 Springer Berlin / Heidelberg." *10th International Conference, ICCHP 2006*, Linz, Austria, July 11–13 (2006).

6. Sanchez, J., and E. Maureira. "Subway mobility assistance tools for blind users." *LNCS* 4397, pp. 386–404 (2007).

7. Jerry, T., H. Goh, and K. Tan. "Accessible bus system: A bluetooth application." *Assistive Technology for Visually Impaired and Blind People*, pp. 363–384 (2008).

8. Noor, M.Z.H., A. Shah, I. Ismail, and M.F. Saaid. "Bus detection device for the blind using RFID application." *5th International Colloquium on Signal Processing & Its Applications*, pp. 247–249 (2009).

9. Quoc, T., M. Kim, H. Lee, and K. Eom. "Wireless sensor network apply for the blind u-bus system." *International Journal of u- and e- Service, Science and Technology* 3, no. 3 (2010): 13–24.

10. Manoj Kumar, R., and Dr. R. Senthil. "Effective control of accidents using routing and tracking system with integrated network of sensors." *International Journal of Advancements in Research & Technology* 2, no. 4 (2013): 69–74.

11. Wang, Shu, Jungwon Min, and Byung K. Yi. "Location based services for mobiles: Technologies and standards." *IEEE International Conference on Communication (ICC)*, Beijing, China (2008).

12. Dai, Jiangpeng Jin Teng, Xiaole Bai, and Zhaohui Shen. "Mobile phone based drunk driving detection pervasive computing technologies for healthcare." *2010, 4th International IEEE Conference*, pp. 1–8, March (2010).

13. Chen, H., Y. Chiang, F. Chang, and H. Wang. "Toward real-time precise point positioning: Differential GPS based on IGS ultra rapid product." *SICE Annual Conference, The Grand Hotel*, Taipei, Taiwan, August 18–21.

14. Kohji Mitsubayashi. "Biochemical gas-sensor (biosniffer) for breath analysis after drinking." *SICE 2004 Annual Conference. Vol. 1. IEEE* (2004).

5 Smart Attendance Monitoring IoT-Based Device Using Cloud Services

Suriya Sundaramoorthy and Gopi Sumanth

CONTENTS

5.1 INTRODUCTION

In our day-to-day lives, the Internet has become of a part of us and there are a lot of things that we want to know and experiment with and of course we need a lot of storage services with a lot security which clearly points to one of the finest technologies of the recent years, Cloud computing. What if we have some application which will make use of both the Internet of Things (IoT) and Cloud computing and which saves a lot of time for us and is also eco-friendly in nature? A lot questions are probably running through your mind. So, let's go into the imperative parts.

5.2 CLOUD

Cloud computing was exquisitely defined by the National Institute of Standards and Technology (NIST):'Cloud computing is a model for enabling ubiquitous, convenient, on-demand network access to a shared pool of configurable computing resources that can be rapidly provisioned and released with minimal management effort or service provider interaction'. Cloud computing has become prevalent due to it being a virtualized resource with a lot of storage capacity, processing expanse, high security, and nature of being low in cost. So, it is very easy and useful for storing and accessing large amounts of data in a budget-friendly way through Cloud computing.

Cloud computing is mainly characterized with its layered architecture and service models. An equivalent service model for the layered architecture is as shown in Figure 5.1. Service models can be mainly divided into three parts and the unique functionalities are related with Cloud computing as follows: Infrastructure as a Service (IaaS), which mainly deals with the storage, processing, and networking; Platform as a Service (PaaS) is liable for maintaining the layers of the platform along

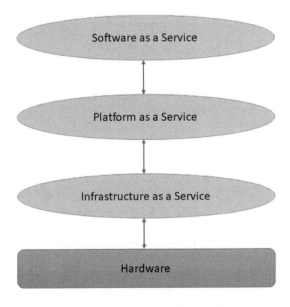

FIGURE 5.1 Layered architecture related with service models.

with the resources; and Software as a Service (SaaS) is at the helm of provisioning the applications.

Depending on the physical availability of resources, Cloud computing can be classified into four major types. A public cloud allows anyone to access public information; a private cloud restricts the accessibility of the Cloud to particular people who owned it; a community Cloud is restricted to a particular community to access the information; and a hybrid Cloud is a combination of any of the three aforementioned architectures.

5.3   SENSORS USING IOT

Internet of Things mainly deals with the Internet and physical devices like sensors and actuators information. Sensors are generally used to make work easy for us by installing them in particular places and collecting the data from the places where we can't get the information easily.

IoT can be described by a layered architecture as shown in Figure 5.2. The perception layer focuses on capturing the data from the sensors and actuators, the network layer deals with the collection and transportation of the captured data from the perception layer, and the application layer will process the data and store it in the Cloud or servers.

Sensors should be installed and maintained in a voltage less than the maximum capacity and they will act as an interaction between the digital world and the physical world. Generally, sensors should always be connected to the electricity and then they will be active and act according to our use case and are also economic in nature.

Takabi et al. [1] has addressed the issues related to security and privacy solutions. Subashini and Kavitha [2] has focused on safety cloud computing. Gubbi et al. [3] has visualized evolution of IoT through sensors and actuators. Wireless Technologies has transformed IoT to fully integrated future Internet. This emerged form of IoT is implemented via Cloud platform Suciu et al. [4] has described the role of distributed

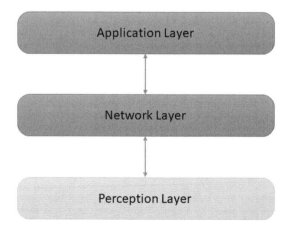

FIGURE 5.2   Internet of Things layered architecture.

nature of cloud computing infrastructures to IoT. The real-time application involving Smart city services are rendered by cloud provisioning effectively. Zhou et al. [5] has focused on CloudThings, to integrate with cloud computing. Soliman et al. [6] has implemented Smart Home applications by integrating Internet of Things and Cloud computing services. The model focuses on sandwiching knowledge into sensors and actuators using Arduino platform, networking using Zigbee technology, Cloud services enabling interactions with smart things, improving data exchange efficiency using JSON data format. Aazam et al. [7,8,9] focuses on the need of integrating IoT and Cloud computing called as "Cloud of Things". The issues that may arise due to this integration is clearly discussed namely protocol support, energy efficiency, resource allocation, identity management, IPv6 deployment, service discovery, quality of service provisioning, location of data storage, security and privacy, and unnecessary communication of data.

5.4 INTEGRATION OF CLOUD AND IOT

There are some specific features of Cloud computing that are required to improve some features of IoT technology. Consider features of Cloud computing like storage capability, services, applications, energy efficiency, and computational capability against the features of IoT, namely smart solutions, renewable smart power grids, remote monitoring of patients, smart home application sensors, and engine monitoring sensors (Table 5.1).

The main reason behind the need for integration of IoT and Cloud technology is developing better solutions in terms of secured and ubiquitous system. The aforementioned features make the convergence towards integration easier. Out of various domains that could benefit from this integration of services, mobile technologies

TABLE 5.1

Mapping Features of Cloud Computing with Features of IoT

	Cloud Services →	Storage	Service	Applications	Energy Efficiency	Computational Capability
IoT Characteristics	Smart Solution	Y	Y	Y	N	Y
	Renewable Smart Power Grids	Y	Y	N	Y	Y
	Remote Monitoring of Patients	N	Y	Y	N	
	Smart Home Application Sensors	Y	Y	Y	Y	Y
	Engine Monitoring Sensors	N	Y	Y	Y	Y

and the health industry stand to benefit first[10]. Other research has focused on voice pathology monitoring of people using IoT and Cloud Services, which has high degree of accuracy in detection[10].

Atlam et al. [11] keenly show the challenges and issues that arise due to the integration of Cloud computing and IoT technologies. The challenges are security, privacy, heterogeneity, big data, performance, legal aspects, monitoring, and the large scale. The issues are standardization, Fog computing, Cloud capabilities, service level agreement (SLA) enforcement, energy efficiency, security, and privacy. Various Cloud–IoT applications [12,13]that could benefit because of this integration are healthcare, smart cities, smart houses, video surveillance, automotive and smart mobility, smart energy and smart grids, smart logistics, and environmental monitoring. Cloud–IoT integration has led to many new service models like SaaS (Sensing as a Service), EaaS (Ethernet as a Service), SAaaS (Sensing and Actuation as a Service), IPMaaS (Identity and Policy Management as a Service), DBaaS (Database as a Service), SEaaS (Sensor Event as a Service), SenaaS (Sensor as a Service), and DaaS (Data as a Service). The various general benefits of Cloud–IoT integration are communication, storage, processing capabilities, and new service models.

When there is loads of data each day and everything has to be stored and maintained constantly, then the perfect solution for the storage and maintenance of the data with good security is Cloud technology. When we want to integrate the two big technologies for better applications, we do have some pros and cons and also some procedures to follow. So, let's see how we can do this.

There are major problems, namely standardization, heterogeneity, context awareness, middleware, IoT node identity, energy management, and fault tolerance with existing cloud based IoT platforms [14]. Bandyopadhyay et al. [15] has addressed heterogeneity issues by enabling effective communication between IoT devices. Chang et al. [16] has focused on technologies and frameworks required for Cloud-based IoT systems and classifications of IoT supporting technologies. Celesti et al. [17] has described the importance of IoT Cloud federation architecture toward new business opportunities.

5.5   CLOUD AND IOT: DRIVERS OF INTEGRATION

IoT devices generally capture information, process it, and then send it to other devices for further manipulations. The Cloud generally stores the information, processes it, and then helps in retrieving the data for users. We have to make sure that the integration is smooth and have to take care that their individual features are not getting affected. Some of the factors that we have to take care of are as shown in Figure 5.3.

5.5.1   PROCESSING SPEED

IoT requires high-end processing speed in order to collect the data continuously, and to process the captured data it should be very fast. Cloud computing requires a high processing speed because of its wide range of accessibility for the stored data. But comparatively, Cloud computing has high processing speed due to its nature to handle large resources.

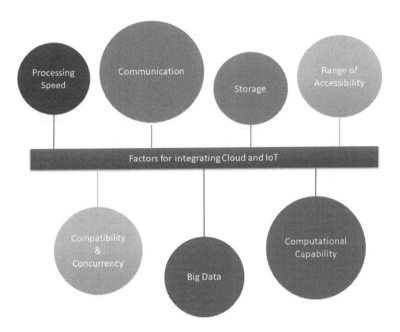

FIGURE 5.3   Factors involving integration of Cloud and IoT.

5.5.2   BIG DATA

The source of data is IoT devices, which collect the data in very large amounts each day. So, IoT has a very large amount of data, but the Cloud does not have data unless something triggers it to store the data. Even though, IoT devices are collecting a lot of data and its incapable nature of storing, there needs to be a technology that stores all the data, which is the Cloud in this case.

5.5.3   RANGE OF ACCESSIBILITY

The Internet of Things has a very limited range of access because of the sensors' capacity to read and process. If we want to increase the range of IoT devices, we have to install more of them. Cloud computing has a wide range of accessibility due to its ubiquitous nature. When integrating, IoT devices will be sending the data to be stored and the Cloud is able to access the data within a very wide range.

5.5.4   COMPATIBILITY AND CONCURRENCY

There will be a lot of IoT devices that will be reading and sending the data to the network gateways. IoT devices only send the information and don't really care about the same data being captured many times. The Cloud will get all the data and during the processing, it should take care of all the issues related to compatibility and concurrency and store it in a way for further processing of data into applications.

5.5.5 COMMUNICATION

IoT devices generally transfer data through a broadband. The Cloud communicates according to its data storage density and processing power. Automation can be used for data collection and storage at a low cost. Due to the low capacity of the broadband, it might not be transferring the data to the same capacity as the Cloud.

5.5.6 STORAGE

The Cloud has a large storage capacity, as it is a virtual resource and can be operated from anywhere. IoT devices have a limited or null storage capacity. So, it is best to integrate both the technologies to overcome the storage crisis in IoT devices.

5.5.7 COMPUTATIONAL CAPABILITY

As IoT is a physical device, it is small in size and has a limited or less computational capability. As Cloud computing is a virtual entity, it has a virtually unlimited computational capability. In order to balance the computational capabilities, we have to integrate IoT with the Cloud for better results in terms of computation.

The drivers for integration of Cloud computing and IoT deals with some of the latest technologies like smart grid, smart home applications, smart solutions, and digital twins, and also deals with all kinds of internal mechanisms, architectures, and classifications involved. These work in a process of collecting the data, aggregating and mapping it with the metadata, and then storing in the Cloud. Drivers for integration mainly involve components as shown in Figure 5.4. These are the main aspects through which the drivers will be streamlined and then the other factors will be supportive for better integration and productivity.

5.6 OPEN ISSUES IN CLOUD-BASED IOT INTEGRATION

The open issues that are encountered during the integration of Cloud-based IoT are standardization, fog computing, cloud capabilities, SLA enforcement, big data, energy efficiency, security, and privacy, as shown in Figure 5.5.

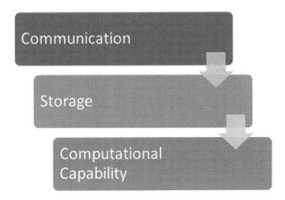

FIGURE 5.4 Drivers for integrating Cloud and IoT.

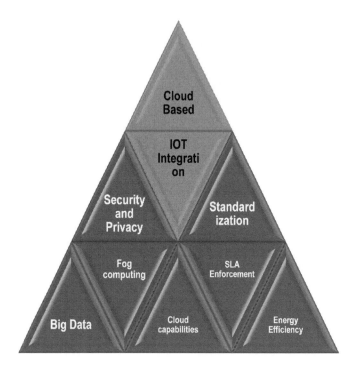

FIGURE 5.5   Open issues in Cloud-based IoT Integration.

Standardization is a critical issue because the lack of standardization affects the performance of a Cloud-based IoT environment. Cloud-based IoT requires interconnection between various IoT devices to enable generation of Cloud services. Hence, standardization plays a key role in defining standards, architectures, and protocols.

Fog computing refers to extension of Cloud services toits edge. It acts as a platform enabling communication between end users and the services they require. It attends to delay over the network, as it is best suited for latency-sensitive applications. It includes features like awareness towards location, edge location, real-time interaction, and mobility.

Cloud capabilities are all about security, since security always remains a major issue with any networking environment. Cloud-based IoT has the threat of attacks from the Cloud and IoT. Threat of attacks of IoT includes attacks related to data integrity, confidentiality, and authenticity.

SLA enforcement is focused towards quality of service in a Cloud-based IoT environment. There is a tendency of Cloud service providers to violate SLAs, which raises issues related to quality of service management.

Energy efficiency represents the issues related with energy consumption caused due to transmission of data from IoT devices to the Cloud. Compression technologies, data transmission, and data caching techniques are used to overcome these energy efficiency issues.

5.7 PLATFORMS

In the present world, integrating IoT devices with Cloud computing isn't an easy task. It is very difficult to properly build a platform like that and also involves huge installation costs as well. So, many industries have created a virtual platform for testing all their applications based on cloud computing and IoT. We can see a lot of paradigms of Cloud computing that support the IoT device information such as SaaS, DaaS, AaaS (Analytics as a Service), XaaS (Everything as a Service), TaaS (Testing as a Service), PaaS, EaaS, and SEaaS. These platforms are so efficient in terms of parameters such as processing speed and time consumption compared to other Cloud servicing platforms. With the use of the virtualized infrastructure, we can customize the platform as per our requirements such as high-level architecture with the use of a lot of open source and commercial platforms. To integrate IoT devices and Cloud computing as a platform, we should have something like an API as a middleware service. The most commonly used platforms are ThingSpeak, IBM BlueMix, Microsoft Azure IoT Hub, Amazon Web Services IoT, Google Cloud's IoT, Cisco IoT Cloud Connect, OpenIoT, and IoT Toolkit.

5.8 OPEN CHALLENGES

Whenever we take an application or technology or whatever it is, it will have a positive side and a negative side, coming to Cloud computing and IoT. We have seen a lot of factors and drivers for integrating Cloud computing and IoT. Now, we will see the main hurdles for integration which we can call open challenges (Figure 5.6).

5.8.1 SECURITY AND PRIVACY

Cloud computing is a virtual kind of storage and if one of the Cloud data centres has been breached, then a lot of data will be lost because the Cloud is distributed system. This will be major concern for the cloud computing. IoT devices will also have issues when the particular device has been tampered with or if it is connected to some device for manipulating the function of the IoT device. If we are integrating them, then we have to take care of the data flow from the IoT device to the Cloud, because through the middleware APIs, the data can be easily stolen and privacy will be lost. This is one of the major challenges in integration.

5.8.2 NETWORK INFRASTRUCTURE

The Cloud's infrastructure is based on its service model and the service model helps us to maintain data easily. Because the Cloud is a distributed system, there is no challenge regarding infrastructure. IoT devices made up of sensors and actuators involve some network infrastructure, and if there is a small change in the infrastructure, then the network becomes a tedious one ; it is better to build a new one instead of fixing the affected one. On integrating both, we might face more problems because the middleware's infrastructure should also be taken care of.

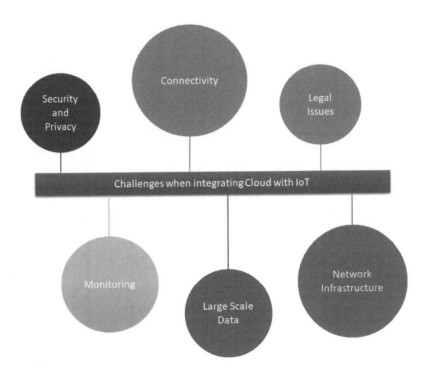

FIGURE 5.6　Open challenges when integrating Cloud with IoT.

5.8.3 Connectivity

The Cloud will be connected to different datacentres across the world and its main advantage is replicating the data across the datacentres, so that if one of the data centres is not working it can connect to another. The data should be safe and it always connected to the Internet for better communication. If an IoT device loses connectivity to the Internet, then the data will be lost. If we lose the data, then the Cloud can't collect or process the data, so the application gives bad performance and results.

5.8.4 Large-Scale Data

The Cloud store large amounts of data across its data centres and IoT devices collect and send a lot of data. As IoT devices can't store data, everything is stored in the Cloud. As we continue storing the data, the storage will keep increasing. After certain period of time, a lot of old data seems to be useless and it won't be worth maintaining huge amounts of data. So scaling the data accordingly has huge maintenance costs and it will be quite a challenge after a certain point of time.

5.8.5 Legal Issues

Cloud computing and IoT devices are still in the development stages, thus no one knows the full practical details of their working. Governments haven't assigned predefined standards for using these technologies, but when integrating them some rules may be created. However, once governments have issued standards, application developers must meet those standards in order to avoid legal issues.

5.8.6 MONITORING

When it comes to monitoring the Cloud and IoT devices, we have to concentrate more on IoT devices because of their wear and tear, managing resources, security, and performance. Though the Cloud just stores the data, we have to monitor whether it is correctly storing the data and according to its service model. In an integrated technology, we have to make sure that the diagnostics, performance, security, and network are maintained and monitored.

These are the major open challenges that are faced with these technologies. We can integrate IoT with the Cloud by overcoming the open challenges, and provide a better application with all the standards and nature of being user-friendly.

5.9   IOT-SUPPORTING TECHNOLOGIES AND CLOUD SERVICE FRAMEWORK

Chang et al. [16] discussed the technologies and framework required for a Cloud-based IoT system. This paper focuses on classification of IoT-supporting technologies into thirteen categories. IoT-enabling technologies are diverse and not interoperable in nature. IoT-supporting technologies are identification technology, IoT architecture technology, communication technology, network technology, network discovery, software and algorithms, hardware, data and signal processing technology, discovery and search engine technologies, relationship network management, power and energy storage technologies, security and privacy technologies, and standardization. A Cloud-based IoT system relies on the Cloud for the processing services required. This brought a strong foundation for a common service framework. This avoided the scenario of implementation of various access methods between different clouds on IoT. Based on IP transport and Session Initiation Protocol, IP Multimedia Subsystem (IMS) has provided a common service framework for Cloud-based IoT systems. The scalability and efficiency of IMS architecture is improved because of Cloud computing technologies. This framework also has few challenges, such as the IoT environment requires redesigning of the Home Subscriber Server (HSS) database schema, the IMS service discovery and search function needs to be improved for the IoT environment, IoT requires an improved original IMS-URI (Universal Resource Identifier) Naming Architecture, and there is a need for improvement of the Service Capability Interaction Manager for service composition and interaction.

5.10   VIRTUALIZATION IN IOT

Virtualization builds services that hide the complexity of the underlying hardware or software. This results in offering guaranteed services with reliability, elasticity, scalability, isolation, and resource optimization. A Cloud-based IoT system relies more on the Linux container virtualization (LCV). This LCV creates containers for different applications in the Linux environment. It is an OS level virtualization technique that helps in creating multiple instances of the typical OS at the user space while sharing the kernel of the host OS. The notable features of LCV are flexibility in setup, configuration, optimization, and management of sensing and actuating capabilities. In general, there are two models for containers: application containers and system containers. An application container is capable of running only a single application inside the container. A system container has flexibility where the container boots an instance of user space. The beauty of this flexibility is simultaneously multiple instances of the user space run inside the container. Each of these instances has its own Init process, process space, file system, and network stack. The best examples of containers are Docker, LXC, lmctfy, and OpenVZ.

Let us discuss about the differences between the traditional hypervisor virtualization technique and Linux container virtualization based on start-up time, dynamic runtime control, speed, isolation, flash memory consumption, communication channels between the virtual and physical environment, dynamic resource assignment or separation, and direct hardware access. Linux container virtualization is favourable with respect to initial start-up time and application start-up time when compared to that of the hypervisor virtualization technique. From the perspective of dynamic runtime control, the hypervisor virtualization technique starts or kills the application in a container through semaphores or virtual network connections or serial connections, whereas Linux container virtualization does it directly from the host. The next comparison parameter is speed: the hypervisor virtualization technique has a reasonably good speed in directly accessing virtual machines because Linux container virtualization accesses the virtual machines through the same driver running on the host system. Latency and throughput are affected by either Linux container virtualization overhead or hypervisor virtualization overhead depending on the scenario under consideration. LCV does not support this concept of isolation on the system level, but isolation has an effective role in the case of hypervisor virtualization since it supports isolation at both the system level and user level. The flash memory consumption aspect is realized in the case of Linux container virtualization by proving its speciality in sharing both the operating system kernel and user space. But hypervisor virtualization puts forth the constraint of storing the virtual machine images separately and does not allows sharing at all. The unity of both Linux container virtualization and hypervisor virtualization is observed in case of a communication channel between virtual and physical environments as they permit communication via serial or networking interfaces. At situations it is also achieved using shared file systems with semaphores. Load management and fail over of a system rely more on dynamic resource assignment or separation. It focuses on assigning additional CPUs to heavily loaded virtual machines or removing CPUs from idle virtual machines or containers. Hypervisor virtualization permits direct access to hardware peripherals from the virtual machines, whereas this is not supported in the Linux container virtualization technique.

5.11   PROBLEMS OF THE EXISTING ATTENDANCE MONITORING SYSTEM

Till now, we have been discussing the technologies that we are going to use and the factors, challenges, and drivers for integrating them. Now, we will talk about the application of attendance monitoring. First, we will discuss some of the native methods of attendance monitoring.

In olden days, when there were fewer people who went to school, teachers noticed when students did not show up. Teachers would directly go to the students' homes to ask them about their absence. They followed this kind of procedure because of the low number of students and the students 'homes were often within the vicinity of the school.

However as time progress, people slowly realized the importance of education and started to join schools and institutions and the number of people joining educational institutions has increased. Also many students travelled from a long distance. Thus the previous system of teachers travelling to students' homes to enquire about absences was no longer feasible. So teachers started maintaining books for attendance for each class. Nowadays, attendance has become a major requirement in some schools and colleges, and students should maintain a minimum percentage of absences. It has become tough for teachers to maintain attendance records without making some errors, even if done so electronically or online. This system is currently followed not only in most educational institutions but also in a lot of private and government offices.

Some colleges have changed this system to a biometric attendance system, which processes attendance based on the eye scan or fingerprint of the student, and then directly processed to the Internet. Some corporate offices use employee access cards to keep track of employees' comings and goings. However, when there are a lot of people waiting in long queues for scanning of access cards or for biometrics, it consumes a lot of time and will be very difficult to keep attendance daily. So, it was not instituted, and most institutions follow the attendance book system. In order to get rid of the aforementioned problems, our Smart Attendance application will be useful and will take attendance of everyone without making any errors and without wasting a lot of time. So, let's see the process step by step.

5.12   HARDWARE SUPPORT FOR SMART ATTENDANCE

We have seen the problems with the current system and now we have to know about the working of the Smart Attendance application; we have to know the hardware and software parts of the system. First, we will learn about the hardware.

When it comes to hardware, the main things that run in our minds are about the installation like where to install, how to install and how to make that work. Along with these, hardware also means that we will have some installation costs as well. The wear and tear of hardware will be relatively less and should make proper use of the hardware.

In the process of making ID cards, we need a proximity sensor radiator, which will generate the signals for the receiver to capture. We also need a circuit that will produce a constant result throughout the card's lifetime. The constant result will help identify the single person from many in a group and then it produces an easy way in the software processing point of view. Before installing the sensor between the magnetic strips of the ID card, we have to test specific information to make sure that it will be assigned to a unique person.

Magnetic strips are mainly useful for transporting the signal and they also contain the encoded number which is the same as that being written on the sensor. The magnetic strips will be deactivated if there is more than one ID card in the nearby vicinity. (This point will be clearly explained later.) We can see the design of the sensor getting integrated with the ID card in Figure 5.7. When we have installed the sensors, we have to make sure that the sensors are properly installed in the ID cards and in the same way we have to make sure of the performance, reliability, and other related factors. The magnetic strips will have some printing on the cards in order to directly identify the ID card.

We have seen the hardware part of generating or sending the information through ID cards. We will need a device that can capture the information generated by each ID card. For this, we will be installing a proximity sensor receiver which will have a limited range of access and can capture only a limited number of data for each second. We will be installing the receivers in a classroom depending on the number of people and the area of the room. These receivers will be connected to the Internet in the background, and each piece of data they receive will be transferred to the storage device for processing of the data.

The hardware part will be the most challenging part because of costs and the circuits involved. The ID cards will be lightweight and just like a credit card or debit card. The receivers will have a camera which will capture the data and transfer the data. The relationship between the receiver and radiator should be very proper and should be very highly secured in order to safeguard from the loss of data or tampering of the data.

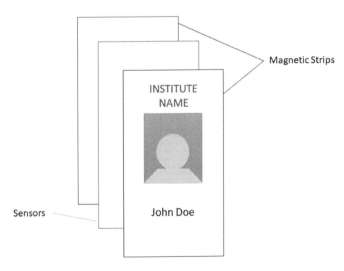

FIGURE 5.7   Sensor integration with ID card design.

5.13 SOFTWARE SUPPORT FOR SMART ATTENDANCE

We have described the hardware setup of the application, now we will cover the software setup that is required for this application. Unlike hardware there won't be much cost involved in software, and there is no need of any circuits. All we need here is storage in the Cloud and a platform to process the data which will be given to the Cloud. We will now review the processing of the data captured and the functionalities included in this application.

After the unique numbers have been assigned to the sensor, that unique number should be mapped to one of the persons in the organization, and the person's details should be printed on the magnetic strips of the ID card. When the ID cards are in the vicinity of the proximity sensor receivers, they will start capturing the data and sending it to receivers. Receivers will capture the data and then it will be transferred to the Cloud through the broadband. The Cloud will store the data and process the data at the same time.

We need to check whether all the data is entered correctly by mapping the unique ID card number with the correct person. The entire process and control flow of the software process is shown in Figure 5.8. All the required data that is to be used in the software process is collected by the proximity sensor receivers.

Processing of the data includes the time for which the student is inside the specified room of the institution such that the software checks the time for which the student is inside the room and will that student is eligible to get the attendance for that particular time. The start and the end times should be fixed by the teacher who manages all the other students in the classroom. According to the timings, it will be known whether the student has spent the minimum amount of time in the class to be credited for attendance.

For better accuracy of the attendance, the teacher who manages the class should give the confirmation of students who are present by checking attendance for the processed data whether it is processed correctly or not.

After all the processing of the data, the data will be updated and stored in a separate place in the Cloud for future use. The processed attendance will also be updated directly into the student's attendance portal. Because of this updating, parents can continuously monitor their children.

The biggest hurdle here is the data that is collected will have a lot of duplicate entries because more than one receiver will be capturing the data. In order to resolve this issue, we have to remove all the duplicates and then sort the data according to the unique ID card number and sort it according to the timestamp for easy calculation of the duration involved and assign the attendance. If more than one ID card is

FIGURE 5.8 Control flow of software process in attendance monitoring.

in the same vicinity, then all the cards will be deactivated and they have to wait till activation of the card ;until then they won't be processed for attendance.

This is how the software is designed and handled by the input coming from the hardware devices. This might have some common problems in designing and they will be further discussed in a later section.

5.14   ARCHITECTURE OF ATTENDANCE MONITORING SYSTEM

The hardware and software parts of the attendance monitoring system have been analyzed separately, and in order to get the working application, we need to integrate the hardware and software parts. IoT technology is used in the initial phase of collecting data, and after collection of data, the Cloud computing technology comes into the picture for the further part of the application. This can be easily explained through an architecture as shown in Figure 5.9.

The Internet of Things physical devices are the first part involved and they have the important task of collecting the data regarding to ID cards. Receivers will be reading the ID card continuously and will be transferring the data to the other Cloud computing technology. Cloud computing technology will be getting all the data and will process all the information and then update in each and every place accordingly. Cloud computing will be an added advantage because of its virtual storage and we can store a very large amount of data.

Proximity sensors are used in IoT devices because of their usefulness to the application in such a way that they will be having a feature of collecting the data wirelessly from a stipulated amount of distance. In the same way, each proximity sensor receiver will be able to read data of only a limited amount of devices at a particular second. So, we have to install the receivers according to the density of people and the area of the room. There will be some devices setup in the college in order to check

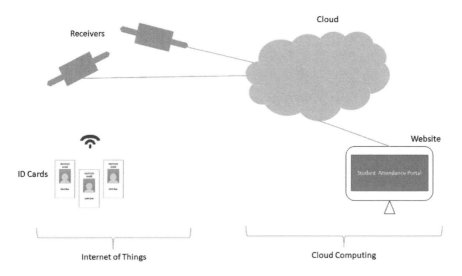

FIGURE 5.9   Attendance monitoring system architecture.

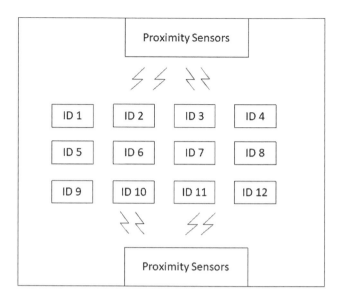

FIGURE 5.10   Proximity sensor receivers collecting data from ID cards.

whether cards are working, and if the card is not working or lost, we can directly report to the college and get a new ID card. The example setup of the sensors is shown in Figure 5.10.

If there is any problem in the attendance updated on the website, the student has to contact the teacher and then report the issue to administration to resolve the problem. There will always be a person who will be managing the data between sensors and the Cloud, such that there will be no flaws and the integration of the technologies will be smooth and easy.

The main problem will be safety and security of the data because the data can be easily tampered when transferring the data to the Cloud, or the sensor information present in the card can be breached and can be made to malfunction. The data stored in the Cloud is not so secure because only some Cloud technologies have good security features. So for the safety and security of the data, we will have to buy the Cloud and install proper security features. The data collected from the sensors will be stored in the Cloud for only one year and after that the data will be dumped because we can conserve storage and in the same way the sensor readings will not be useful in any way, just the consolidated attendance will be fine.

With the help of the Smart Attendance Monitoring System, we can put a full stop to the original methods of writing it in an attendance register. By using this system, we can save paper and a lot of trees can be safeguarded as well. The only thing that is required in this application is some installation cost for receivers and some cost for making the ID cards for every student in the class. Software costs will be less than hardware costs, because for software there will be some resources to get access to the Cloud and broadband for capturing the data of the sensors.

This is how the Smart Attendance Monitoring System is making use of the two vast technologies and integrating them for the purpose of the application, and even though there are significant hurdles and difficulties, we can easily overcome them and make the application work with much ease. There might be some difficulty in making ID cards because they are the most used and through parts, so in order to avoid them from the unique sensor number, they can be mapped to ID cards for new persons and we can reuse the data for another person. By doing this way we can reduce economic costs and reduce e-waste.

5.15   CHALLENGES FACED WITH SMART ATTENDANCE

The main challenge we can face in the colleges is proxy. Using this Smart Attendance application, there will be no chance of such scenarios. There are also some other concerns:

- What if a single person wears more than one card at a time?
  - If two ID cards are in contact with each other, they both will become deactivated. This happens due to the magnetic strips present on the manufacture of ID cards, such that if two magnetic cards come in close contact, then the cards will become deactivated.
- What if we keep the identity cards in bags or desks?
  - If that's the case, then the receivers won't be receiving the signals from the identity cards because of the obstacles occurring between the radiators and receiver. So, one person cannot get attendance for more than one person.

REFERENCES

1. Takabi, Hassan, and James B. D.Joshi. "Security and privacy challenges in cloud computing environments." *IEEE Computer And Reliability Societies*, pp. 24–31, 2010.
2. Subashini, S., and V.Kavitha. "A survey on security issues in service delivery models of cloud computing." *Journal of Network and Computer Applications*, Volume no. 34, Issue 1, pp. 1–11, 2010.
3. Gubbi, Jayavardhana et al, "Internet of Things (IoT): A vision, architectural elements, and future directions." *Future Generation Computer Systems*, Volume 29, Issue 7, pp. 1645–1660, 2013.
4. Suciu, George et al, "Smart cities built on resilient cloud computing and secure Internet of things." 19th International Conference on Control Systems and Computer Science, Bucharest, 2013.
5. Zhou, Jiehan et al, "CloudThings: A common architecture for integrating the Internet of things with cloud computing." Huazhong University of Science and Technology, Wuhan, In *Proceedings of the 2013 IEEE 17th International Conference on Computer Supported Cooperative Work in Design (CSCWD)*, 2013.
6. Soliman, Moataz et al, "Smart home: Integrating Internet of things with web services and cloud computing." IEEE International Conference on Cloud Computing Technology and Science, Oulu, 2013.
7. Aazam, Mohammad et al, "Cloud of things: Integrating Internet of things and cloud computing and the issues involved." 11th International Bhurban Conference on Applied Sciences & Technology (IBCAST), Islamabad, 2014.

8. Aazam, Mohammad et al, "Smart gateway based communication for cloud of things." IEEE 9th International Conference on Intelligent Sensors, Sensor Networks and Information Processing (ISSNIP)Symposium on Public Internet of Things, Singapore, 2014.

9. Aazam, Mohammad et al, "Cloud of things: Integration of IoT with cloud computing."Springer International Publishing, Book "Robots and Sensor Clouds" pp. 77–94, 2016.

10. Muhammad, Ghulam, SK Md Mizanur Rahman, Abdulhameed Alelaiwi, and Atif Alamri. "Smart health solution integrating IoT and Cloud: A case study of voice pathology monitoring." *IEEE Communications Magazine*, Volume 55, Issue 1, pp. 69–73, 2017.

11. Atlam, Hany F., AhmedAlenezi, Abdulrahman Alharthi, Robert J. Walters, and Gary B. Wills. "Integration of cloud computing with Internet of things: Challenges and open issues." IEEE International Conference on Internet of Things (iThings) and IEEE Green Computing and Communications (GreenCom) and IEEE Cyber, Physical and Social Computing (CPSCom) and IEEE Smart Data (SmartData), pp. 670–675, June 2017.

12. LeTu'n, A., H. N. M. Quoc, M. Serrano, M. Hauswirth, J. Soldatos, T. Papaioannou, and K. Aberer. "Global sensor modeling and constrained application methods enabling cloud-based open space smart services." Ubiquitous Intelligence & Computing and 9th International Conference on Autonomic & Trusted Computing (UIC/ATC), pp. 196–203, 2012.

13. Irwin, David, Navin Sharma, Prashant Shenoy, and Michael Zink. "Towards a virtualized sensing environment." International Conference on Testbeds and Research Infrastructures, Springer, Berlin, Heidelberg, pp. 133–142, 2010.

14. Ray, Partha Pratim. "A survey of IoT cloud platforms." *Future Computing and Informatics Journal* 1, no. 1–2, pp. 35–46, 2016.

15. Bandyopadhyay, D., and J. Sen. "Internet of things: Applications and challenges in technology and standardization." *Wireless Personal Communications*58, no. 1, pp. 49–69, 2011.

16. Chang, K. D., C. Y. Chen, J. L. Chen and H. C. Chao. "Internet of things and cloud computing for future Internet." International Conference on Security-Enriched Urban Computing and Smart Grid, Springer, Berlin, Heidelberg, pp. 1–10, September, 2011.

17. Celesti, A., M. Fazio, M. Giacobbe, A. Puliafito, and M. Villari. "Characterizing cloud federation in IoT." 30th IEEE International Conference on Advanced Information Networking and Applications Workshops (WAINA), pp. 93–98, March 2016.

WEB REFERENCES

1. https://www.webopedia.com/TERM/C/cloud_computing.html
2. https://www.researchgate.net/publication/283236612_Integration_of_Cloud_comp uting_and_Internet_of_Things_A_survey
3. https://www.researchgate.net/publication/317585066_Integration_of_Cloud_Com puting_with_Internet_of_Things_Challenges_and_Open_Issues

6 Encryption of Data in Cloud-Based Industrial IoT Devices

Ambika N.

CONTENTS

6.1 INTRODUCTION

The Internet of Things (IoT)has become a necessity in today's world. The technology aims to provide a common platform. These devices of different capabilities can communicate with each other. The devices communicate through the Internet as its medium. These devices can be located and addressable. Availing this facility, the data can be accessed from anywhere and anytime. Many applications (Babar et al. 2011) like healthcare (C. Doukas, et al. 2012) (Doukas and Maglogiannis 2012; Bhatt, Dey, and Ashour 2017), smart homes (Arabo 2014; Harper 2006), industry (Huang and Sun 2018) and agriculture (TongKe 2013; Mukhopadhyay 2012) use these devices. As the usage of them is increasing, the data storage is also becoming bulkier day-by-day. This is where cloud computing is providing its solution. The technology is aiding IoT in providing huge data storage (Jiang et al. 2014).

Cloud computing is featured with computing, services, and applications over the Internet. The integration (Hasan, Hossain, and Khan 2015) of technology of the cloud with devices requires some modifications. These modifications are in terms of storage, energy (Bahrami, Khan, and Singhal 2016), computational power, and context awareness (Arabo 2014). Some features of the cloud are:

- Storing over the Internet requires the Transmission Control Protocol (Rojviboonchai and Aida 2004) or Internet Protocol (Huitema 1998) platform. The servers and storage devices are linked to bring the technology into play.
- Services are provided to customers with abundant efficiency and speed.
- The applications are made to run by executing the programme through the Internet.
- Managing the energy and restraining the growth of the same is a challenge. Delivering the services to the same energy input or the same services to decrease energy input can serve the purpose. The system is found to be computationally capable if it meets the requirements.

Singh et al. (2016) have suggested security conditions to be considered when integrating IoT with the cloud. The conditions are as follows:

- Secure communication aids in preventing eavesdropping (Zh, et al. 2011) or data leakage and data from corruption. The work suggests using Transport Layer Security (TLS) to bring in security. TLS employs a certificate-based model.
- Access controls (Wang, Cardo, and Guan 2005) govern the actions taken on the objects. They can vary from accessing data to issuing a query to performing a computation. Authentication (Lee, Hwang, and Liao 2006) and authorization are two stages that avail rights to the users.
- The data being communicated contains some sensitive information. Confidentiality has to be maintained in the system. Using applications that provide location information will help in providing better security. The cloud architecture avails being public, private, and hybrid. The customer should understand the requirement of the system before availing the service.
- The cloud offers data sharing. The customer should avail the facility after going through the policies offered. Personal data should be assured to be protected and should be shared on user acceptance.
- To protect the data from misuse, the users have to be availed the service of encryption (Boneh, Dan and Franklin 2001). The customers should be availed the option to authorize, revoke, and issue a new set of keys.
- The application taking care of personal credentials should raise some security concerns (Narayanan and Shmatikov 2010). The concerns relate to the trade-offs between the benefits and revealing data. The functional benefits of combining data should not reveal the credentials.
- An issue related to identity concerns has to be addressed. The IoT provider should have complete details of the user belongings. Trusted platform models (Morris 2011) aid in the process.
- A manipulation, creation, or update has to identify the provider. The cloud services should be accountable for the coordination and distribution of things.

- The cloud is designed to provide support to store the data. The scalability is one of the prior properties that the cloud should possess.
- The cloud providers should ensure law and enforcement in the system. The mentioned rules should be obliged by all customers.
- The cloud is into sharing when the users authorize it to do so. It also provides various types of access and controls to its users. This has to be looked after even in case of different kinds of attacks.
- The cloud has its policies to preserve integrity (Krutz and Vines 2010) to the data. The invalid things are disowned by the system.

The Internet of Things is used in industry setups. They provide a platform to increase efficiency and accelerate the procedure. IIoT (Hossain and Muhammad 2016) with cloud technology connects machines, controllers, and drivers. New emerging technologies, remote controlling, and automatic production are the boons to IIoT-cloud. The technology suffers from many security issues (Roman, Zhou, and Lopez 2013). Huang and Sun (2018) explain the security issues. The work suggests the analytic hierarchy process. This is a structured methodology for decision-making problems. Planning, designing, and risk assessing (Zhao et al. 2005; Singh et al. 2016; Narayanan and Shmatikov 2010) are some of the problems encountered. The risk encountered can breach confidentiality, integrity, and availability to data. As the devices are wireless, the security issues (Roman, Zhou, and Lopez 2013; Aazam et al. 2014; Liu et al. 2015) are large. Backdoor (Diksha and Shubham 2006), key cracking (Dagon, Lee, and Lipton 2005), Sybil, DoS, Monitor (Chowdhury and Panda 2017), and access right attacks are found in these networks. Hence, providing a solution against them becomes essential.

The proposed work aids in bringing better security to the system. The system uses location keys to encrypt its data. The AES (Advanced Encryption Standard) and RSA algorithms are used in the work. The sensor's location is concatenated with the location of the object. This key is used to encrypt the data using the AES algorithm. A hash code (Maurer 1968) is derived from the concatenated key. This hash code is encrypted using the user location key. The user will be able to generate private and public keys using his location details. The public key is shared with the cloud. The approach aids in bringing reliability (Bauer and Adams 2012) to the system. It also minimizes attacks.

6.2   LITERATURE SURVEY

Security becomes essential for wireless devices (Carcelle, Dang, and Devic 2006). The IoT-cloud interface is used to store and retrieve data as a convenience. The confidential data being transmitted requires high security. Authentication and encryption (Mao, et al. 2016) are some of the measures taken to safeguard data. Authentication being a prevention measure verifies the communicating parties. Cryptography (Koblitz, Neal and Menezes 2005) is a methodology adopted to secure data transmitted. In this section different attempts (Alassaf, Alkazemi, and Gutub 2017) towards securing the interface are narrated.

Stergiou et al. (2018) considered a wireless sensor setup with a source–destination pair. The assumptions also consist of trusted relays and eavesdroppers. The study considers only the source message to be confidential. The eavesdropper channel, cooperative protocol, source encoding, and decoding scheme are assumed to be public. The cooperative protocol is limited to forwarding after decoding and forwarding after amplifying. The source broadcasts the encoded message using the first transmission slot. Trusted relays use the second slot for its transmission. It decodes the message, recodes it, and cooperatively transmits. Using the weighted version of the recoded symbol, eavesdropping is detected. The RSA algorithm is used in assistance with the trusted platform module.

A secure monitoring framework is proposed by Muhammad et al. (2018). The methodology is an energy-effective methodology. Video sum-up and image encryption are combined to bring security to the system. The video summarization methodology is utilized to obtain informative frames. The image encryption methodology is suggested that is map employed and cryptosystem derived. The cypher is derived from the embedded random bits of plain images. These images are undistinguished from the accidental noise. The processing capabilities of visual sensors are used to obtain the frames. After detecting an event, an alert message is sent to the respective authorities. The final decision is made based on many factors including the extracted keyframes and their alterations during some attacks.

Belguith, Kaaniche, and Russello (2018) have designed a lightweight attribute-based encryption. The scheme supports bandwidth-limited applications. PU-ABE is a new variant key attribute providing its assistance in the work. The system aims to access policy updates efficiently. The two communicating parties are the cloud and the data owners. The access policies can be updated without revealing secret keys among them. The encrypting entity generates the cyphertext. This is attached to the encrypted data and transmitted to the cloud. The cloud server will update upon receiving the data. The system ensures to preserve privacy with good access control to data that is outsourced. The cyphertext remains to be of constant size. It also remains independent of other attributes.

The paper by Rahman, Daud, and Mohamad 2016) considers a five-layer approach to the ecosystem (Prentice 2014). IoT sensor nodes and devices make up the first layer. Communication and network make up the second layer. Layer three is made up of computing and storage. Applications and services are inlayer four. Analytics make up layer five. A secure framework is suggested in this work.

The authentication system model is suggested by Barreto et al. (2015). Different user cases are discussed in the work. The work explains how the work is effective in the provider/service provider model.

Two kinds of requests are considered in the work. Direct users request to IoT devices and user requests performed by the cloud are considered. The model supports single-sign-on and single-logout tasks. Two types of authentication are proposed. Basic user authentication is capable of accessing services and resources through the cloud. The IoT devices are transparent to their users. The authentication phase is the relying party in the first phase. This approach eases the users' access to the cloud. The authentication on the IoT device is performed in the second phase. This step is performed on behalf of the basic user. The second kind of authentication considered in the work is advanced user authentication. In this approach, the admin user, cloud platform, and manufacturer can directly access the IoT device.

Cloud-centric multi-level authentication is proposed by Burton et al. (2016). It addresses scalability and time constraints. The model uses the hierarchical approach. The device can be carried or worn by public safety respondents. It offloads continuous authentication. It is an enhanced work of two-level authentication (I. W. Butun 2012). In the work, the authors have to authenticate through the cloud service provider. The user, wearable node, wearable network coordinator and cloud service provider are four entities in the model. The scheme runs in three stages: initialization, registration, and authentication. In the initialization phase, key agreement and distribution takes place. The system adopts the elliptical curve cryptography. This algorithm is used for generation of the signature digitally and for scrutinization. The verification is done between the users and the cloud service provider. The elliptical curve digital signature is used during the registration phase. The algorithm is used to swap the secret message authorization code. The scheme also uses the elliptic curve Diffie-Hellman key exchange algorithm. This algorithm is used between the cloud service provider and users. Wearable devices (Di Rienzo et al. 2006) are accessed through wearable network coordination after the authentication phase.

The work by Gehrmann and Abdelraheem (2016) tackles the problem of securing distributed IoT devices. A novel scheme is designed to tackle the problem. The machine is highly available and attack resistant. The IoT device is designed with a suitable high-level application model. This model aids in execution of the system. The activities are reflected in a cloud-executed machine. The machine accepts all directly addressed issues. The system hinders any direct communication with the IoT devices. A dedicated synchronization protocol is used to communicate with the machine. The state information and state alterations are transmitted using this protocol. The IoT device initiates the synchronization procedure. The device is given the privilege to turn off the entire network interfaces.

Chandu et al. (2017)suggested using hybrid encryption mechanism. The algorithm uses the AES encryption (Rijmen and Daemen 2001)algorithm and the RSA algorithm (Agrawal 2010). The data gathered by the sensor is provided to the microcontroller. The device hardware is AES encryption supported. This data is cyphered using the AES encryption algorithm. The key to the encryption is saved for later use. This encrypted data with the key is fed to the cloud. The cloud provides security to the received message. The transmitter is provided with a password to access the objects. If any device other than the transmitter requests the data, a message is sent to the transmitter. The transmitter uses the public key of the requester and encrypts the key of the data. This message is shared with the requester. The requester can decrypt the key using his private key. Using the decrypted key, the data is availed by the requester.

Bokefodea et al. (2016)considered the role-based scenario. The access control is provided based on the roles the individual holds or is assigned. Both AES and RSA algorithms are used in the work. The data stocked in the cloud is encrypted using a key. The secret key used for encryption is created using the AES algorithm. This credential is used to encrypt the data. For every role, a public and private key is generated using the RSA algorithm. The public key is used to encrypt the secret key by the cloud. The user raises a request for the data. The user uses the private key to decrypt the secret key. This key is used to decrypt the data provided by the cloud.

The Message Queue Telemetry Transport protocol is suggested by Singh et al. (2016). The algorithm is a derivative of key/cyphertext policy-attribute-based encryption methodology. The lightweight elliptic curve cryptography is used in the work. In the suggestion, the transmitter encrypts the data based on a set of terms. These terms give access to the policy. The receiver will be able to decrypt the received message if it satisfies the access policy. The user attributes are used to make the access policy. The algorithm commences with the public key generator (PKG) making the master secret key and the access policy. The devices in the system have to register themselves with the public key generator. During this procedure, the attributes provided by the devices are verified by the PKG. The generator transmits the public parameters to the respective device. These parameters are used by the device for encryption. The private key is transmitted to the receiver. This key is used to decrypt the attributes encapsulated in the access policy.

A lightweight encryption methodology (Alassaf, Alkazemi, and Gutub 2017) is suggested by Al Salami et al. (2016). The scheme applies to smart homes. It promises confidentiality service without an increase in overhead with reference to computation and communication. The proposal is an identity-based scheme (Bao, Hou, and Choo 2016) with a combination of the Diffie-Hellman encryption mechanism (Kocher 1996). The system supports fast random encryption standard. Two algorithms are used in the proposal. One is used to encrypt the session key. The second one is used to encrypt the messages using the selected key.

Rahulamathavan et al. (2017) work supports confidentiality and access control. The system adopts a single encryption methodology. Four entities are involved in the procedure: data owners/cluster heads, blockchain miners, attribute authorities, and distributed ledgers. The system generates the private and public keys using the attributes (Ambrosin et al. 2016). The miners interact with some key-issuing scheme. The respective authority dispatches the decrypting material to the miners. The cluster head is responsible to aggregate the data from the various sensors. The data undergo encryption before transmitting them to the respective ones. The encryption done can be verified by the blockchain miners using appropriate attributes. Using the right decryption key by the miners will provide them the sensor readings. The blockchain is appended in the transaction to ensure only the right individuals use the encrypted data. The system aids in providing data privacy through fine-grained access control.

A multi-attribute authority attribute methodology is proposed by Belguith, Kaaniche, Laurent, et al. 2018). The methodology is multi-folded. The CP-ABE scheme (Lewko and Waters 2011) is utilized in the work. The system delegates a part of the decryption phase to the cloud server. The scheme provides provisions for the adversary to query for the secret keys. The challenger in the methodology commences by initializing an empty set and a table. The adversary is provisioned to query the secret keys related to the set of attributes. The authority acknowledges back with the corresponding secret keys. During the transformation key query stage, the adversary is provisioned to query the secret keys. The challenger acknowledges with the keys if it finds the keys during the search operation. If not available it runs the algorithm to generate a set of transformation keys. The methodology supports a hidden access policy. The system proves to be feasible, secure, and privacy-preserving. Homographic encryption is used in the work of Talpur, Bhuiyan, and Wang (2015).

The methodology is used for health monitoring. The services are shared among the users to reduce the cost. An Internet Protocol–based methodology is proposed to secure the system. The system can configure the system dynamically. The combined methodology is used in the proposed work – single-user node and shared nodes. The single-user node behaves as the identity node in the system. It validates the user access in the network. It also is responsible to validate the data leaving the network. The shared node validation is also the responsibility of the single node.

The AES (Feldhofer, Dominikus, and Wolkerstorfer 2004) and generic algorithms (Masood, Rattanawong, and Iovenitti 2003) are used by Aljawarneh and Yassein (2017) to increase security. The proposed system cuts down the key distribution and key updation procedure. The keys are generated utilizing the data in the proposed work. The input file is read in a multi-size block during the encryption procedure. The block is divided into two divisions – plaintext and key. The key is encrypted using the Feistel encryption methodology (Bellare, Hoang, and Tessaro 2016). The plaintext is encrypted using AES. The encrypted text and key are further aggregated using the generic algorithm. Table 6.1.

6.3   PRELIMINARIES

Chandu et al. (2017)suggested using the hybrid encryption mechanism. AES encryption (Rijmen and Daemen 2001)and the RSA algorithm (Agrawal 2010)are used to bring security to the data. The data collected by the sensors are provided to the microcontroller. The device hardware is AES encryption supported. This data is encrypted using the AES encryption algorithm. The key to the encryption is saved for later use. This encrypted data with the key is fed to the cloud. The cloud provides security to the received message. The transmitter is provided with a password to access the objects. If any device other than the transmitter requests for the data, a message is sent to the transmitter. The transmitter uses the public key of the requester and encrypts the key of the data. This message is shared with the requester. The requester will be able to decrypt the key using his private key. Using the decrypted key the data is availed by the requester.

Let T_i be the transmitter and R_i be the receiver. Let D_i be the data to be transmitted. Let K_A be the key used by the AES algorithm to encrypt them. Let D be the concatenation of D_i and key K_A. Let K_R be the key used by RSA algorithm to encrypt the key K_A. From Equation 6.1, the transmitter T_i is encrypting data. D_i is the data to be transmitted to the requester R_i. K_A is the encryption key used. After encryption the encrypted data and key K_A are concatenated.

$$T_i \rightarrow K_A(D_i) \quad K_{A_n^*} \tag{6.1}$$

$$K_A(D_i) \parallel K_R(K_A) \rightarrow R_i \tag{6.2}$$

In Equation 6.2, the key K_R is used to encrypt K_A using the RSA algorithm. The resultant data obtained by concatenating is dispatched to the receiver R_i.

Bokefodea et al. (2016) considered the role-based scenario. The access control is provided based on the roles the individual holds or is assigned. Both AES and RSA

TABLE 6.1

Characteristics, Advantages, and Space and Time Complexity of Various Studies

Contribution	Characteristics	Advantages	Space Complexity	Time Complexity
Carcelle, Dang, and Devic (2006)	The IoT-cloud interface is used to store and retrieve data as a convenience. The confidential data being transmitted requires high security.	Authentication and encryption are some of the measures taken to safeguard data.	$O(n \log n^2)$	$O(\log n)$
Stergiou et al. (2018)	The study considers only the source message to be confidential. The eavesdropper channel, cooperative protocol, source encoding, and decoding scheme are assumed to be public.	RSA algorithm is used to make data secure.	$O(n^3)$	$O(\log n/2)$
Belguith, Kaaniche, and Russello (2018)	The system uses lightweight attribute-based encryption. The scheme supports bandwidth-limited applications. PU-ABE is a new variant key attribute providing its assistance in the work.	The system aims to access policy updates efficiently.	$O(n^2)$	$O(2^n)$
Rahman, Daud, and Mohamad 2016)	The ecosystem consists of five layers. IoT sensor nodes and devices make up the first layer. Communication and network make up the second layer. Layer three is made up of computing and storage. Applications and services are in layer four. Analytics make up layer five.	A secure framework is suggested in this work.	$O(n^2)$	$O(n^4)$
Barreto et al. (2015)	Direct users request to IoT devices and user requests performed by the cloud are considered.	The model supports single-sign-on and single-logout tasks.	$O(2^n)$	$O(n!)$

(Continued)

TABLE 6.1 (CONTINUED)

Characteristics, Advantages, and Space and Time Complexity of Various Studies

Contribution	Characteristics	Advantages	Space Complexity	Time Complexity
Butun et al. (2016)	The model uses the hierarchical approach. The device can be carried or worn by public safety respondents.	It offloads continuous authentication.	$O(n \log n)$	$O(n^2)$
Gehrmann and Abdelraheem (2016)	The IoT device execution state is devised with a suitable high-level model. The activities are reflected in a cloud-executed machine. The machine accepts all requests targeting the devices.	The machine is highly available and attack resistant.	$O(n^2 \log n)$	$O(n!)$
Chandu et al. (2017)	The data gathered by the sensor is provided to the microcontroller. This data is encrypted using the AES encryption algorithm. The key to the encryption is saved for later use. This encrypted data with the key is fed to the cloud. The cloud provides security to the received message.	The device hardware is AES encryption supported.	$O(n)$	$O(\log n)$
Singh et al. (2015)	The algorithm is a derivative of key/cyphertext policy-attribute-based encryption methodology. The lightweight elliptic curve cryptography is used in the work.	The sender is connected to the receiver using a bridge to envisage Internetworking.	$O(n^m \log n)$ m is the number of attributes used.	$O(n \log n)$
Al Salami et al. (2016)	The proposal is an identity-based scheme with a combination of Diffie-Hellman encryption mechanism.	The scheme reduces computing exponentiations. A fast random encryption standard is adopted.	$O(n^3)$	$O(n)$

(Continued)

TABLE 6.1 (CONTINUED)

Characteristics, Advantages, and Space and Time Complexity of Various Studies

Contribution	Characteristics	Advantages	Space Complexity	Time Complexity
Muhammad et al. (2018)	Video sum-up and image encryption are combined to bring security to the system. The video summarization methodology is used to obtain informative frames. The image encryption methodology is suggested that is map employed and cryptosystem derived. The cypher is derived from the embedded random bits of plain images.	The system is energy-effective methodology.	$O(n!)$	$O(n^3)$
Rahulamathavan et al. (2017)	The cluster head is responsible to aggregate the data from the various sensors. The data undergo encryption before transmitting them to the respective ones. The encryption done can be verified by the blockchain miners using appropriate attributes.	The system aids in providing privacy to data through access control.	$O(\log n)$	$O(n \log n)$
Belguith, Kaaniche, Laurent, et al. 2018)	The adversary is provisioned to query the secret keys related to the set of attributes. The authority acknowledges back with the corresponding secret keys. During the transformation key query stage, the adversary is provisioned to query the secret keys. The challenger acknowledges with the keys if it finds the keys during the search operation. If not available it runs the algorithm to generate a set of transformation keys.	The system supports the multi-attribute methodology. A partial decryption procedure is delegated to the semi-trusted cloud server. The methodology supports a hidden access policy. The system proves to be feasible, secure, and privacy-preserving.	$O(n^2 \log n)$	$O(2^n * n)$

(*Continued*)

TABLE 6.1 (CONTINUED)

Characteristics, Advantages, and Space and Time Complexity of Various Studies

Contribution	Characteristics	Advantages	Space Complexity	Time Complexity
Talpur, Bhuiyan, and Wand (2015)	Homographic encryption is used in the work. The methodology is used for health monitoring. An Internet Protocol–based methodology is proposed to secure the system. The system can configure the system dynamically. The combined methodology is used in the proposed work – single-user node and shared nodes. The single-user node behaves as the identity node in the system. It examines the user access in the network. It also is responsible to scrutinize the data leaving the network. The shared node validation is also the responsibility of the single node.	The services are shared among the users to reduce the cost.	$O(n^m)$	$O(n^4)$
Aljawarneh and Yassein (2017)	The input file is read in a multi-size block during the encryption procedure. The block is divided into two divisions – plaintext and key. The key is encrypted using the Feistel encryption methodology. The plaintext is encrypted using AES. The encrypted text and key are further aggregated using the generic algorithm.	The proposed system cuts down the key distribution and key updating procedure. The keys are generated utilizing the data in the proposed work.	$O(n^3)$	$O(\log n)$

algorithms are used in the work. The data stocked in the cloud is encrypted using a key. The secret key used for encryption is created using the AES algorithm. This credential is used to encrypt the data. For every role, a public and private key is generated using the RSA algorithm. The public key is used to encrypt the secret key by the cloud. The user raises a request for the data. The user uses the private key to decrypt the secret key. This key is used to decrypt the data provided by the cloud.

6.4   PRINCIPLE OF THE SYSTEM

The Internet of Things aids in connecting devices. The devices are provided with a platform to communicate. The cloud is a platform that provides a storage location to the sensed data. The collaboration of both in the industrial platform is considered in the work.

6.4.1   Assumptions Made in the Study

The following assumptions are made in the study:

- The cloud provider is considered to be the most reliable resource.
- The cloud is capable of generating the hash code using location details.
- The devices are able to generate the encryption key from the hash code provided by the cloud.
- The system is liable to get compromised.

6.4.2   Notations Used in the Study

See Table 6.2.

6.4.3   Workflow of the System

The purpose of the Industrial Internet of Things (IIoT) with the cloud is to provide data anytime anywhere. This implementation can aid in bringing better

TABLE 6.2
Notations Used in the Work

Notations	Description
U_i	ith user
C_i	Cloud provider
L_i	User location
E_i	ith encryption key
H_i	Hash key
U_N	Username
U_P	Password
R_i	Request from ith user
Q_i	Query transmitted by ith user
N_i	Network in consideration
S_l	Location details of the sensor
O_l	Location details of the observed object
K_A	Encryption key used to encrypt data using AES algorithm
D_i	Data transmitted by the sensor
D_N	Encrypted data
K_H	Encrypted form of hash code
D_K	Decryption key

planning and design. They can aid in providing solutions in emergencies. The proposed work aims to bring in better reliability to the data. The encryption keys are generated by the combination of the location of the device and location of the object address. Using the location of the device provides authentication of the device. Location of the object addressing ensures the information sent. The concatenated key is used by the AES algorithm to cypher the data. The generated code is derived from the encryption key and attached to the data before transmission.

The procedure is divided into three stages. The process of generating the encrypted data and decrypting the data is as follows:

Stage 1

1. The user U_i authenticates using the key credentials. The username and password act as the credentials. The user shares his location details with the cloud. In Equation 6.3, the user U_i is transmitting username U_N, password U_P, and location L_i.

$$U_i \rightarrow U_N \| U_P \| L_i \tag{6.3}$$

2. After positive confirmation, the user is allowed to input his request. The device generates the public key and private key using location details. The public key is shared with the cloud. In Equation 6.4 user U_i is generating public key PUB_i using location key L_i. In Equation 6.5 private key PR_i is generated using the location key L_i. In Equation 6.6 the user U_i is transmitting public key PUB_i to the cloud C_i.

$$U_i : PUB_i \rightarrow algo1(L_i) \tag{6.4}$$

$$U_i : PR_i \rightarrow algo2(L_i) \tag{6.5}$$

$$U_i : PUB_i \rightarrow C_i \tag{6.6}$$

3. The user inputs his query to the cloud. In Equation 6.7 the user U_i is sending a query Q_i to the cloud C_i. The steps of stage1 are portrayed in Figure 6.1.

$$U_i : Q_i \rightarrow C_i \tag{6.7}$$

Stage 2

4. The cloud in turn forwards the query to the sensor network. In Equation 6.8 the cloud C_i forwards the same to the network N_i.

$$C_i : Q_i \rightarrow N_i \tag{6.8}$$

5. The sensor responds with data. It transmits its location details and location details of the observed object. In Equation 6.9 the user U_i is transmitting the

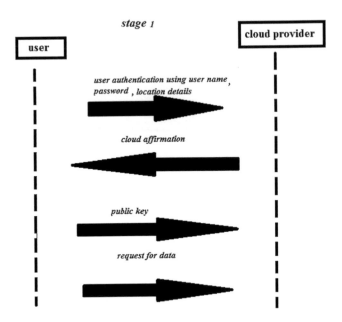

FIGURE 6.1   Steps followed in stage 1 of the proposed work.

requested data D_i, its location S_l, and location of observed object O_l to the cloud C_i.

$$U_i : (D_i \| S_l \| O_l) \rightarrow C_i \qquad (6.9)$$

6. The cloud generates the encryption key using these details. The keys of location details received by the cloud are generated and concatenated. In Equation 6.10 the cloud C_i generates the key K_A from the location details received. In the equation, S_l is the location of the sensor and O_l is the location of the object of interest. The moves of the stage2 methodology followed are depicted in Figure 6.2.

$$C_i : K_A \rightarrow algo3(S_l \| O_l) \qquad (6.10)$$

Stage 3

7. The cloud uses this as an encryption key to the AES algorithm. The data is encrypted using the key. A hash code is generated using the concatenated encryption key. In Equation 6.11 encrypted data D_N is generated using the encryption algorithm AES and key K_A on the data D_i. In Equation 6.12 the hash code H_i is derived from the encryption key.

$$C_i : D_N \rightarrow K_A(D_i) \qquad (6.11)$$

$$C_i : H_i \rightarrow algo4(K_A) \qquad (6.12)$$

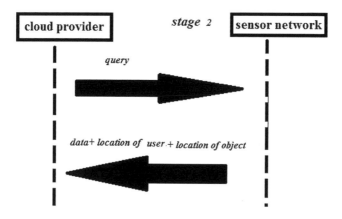

FIGURE 6.2   Steps followed in stage 2 of the proposed work.

8. The cloud uses the public key to encrypt hash code using the RSA algorithm. In Equation 6.13 the cloud C_i is using the public key PUB_i on the hash code H_i to generate its encrypted form K_H. In Equation 6.14 the encrypted data D_N is concatenated with the encrypted hash code K_H and dispatched to user U_i by the cloud C_i.

$$C_i : K_H \rightarrow PUB_i\left(H_i\right) \tag{6.13}$$

$$C_i : (D_N \parallel K_H) \rightarrow U_i \tag{6.14}$$

After receiving the data, the device at the user end decrypts the hash code using the private key. In Equation 6.15 the user obtains the hash code H_i by decrypting using the private key PR_i on key K_H.

$$U_i : H_i \rightarrow PR_i\left(K_H\right) \tag{6.15}$$

9. The device uses the hash code to create the decryption key and decrypts the data. In Equation 6.16 the user U_i is creating the decryption key D_K using the hash code H_i. In Equation 6.17 the user can obtain the data D_i by using the decryption key D_K on encrypted data D_N. The procedure is depicted in Figure 6.3.

$$U_i : D_K \rightarrow algo5\left(H_i\right) \tag{6.16}$$

$$U_i : D_i \rightarrow D_K\left(D_N\right) \tag{6.17}$$

6.5   ANALYSIS OF THE WORK

The proposed work is used to bring in reliability to the system. The location details of the sensors and the object of interest increases reliability. The combination of the

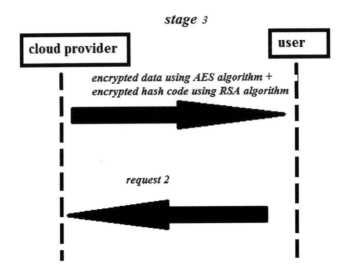

FIGURE 6.3   Steps followed in stage 3 of the proposed work.

key is used to derive the public and private keys. The AES and RSA algorithms are used to bring in better security to the network.

The proposed work uses location information to encrypt the data. The work is divided into three stages. In the first stage, the user logs in to the system using his username and password. The user requests the required data. In the second stage, the cloud queries the network with the request. The respective sensors transmit the data along with the location details. The encryption key is derived by concatenating the location details. The cloud uses this key to encrypt the data received. The encryption is done using the AES algorithm. In the third stage, it generates a hash code from the encryption key. The encrypted form is generated by using the public key on the hash code. The cyphered hash code and the encrypted data are transmitted to the user by the cloud. The user on the other end decrypts the hash code using his private key. The obtained hash code is used to regenerate the decryption key and decrypt the data received.

6.5.1   RELIABILITY TO DATA

Using the location information of the user and the object of interest brings in better reliability to the system. The combination of the key aids in making the system secure. The proposed work is compared to the algorithm using AES with the RSA algorithm. The proposed work proves to be 10.05% more reliable. The same is depicted in Figure 6.4.

6.5.2   COMPUTATIONAL TIME

The proposed work uses different key lengths to encrypt data using the AES algorithm, and 128 bits, 292 bits, and 256 bits are used. To encrypt data AES-128 requires

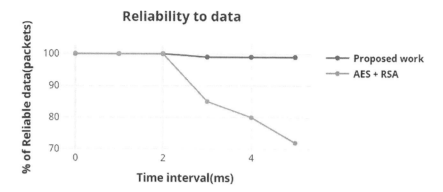

FIGURE 6.4   Reliability to data.

10 rounds to complete its computation. AES-192 bits require 12 rounds and AES-256 bits require 14 rounds to complete their computation. The data size of 32 bits, 64 bits, and 128 bits are used. The AES algorithm of 128 bits, 192 bits, and 256 bits are considered. The times required to perform the computation are shown in Figure 6.5.

6.6   INFERENCE

The Internet of Things provides a platform for devices to communicate. The cloud is used to store abundant data. The technology can be used by the user from any geographical location. The scenario considered is where the sensors are deployed in the network. They provide data to the cloud when queried. In the proposed work, a different encryption standard is followed to enhance reliability to data. The user logs into his cloud account. The user provides a username and password. A query is sent by the user to the cloud. The same forwards the request to the network. The sensors provide the queried data with location details. The location details consist of a sensor and observed object location. The cloud uses location details to obtain the key for the AES algorithm. Using the AES algorithm, the data is encrypted. A hash code is derived from the encryption key. The user uses his location details to create

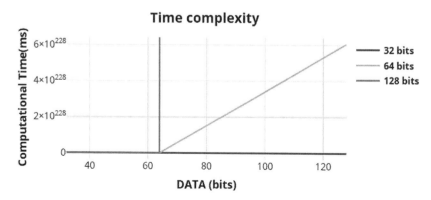

FIGURE 6.5   Encrypting variable size data using AES algorithm.

the public and private key. The public key is shared with the cloud. The cloud uses the key to cypher the hash code. The hash code is encrypted using the public key. It concatenates the same and encrypts the data. This data is transmitted to the user. The user uses his private key derived from his location to decrypt the hash code. The cyphered key is generated from the hash code. Using the cyphered key the data is decrypted. The work provides 10.05% more reliability to data compared to regular encryption methods.

REFERENCES

Aazam, M., I. Khan, A. A. Alsaffar, and E. N. Huh. 2014. "Cloud of things: Integrating Internet of things with cloud computing and the issues involved." *International Bhurban Conference on Applied Sciences & Technology*, Islamabad, Pakistan: IEEE.

Agrawal, H.2010. "Matlab implementation, analysis & comparison of some RSA family cryptosystems." *IEEE International Conference on Computational Intelligence and Computing Research*, Coimbatore, India, IEEE, 1–3.

Al Salami, S., J. Baek, K. Salah, and E. Damiani. 2016. "Lightweight encryption for smart home."*11th International Conference on Availability, Reliability and Security (ARES)*, Salzburg, Austria: IEEE, 382–388.

Alassaf, N., B. Alkazemi, and A. Gutub. 2017. "Applicable light-weight cryptography to secure medical data in IoT systems." *Journal of Research in Engineering and Applied Sciences (JREAS)*2(2): 50–58.

Aljawarneh, S., and M. B. Yassein. 2017. "A resource-efficient encryption algorithm for multimedia big data." *Multimedia Tools and Applications*76(21): 22703–22724.

Ambrosin, M., A. Anzanpour, M. Conti, T. Dargahi, S. R. Moosavi, Rahmani, A. M., and P. Liljeberg. 2016. "On the feasibility of attribute-based encryption on Internet of things devices." *IEEE Micro*36(6): 25–35.

Arabo, A. 2014. "Privacy-aware IoT cloud survivability for future connected home ecosystem."*11 International Conference on Computer Systems and Applications*, Doha, Qatar: IEEE, 803–809.

Babar, S., A. Stango, N. Prasad, J. Sen, and R. Prasad. 2011. "Proposed embedded security framework for Internet of Things (IoT)."*2nd International Conference on Wireless Communication, Vehicular Technology, Information Theory and Aerospace & Electronic Systems Technology (Wireless VITAE)*, Chennai, India: IEEE, 1–5.

Bahrami, M., A. Khan, and M. Singhal. 2016. "An energy efficient data privacy scheme for IoT devices in mobile cloud computing." *IEEE International Conference on Mobile Services*, San Francisco, CA: IEEE, 190–195.

Bao, Q., M. Hou, and K. Kwang R. Choo. 2016. "A one-pass identity-based authentication and key agreement protocol for wireless roaming." *Sixth International Conference on Information Science and Technology (ICIST)*, Dalian, China: IEEE, 443–447.

Barreto, L., A. Celesti, M. Villari, M. Fazio, and A. Puliafito. 2015. "An authentication model for IoT clouds." *IEEE/ACM International Conference on Advances in Social Networks Analysis and Mining*, Paris, France: IEEE, 1032–1035.

Bauer, E., and R. Adams. 2012. *Reliability and Availability of Cloud Computing*, Piscatway, NJ, John Wiley & Sons.

Belguith, S., N. Kaaniche, and G. Russello. 2018. "PU-ABE: Lightweight attribute-based encryption supporting access policy update for cloud assisted IoT."*11th International Conference on Cloud Computing*, San Francisco, CA, IEEE, 924–927.

Belguith, S., N. Kaaniche, M. Laurent, A. Jemai, and R. Attia. 2018. "Phoabe: Securely outsourcing multi-authority attribute based encryption with policy hidden for cloud assisted iot." *Computer Networks*133: 141–156.

Bellare, M., V. T. Hoang, and S. Tessaro. 2016. "Message-recovery attacks on Feistel-based format preserving encryption." *ACM SIGSAC Conference on Computer and Communications Security*, Vienna, ACM, 444–455.

Bhatt, C., N. Dey, and A. S. Ashour. 2017. *Internet of Things and Big Data Technologies for Next Generation Healthcare*. Turkey, Springer.

Bokefodea, J. D., A. S. Bhiseb, P. A. Satarkara, and D. G. Modani. 2016. "Developing a secure cloud storage system for storing IoT data by applying role based encryption." *Twelfth International Multi-Conference on Information Processing*, Bangalore, India, 43–50.

Boneh, D., and M. Franklin. 2001. "Identity-based encryption from the Weil pairing." *Annual International Cryptology Conference*, Berlin, Heidelberg: Springer, 213–229.

Butun, I., M. Erol-Kantarci, B. Kantarci, and H. Song. 2016. "Cloud-centric multi-level authentication as a service for secure public safety device networks." *IEEE Communications Magazine*, 54, 4 ,47–53.

Butun, I., Y. Wang, Y. S. Lee, and R. Sankar. 2012. "Intrusion prevention with two-level user authentication." *International Journal of Security and Networks*7(2): 107–121.

Carcelle, X., T. Dang, and C. Devic. 2006. *Wireless Networks in Industrial Environments: State of the Art and Issues. In Ad-Hoc Networking*. Boston, MA: Springer.

Chandu, Y., K. R. Kumar, N. V. Prabhukhanolkar, A. N. Anish, and S. Rawal. 2017. "Design and implementation of hybrid encryption for security of IOT data." *International Conference on Smart Technologies for Smart Nation*, Bangalore, India: IEEE, 1–4.

Chowdhury, A., and R. Panda. 2017. "Multi-view surveillance video summarization via joint embedding and sparse optimization." *IEEE Transactions on Multimedia*, 19(9): 2010–2021.

Dagon, D., W. Lee, and R. Lipton. 2005. "Protecting secret data from insider attacks." *International Conference on Financial Cryptography and Data Security*, Berlin, Heidelberg: Springer, 16–30.

Di Rienzo, M., F. Rizzo, G. Parati, G. Brambilla, M. Ferratini, and P. Castiglioni. 2006. "MagIC system: A new textile-based wearable device for biological signal monitoring Applicability in daily life and clinical setting." IEEE Engineering in Medicine and Biology, Shanghai, China, IEEE, 7167–7169.

Diksha, N., and A.Shubham. 2006. "Backdoor Intrusion in Wireless Networks-problems and solutions." *International Conference on Communication Technology*, Guilin, China, IEEE, 1–4.

Doukas, C., and I. Maglogiannis. 2012. "Bringing IoT and cloud computing towards pervasive healthcare."*6 International Conference on Innovative Mobile and Internet Services in Ubiquitous Computing*, Italy: Palermo.

Doukas, C., I. Maglogiannis, V. Koufi, F. Malamateniou, and G. Vassilacopoulos. 2012. "Enabling data protection through PKI encryption in IoT m-Health devices."*12th International Conference on Bioinformatics & Bioengineering (BIBE)*, Larnaca, Cyprus: IEEE, 25–29.

Feldhofer, M., S. Dominikus, and J. Wolkerstorfer. 2004. "Strong authentication for RFID systems using the AES algorithm." *International Workshop on Cryptographic Hardware and Embedded Systems*, Berlin, Heidelberg: Springer, 357–370.

Gehrmann, C., and M. A. Abdelraheem. 2016. "IoT protection through device to cloud synchronization." *IEEE 8th International Conference on Cloud Computing Technology and Science*, Luxembourg City, Luxembourg: IEEE, 527–532.

Harper, R.2006. *Inside the Smart Home*. Springer Science & Business Media, Berlin, Heidelberg.

Hasan, R., M. M. Hossain, and R. Khan. 2015. "Aura: An IoT based cloud infrastructure for localized mobile computation outsourcing."*3rd IEEE International Conference on Mobile Cloud Computing, Services and Engineering*, San Francisco, CA: IEEE.

Hossain, M. S., and G. Muhammad. 2016. "Cloud-assisted Industrial Internet of Things (IIOT)–enabled framework for health monitoring." *Computer Networks*, 101,192–202.

Huang, Y. L., and W. L. Sun. 2018. "An AHP-based risk assessment for an industrial IoT cloud." *IEEE International Conference on Software Quality, Reliability and Security Companion*, Lisbon, Portugal, IEEE, 637–638.

Huitema, C.1998. *IPv6: The New Internet Protocol*. NJ, Prentice Hall PTR.

Jiang, L., Da Xu, L., Cai, H., Jiang, Z., Bu, F., and Xu, B.2014. "An IoT-oriented data storage framework in cloud computing platform." *IEEE Transactions on Industrial Informatics*10(2): 1443–1451.

Koblitz, N., and A. Menezes. 2005. "Pairing-based cryptography at high security levels." *IMA International Conference on Cryptography and Coding*, Berlin, Heidelberg: Springer, 13–36.

Kocher, P. C.1996. "Timing attacks on implementations of Diffie-Hellman, RSA, DSS, and other systems." *Annual International Cryptology Conference*, Berlin, Heidelberg: Springer, 104–113.

Krutz, R. L., and R. D. Vines. 2010. *Cloud Security: A Comprehensive Guide to Secure Cloud Computing*. Tokyo, Wiley Publishing.

Lee, C. C., M. S. Hwang, and I. E. Liao. 2006. "Security enhancement on a new authentication scheme with anonymity for wireless environments." *IEEE Transactions on Industrial Electronics*53(5): 1683–1687.

Lewko, A., and B. Waters. 2011. "Decentralizing attribute-based encryption." *Annual International Conference on the Theory and Applications of Cryptographic Techniques*, Berlin, Heidelberg: Springer, 568–588.

Liu, C., C. Yang, X. Zhang, and J. Chen. 2015. "External integrity verification for outsourced big data in cloud and IoT: A big picture." *Future Generation Computer Systems*49: 58–67.

Mao, Y., J Li, M.-R. Chen, J. Liu, C. Xie, and Y. Zhan. 2016. "Fully secure fuzzy identity-based encryption for secure IoT communications." *Computer Standards & Interfaces*44: 117–121.

Masood, S. H., W. Rattanawong, and P. Iovenitti. 2003. "A generic algorithm for a best part orientation system for complex parts in rapid prototyping." *Journal of materials processing technology*139(1–3): 110–116.

Maurer, W. D.1968. "Programming technique: An improved hash code for scatter storage." *Communications of the ACM*11(1): 35–38.

Morris, T.2011. "Trusted platform module." In *Encyclopedia of Cryptography*, H.C. A. Van Tiborg and S. Jajodia, Eds. Boston, MA, Springer, 1332–1335.

Muhammad, K., R. Hamza, J. Ahmad, J. Lloret, H. Wang, and S. Wook Baik. 2018. "Secure surveillance framework for IoT systems using probabilistic image encryption." *IEEE Transactions on Industrial Informatics*14(8): 3679–3689.

Mukhopadhyay, S. C.2012. *Smart Sensing Technology for Agriculture and Environmental Monitoring*. Berlin Heidelberg: Springer.

Narayanan, A., and V. Shmatikov. 2010. "Myths and fallacies of personally identifiable information." Communications of the *ACM*53: 24–26.

Oliveira, L., B., Diego F. Aranha, E. Morais, F. Daguano, J. López, and R. Dahab. 2017. "Tinytate: Computing the tate pairing in resource-constrained sensor nodes." *Sixth IEEE International Symposium on Network Computing and Applications (NCA 2007)*, Cambridge, MA: IEEE, 318–323.

Prentice, S.2014. *The Five SMART Technologies to Watch*. Gartner.

Rahman, A. F. A., M. Daud, and M. Z. Mohamad. 2016. "Securing sensor to cloud ecosystem using Internet of Things (IoT) security framework." *International Conference on Internet of Things and Cloud Computing*, Cambridge, UK: ACM, 779.

Rahulamathavan, Y., R. C. W. Phan M. Rajarajan, S. Misra, and A. Kondoz. 2017. "Privacy-preserving blockchain based IoT ecosystem using attribute-based encryption." *IEEE International Conference on Advanced Networks and Telecommunications Systems (ANTS)*, Bhubaneswar, India: IEEE, 1–6.

Rijmen, V., and J. Daemen. 2001. "Advanced encryption standard." *Proceedings of Federal Information Processing Standards Publications*, National Institute of Standards and Technology, San Mateo, CA: Morgan, 19–22.

Rojviboonchai, K., and H. Aida. 2004. "An evaluation of multi-path transmission control protocol (M/TCP) with robust acknowledgement schemes." *IEICE Transactions on Communications*87(9): 2699–2707.

Roman, R., J. Zhou, and J. Lopez. 2013. "On the features and challenges of security and privacy in distributed Internet of things." *Computer Networks*, 57, 10,1389–1286.

Singh, J., T. Pasquier, J. Bacon, H. Ko, and D. Eyers. 2016. "Twenty security considerations for cloud-supported Internet of things." *Internet of Thing, 3(3)*,1–17.

Singh, M., M. A. Rajan, V. L. Shivraj, and P. Balamuralidhar. 2015. "Secure mqtt for Internet of Things (IoT)." *Fifth International Conference on Communication Systems and Network Technologies*, Gwalior, India: IEEE, 746–751.

Stergiou, C., K. E. Psannis, B. G. Kim, and B. Gupta. 2018. "Secure integration of IoT and Cloud Computing." *Future Generation Computer Systems*78: 964–975.

Talpur, M. S. H., M. Z. A. Bhuiyan, and G. Wang. 2015. "Shared–node IoT network architecture with ubiquitous homomorphic encryption for healthcare monitoring." *International Journal of Embedded Systems*7(1): 43–54.

Tong K., F.2013. "Smart agriculture based on cloud computing and IOT." *Journal of Convergence Information Technology*8(2), 1–7.

Wang, H., J. Cardo, and Y. Guan. 2005. "Shepherd: A lightweight statistical authentication protocol for access control in wireless LANs." *Computer Communications*28(14): 1618–1630.

Zhao, D. M., J. H. Wang, J. Wu, and J. F. Ma. 2005. "Using fuzzy logic and entropy theory to risk assessment of the information security."*4th International Conference on Machine Learning and Cybernetics*, Guangzhou, China, 2248–2253.

Zhu, Q., W. Saad, Z. Han, H. V. Poor, and T. Başar. 2011. "Eavesdropping and jamming in next-generation wireless networks: A game-theoretic approach." *Military Communications Conference*, Baltimore, MD, IEEE, 119–124.

7 Cyber Attack Analysis and Attack Patterns in IoT-Enabled Technologies

Siddhant Banyal, Kartik Krishna Bhardwaj, and Deepak Kumar Sharma

CONTENTS

7.1   INTRODUCTION

Cybersecurity is a crucial issue within the ambit of global peace and security that pertains to protection and preservation of critical infrastructure, ranging from intelligence and technology to private communications and metadata. It is a vexing quandary that has surfaced as a key area of consideration as technology evolves, and states and non-state actors continue to invest in a developing trend of asymmetric cyberwarfare in order to effect as well as preserve strategic gains. Consequently, the community faces a fractious and divisive challenge on this account, as the issue at hand brings with it a litany of socio-political and technological ramifications, with matters such as establishment of a holistic understanding of cyber attacks for development of reasonable standards for and halting the growth of belligerent actions, all key components to a safer world.

7.1.1   IoT-Based Networks and Related Security Issues

Internet of Things (IoT) has asserted itself as the engine that runs our modern technological endeavours under the ambit of public and private domains. Industries along with business institutions have evolved to be more reliant on IoT to perform essential missions and function to improve productivity. On one hand the technology has brought agility to our environment but has induced fragility, which has the potential to cause a great degree of abrasion to our social and economic constitution. As of now there are over 23 billion IoT devices across the globe, and this figure is expected to increase and reach staggering amount of 60 billion in just half a decade. In conjunction with improving productivity this has led to huge economic growth with the expected market size in Europe to shoot up to €242,222 million by end of next year [1]. This underscores the pivotal role IoT and associated technologies, but like a double-edged sword this has opened a myriad of challenges and security related issues. Data encryption, authentication, privacy, side channel attacks, botnets aiming at cryptocurrency, remote access, and untrustworthy communication are few of the contemporaneous challenges the community faces [2].

A plethora of devices have limited processing capacity and memory and run on lower power such as battery-based devices. Current security relies on strong encryption that is unfair for these devices because these constraints render them incapable to perform complex encryption–decryption to transmit data. There are a number of vulnerabilities and potential areas of failures in an IoT system that makes authorisation and verification critical from a security standpoint. Prior to accessing gateways the device must establish its identity. In order to incorporate and manage device updates available throughout the distributed environment with dissimilar devices that communicate via a myriad of unique protocols and track of updates must be kept. Many devices do not facilitate "over-the-air" updation, hence the device must be removed from production to apply updates. Further, data privacy and integrity are essential parts of current discourse on challenges in IoT. Implementation of data privacy encompasses redaction of sensitive data, decoupling personal information, and safe disposal of outdated data. In addition to this, ensuring high availability is essential to daily functioning. Potential disruptions due to device failure, connectivity outages, or denial of service attacks can lead to economic loss, damage to devices, and in severe cases loss of life.

7.1.2   Need for Threat Detection Security Systems in Cyber Networks

We have witnessed technological advances throughout history, and with new opportunities there will always exist a threat from those that exploit them for their own gain. Cyber actors and groups are networking, researching modes to disrupt operations; and testing new tactics, techniques, and procedures. Network security encompasses activities deployed to ensure efficacy, dependability, and security of the network infrastructure and associated information. This focuses on a multitude of threats and impedes their penetration infecting the network.

The initial events in the history of threats security can be dated back to the days of telephony. Electrical signals were sent through copper wires that potentially could be exploited and telephonic talks could be heard. The implementation of cybersecurity involving governance risk and compliance developed separately from the history of computer security software. Threats like network breaches and malware were present during the early times. As per many reports [3] Russians were among the first to employ cyberpower as a cyber weapon. During the year 1986, the first worm by the name of Morris was identified, which was the first one to exploit a global Internet network and disseminate around computers primarily in the United States. During this period in the history of cybersecurity the viruses evolved from an academic prank to a serious threat. The Morris worm employed vulnerability present in the UNIX system and copied itself progressively, thereby maliciously slowing the operations in systems, hence, rendering them useless. The worm was designed by Robert Tapan, who claimed that his actions were an attempt so as to identify the size of the Internet. Later, he was the first entity to be declared guilty under the Computer Fraud and Abuse Act (CFAA) of the United States and currently is a professor at the Massachusetts Institute of Technology. From the invention of the Advanced Research Projects Agency Network, which was designed for the scientific community to exchange information and access remote systems, a new form of system break into networks started. Email applications enabled collaboration on research work and developments over the web. This became an avenue for hackers to exploit and access confidential information such as but not limited to credit card information, passwords, and trade secrets.

Advent of the Internet has augmented the way we live with more than 3 billion active Internet users. Email communication was primarily a means of communication, but now it is being used for purchasing, selling, marketing, advertising, channel for B2B and B2C, etc. Essentially, it created a platform where intruders can exploit by means such as but not limited to probing (observe the network along with its users by using tools), scanning (analyzing the network along with the associated devices for vulnerabilities), malicious code, distributed denial of service, and gaining unauthorized access to a network and their resources.

7.1.3   Managing Threats to Cyber Networks

As discussed in the previous sections, the advent of the World Wide Web resulted into a huge call for controlling data, information, and knowledge. With the current breakthroughs in information system technologies, many applications in various business areas have been computerized. Data has become an important resource to a majority of business organisations and consequently efforts have been made to integrate data sources scattered across various sites using IoT. Further, developments have been made to find information from the data gathered through assessing the patterns and trends. Sources of data may be databases operated by a DBMS and can be warehoused in a repository.

As the need for data and information is on the rise, there is a rise in demand for need of managing the security of the sources, systems, and utilities managing this data. It is essential that the data is safeguarded from unapproved access and malicious altercations. With the rise of Internet and cyber infrastructure it was even more pivotal to safeguard the data and information, as now a multitude of people have the ability or means to gain access to protected confidential information. Therefore, there is a need for appropriate security instruments and infrastructure in cyberspace.

This chapter reviews various threats in cyberspace with a special attention to threats to cybersecurity in IoT-enabled devices. Further, we have addressed a few remedies to these threats and associated technologies. The threats include integrity breaches, access control breaches, unapproved intervention, and sabotage/espionage. The aforementioned solutions encompass cryptographical methods, data-mining methods, and fault tolerance processing methods, and many more. In the second section of this chapter we perform a taxonomic assessment of the terms used in this context from a technological and legal standpoint. Section 7.3 highlights various modelling techniques and paradigms for cyber intrusions and perform an impact assessment for assessing IoT-enabled cyber attacks. Sections 7.4 and 7.5 we highlight existing cyber attack detection software countermeasures to cyberthreats and assess cyber vulnerabilities in various sectors. In addition, we analyze some case studies pertaining to the above. In Section 7.6 we aim to discuss potential solutions to regulatory problems and further developments.

7.2   CLASSIFICATION AND TAXONOMY OF CYBER ATTACKS

Post Industrial Revolution there has been various technological breakthroughs that have significantly ameliorated the way we live. The roles that computers and the Internet plays are well recognized in our society, and this has created a virtual domain for exchanging information known as cyberspace which has been increasing in size steadily. Cyberspace has pervaded all aspects of human life including but not limited to hospitals, banking, education, emergency services, and the military. Threats have been increasing ever since to the same where attacks are employed to disseminate disinformation, impede tactical services, access sensitive data, espionage, data theft, and financial losses. The nature, severity, and complexity of these intrusions are aggregating over time and there exists a gap in the understanding of the modus operandi of these attacks. Figure 7.1 illustrates all the categories and subcategories of cyber attack classification.

7.2.1   BASED ON PURPOSE

Purpose-based attacks are classified as follows.

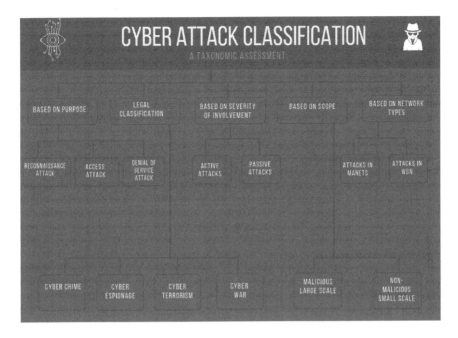

FIGURE 7.1 Classifications of cyber attacks.

7.2.1.1 Reconnaissance Attacks

Reconnaissance attacks entail unsanctioned detection and system charting, and are homogeneous to theft incidents in a neighbourhood, comprising of the following types:

Packet analyzers: A tool used to intercept data flowing in a network. These are employed to monitor traffic between linked computer systems. It commands the network interface card to analyze communication between a particular segment of the computer network. The packet sniffer may be hardware or software in nature.

Port scanning: Scanning of port is performed in order to probe a server or host for open ports. The attackers use this to identify network services in use and exploit vulnerabilities.

Ping sweep: Ping sweep is employed to map the range of IP addresses linked to a live host. Pings may be detected by protocol loggers such as but not limited to ippl.

Queries regarding Internet information: The perpetrator can employ distributed networkservices queries to gain knowledge of who hold a domain and the addresses that have been delegated to that domain.

7.2.1.2 Access-Based Attacks

In access attacks the intruder establishes ingress to an apparatus where the attacker has no authorization for system credentials or authorization. The entity who does not possess the authority to legitimately access the data will try to employ malicious means such as hacking or create a tool which takes advantage of the inherent fragility of the system.

File Transfer Protocol, authentication tools, and Internet applications will be used through exploitation of system weak points or vulnerabilities to gain unauthorised access to online accounts, confidential repositories, and other privileged information. This category of attacks encompass:

Attacks on secret code: This comprises of attempts to crack passwords using all possible combinations. In common parlance it is known as dictionary attacks with two types: guessing and resetting.

Utilization of trust port: Here the attacker employs the trust port of the network to pose as a trusted host, which then is employed to attack the host.

Port redirection: In this type of access attack the network firewall is bypassed and protected hosts are attacked, as the attacker uses a trusted host for this.

Man-in-the-middle attacks: In this, the perpetrator makes independent links with the hosts and transfers messages between them in a deceptive manner in order to establish trust and make them believe the conversation is private.

Social engineering: Here a malicious SQL-based code is used which either changes the content on these websites or infect the users entering them.

Phishing: Phishing is the activity of sending deceptive mail in order to mislead the users while masquerading as a legitimate entity. This allows the perpetrator to gain access to confidential information that can be used for identity theft.

DoS attacks: This type of attack aims to make a system or network unavailable for its designated use for an interim or indefinite period rendering the system capability useless [6].

7.2.2 BASED ON SEVERITY OF INVOLVEMENT

In the case of a cyber attack, it enables the perpetrator to relay information to all the parties, or block the data transmission in one or multiple directions, and they can aim to abort the data sent by the parties in the network as the attacker is located between the communicating nodes.

7.2.2.1 Passive Attacks

Passive cyber attacks employ a non-disruptive means to covertly gain access to a system or network where perpetrators aim to collect information without detection. Primarily, passive attacks are data-gathering operations in which perpetrators use malware or hack system information. This includes identity theft, stealing credit card information, and other forms of privacy/data breaches.

7.2.2.2 Active Attacks

This type of cyber attack aims to change system resources and influence the system's operational capabilities. In this type of attack, the intruder changes the data stream or may introduce new data in the system.

7.2.3 LEGAL CLASSIFICATION

Legal classification entails the following key terms:

1. *Cyber-operation*: Action taken through cyberspace by a state, non-state actors, and/or an individual in concert with specified supporting and related capabilities, which shall be split into cyber-espionage and cyber-offence.
2. *Cyber-espionage*: A type of a cyber-operation constituted by:
    a. The intent to acquire information and data relating to the target 'object'.
    b. The illicit gathering of sensitive information from a government, collective entities (i.e. multinational corporations or financial institutions), or individual entity, as performed by a state or non-state actor.
    c. Distinguishes between the following classifications of spying by means of intelligence gathering in cyberspace for the purpose of identifying the level of adherence to the regulations, by categorizing between spying as a means of:
        i. Maintaining national security, as a function of national security organizations and not as an act of war or use of force.
        ii. Advancing economic or strategic interests irrelevant to the direct security of the state, as being a violation of international norms, but is not to be identified as an act of war.
3. *Cyber-offence*: Any action in cyberspace which alters, corrupts, deceives, degrades, destroys, disrupts, affects, or influences the functioning of/or gains access into a computer network and thereby undermining the related/dependent critical infrastructure, economy, polity, and/or security.
    a. Understanding that:
        i. 'Any action in cyberspace ...' means a cyber-offence includes any action carried out in cyberspace, referring to all computer networks, public or private, connected or controlled via cable, fibre optics or wireless.
        ii. '... which alters, disrupts, deceives, degrades, destroys or affects the functioning of/or gains access into a computer network ...' are all potential elements of a cyber-offence.
        iii. '... and thereby undermining related/dependant ...' refers to the undermining of a computer network by altering, corrupting, deceiving, degrading, destroying, disrupting, affecting or influencing the computer network.
        iv. '... critical infrastructure, economy, polity, and/or security ...' is defined by:
            1. *Critical infrastructure* meaning any system dependent on cyber or electronic structures that is essential for society to maintain its status quo of functionality such as: telecommunications infrastructures, electrical grids, energy sources, water distribution systems, financial organizations, public transportation, information database, health facilities or systems.

2. *Economy* as a state's economic institutions and services.
3. *Polity* as a state's political activity, bodies, entities, and institutions.
4. *Security*: The term 'security' refers to both internal and external security.

v. '... can be termed as a cyber-offence' indicates any action that fulfils all or some of the aforementioned criteria, without ambiguity, will constitute a cyber-offence.

4. *Cyber weapon*: Packet(s) of binary computer code engineered to accomplish the objective of a cyber-operation through software-operated machines, including, but not limited to, computers, servers, routers, mobile phones, or industrial equipments, understanding that:

a. 'Packet(s) of binary computer code' implies code is the elementary makeup of the cyber weapon or the tool used to accomplish the attack. A binary code represents text, computer processor instructions, or any other data used to operate a computer system in most of the computer networks, as well as more complicated algorithms designed to find vulnerabilities in target computer networks such as randomized tree models and other classifiers, which, in practice, are self-changing packets of code that evolve to be more effective by processing data over time.

b. '... the objective of ...' means the purpose and intent of initiating the given cyber-operation, which may have intent to conduct reconnaissance as defined under cyber-espionage, or may have intent to alter, disrupt, deceive, degrade, destroy or affect the functioning of as defined under cyber-offence.

c. 'Engineered to accomplish' refers to code that is 'engineered' or consciously generated either by hand or by using automated tools, subjective to the sophistication, skill, and resources of the attacking entity.

d. 'Through software-operated machines, including, but not limited to, computers, servers, routers, mobile phones, or industrial equipments' refers to software-operated machines and machinery that operate on computer software, which may be manipulated by the use of computer codes to change their operation.

7.2.4 BASED ON SCOPE

7.2.4.1 Large Scale or Malicious

The term 'malicious' means anything characterized by malice or with motive to cause harm or result in some damage. The attacks in this category are large in scale using a multitude of computer systems, which results in a crash of the system worldwide accompanied by a loss of huge volumes of data.

7.2.4.2 Small Scale or Non-Malicious

Non-malicious or small-scale attacks are essentially attacks accidental in nature or damage caused because of mistakes or operational error attributed to a badly trained person which results into data loss or system crash. In this category only a limited number of systems are breached/compromised, and information is usually recoverable, and the recovery corresponds to minor cost.

7.2.5   BASED ON TYPE OF NETWORK

The attacks in cyberspace may be based on the network type on which the attack happened. This includes attacks on mobile ad hoc networks (MANETs) and wireless sensor networks (WSNs) [7].

7.2.5.1   Attacks on MANETs

7.2.5.1.1   Byzantine Attacks

Byzantine attacks happen primarily on MANETs. Here, a device or a group of devices that issue authorization/security is compromised due to a data breach, which causes a network breach. This renders the host unable to distinguish between a hostile user and an authenticated user.

7.2.5.1.2   Black Hole Attacks

In a computer network, a node employs the network routing protocol to assert itself as a desirable candidate for data transmission based on a myriad of factors such as shortest distance (based on the protocol). In a black hole attack, the node then broadcasts its availability of routes irrespective of the information on the routing table. In a back hole attack, the node always has the ability to reply to a route request and pose as a desirable node and drop data packet later [8]. In this type of attack, route request (RREQ) and route reply (RREP) protocols are used.

7.2.5.1.3   Flood Rushing Attacks

The attack this category is essentially altering the routing scheme of the network, in which one node or multiple nodes alter, capture or, fabricate the data packets. They may form circular loops where they drop, delay, or route packets selectively. This leads to an unnatural delay or routing of data packets in non-optimal paths and falsify routing information [9].

7.2.5.1.4   Byzantine Wormhole Attacks

In a Byzantine wormhole attack, the perpetrator nodes gain the ability to transfer packets between them so as to create a shortcut in the network circumventing the desired route. This category of attack is severe but it requires a minimum of two compromised nodes to happen.

7.2.5.2   Attacks on WSNs

Wireless sensor networks are susceptible to two types of attacks, cryptography and non-cryptography, that are dependent on network layers. Attacks on WSNs are categorized based on layers, methods, and area of attacks. Figure 7.2 illustrates examples of cyber attacks in a categorized manner.

NAME OF ATTACKS	EXAMPLES
EXAMPLES OF CYBER ATTACKS	
RECONNAISSANCE ATTACKS	PACKET SNIFFERS, PORT SCANNING, PING SWEEPS AND DNS(DISTRIBUTED NETWORK SERVICES) QUERIES
ACCESS ATTACKS	PORT TRUST UTILIZATION, PORT REDIRECTION, DICTIONARY ATTACKS, MAN-IN-THE-MIDDLE ATTACKS, SOCIAL ENGINEERING ATTACKS AND PHISHING
DENIAL OF SERVICE	SMURF, SYN FLOOD, DNS ATTACKS, DDOS(DISTRIBUTED DENIAL OF SERVICES)
CYBER CRIME	IDENTITY THEFT, CREDIT CARD FRAUD
CYBER ESPIONAGE	TRACKING COOKIES, RAT CONTROLLABLE
CYBER TERRORISM	CRASHING THE POWER GRIDS BY AL-QAEDA VIA A NETWORK, POISONING OF THE WATER SUPPLY
CYBERWAR	RUSSIA'S WAR ON ESTONIA(2007) AND GEORGIA(2008)
ACTIVE ATTACKS	MASQUERADE, REPLY, MODIFICATION OF MESSAGE
PASSIVE ATTACKS	TRAFFIC ANALYSIS, RELEASE OF MESSAGE CONTENTS
MALICIOUS ATTACKS	SASSER ATTACK
NON MALICIOUS ATTACKS	REGISTRY CORRUPTION, ACCIDENTAL ERASING OF HARDDISK
ATTACKS IN MANET	BYZANTINE ATTACK, BLACK HOLE ATTACK, FLOOD RUSHING ATTACK, BYZANTINE WORMHOLE ATTACK
ATTACKS ON WSN	APPLICATION LAYER ATTACKS, TRANSPORT LAYER ATTACKS, NETWORK LAYER ATTACKS, MULTI LAYER ATTACKS

FIGURE 7.2   Examples of cyber attacks.

7.3   MODELLING TECHNIQUES AND PARADIGMS FOR CYBER INTRUSIONS

When UN Secretary-General António Guterres on February 19, 2018, responded to a question pertaining to the nature and the future of modern warfare, he said that war in the future will emerge in the digital domain with cyber attacks paralyzing military capabilities and impeding critical infrastructure [10].

There has been a rise in the demand for online web services, financial transactions encompassing highly sensitive financial, and personal information. Additionally, our cyber infrastructure is now connected to more IoT devices consisting of a myriad of small-scale hardware used to protect critical infrastructure. This makes them even more susceptible to cyber attacks and hence there is ample evidentiary support of attacks on online service providers like Amazon, eBay, Sony, and Yahoo. It has been estimated that every year these attacks cost our global economy about $1 trillion. There is a significant number of researchers and experts engaged in threat modelling to estimate the model of cyber intrusion/attack for any computer network and provide groundwork for future defence systems in cyberspace. The defence systems essentially depend on the understanding of their own network, the rationale behind the attack, the method of attack, and security vulnerabilities.

Analysts uses a myriad of modelling technique to assess instances of cyber attacks. These include but are not limited to:

Tree/attack graph [11]
Kill Chain [16]
Attack surface [13]
Diamond model [14]
Attack vector approach [12]
Open Web Application Security Project (OWASP) threat model [15]

This chapter studies three techniques for cyber attack modelling from the list.

7.3.1 DIAMOND MODELLING

The diamond modelling technique is a unique model of cyber attack analysis where the perpetrator attacks the system based on two pivotal goals instead of employing a designated series of steps such as in attack graph techniques or kill chain modelling. The diamond model comprises four fundamental components that are 'adversary' (perpetrator), 'capability', 'infrastructure', and 'victim'. The adversary is defined as an individual actor or a group of actors who attacks a victim after assessing their 'capability' against the 'victim'.

Firstly, the perpetrator initiates in absence of information of the potential of the victim. Post inspecting the potential of a victim, the perpetrator could conclude that he/she has more capability than the victim to attack or not. The diamond model is pivotal while ecombacting with advanced attackers like those who currently have gained some degree of command or management over the network. Further, the perpetrator additionally assesses the cyber infrastructure of his/her technical and logical ability to direct and manage any of the host's network.

Additionally, features like 'Phases', 'Time Stamp', 'Directions', 'Methodology' and 'Resources' are attributed to the given model to impart additional details. During a breach, the diamond model identifies phases during a time stamp. Elements of the model can be located within Figure 7.3 that shows that the perpetrator searches for a chance to attack a host based on 'capability' or the 'infrastructure'.

7.3.2 KILL CHAIN MODELLING

The given model in Figure 7.4 employs a common attack modelling technique for intrusion, which interprets an attack as a series or an ordered chain of action. It is an ordered attack, that is, a chain of events are followed up by the assailant, progressing as per the plan. The US Department of Defense has defined the Kill Chain technique for attack on a target, outlining the Kill Chain into stages as follows: 'find, fix, track, target and assess'. The Kill Chain has been applied in alternative areas including cyber security where it is used to describe attack levels within a countermeasure system. The analysis through the above technique leads to a description of the Kill Chain as a seven-step operation, which might be represented as:

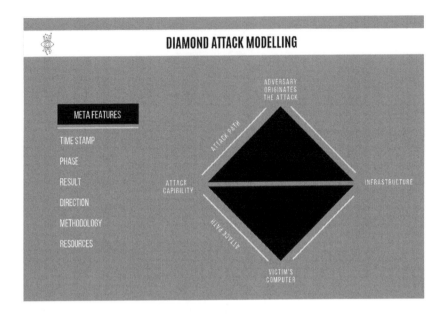

FIGURE 7.3   Diamond attack modelling.

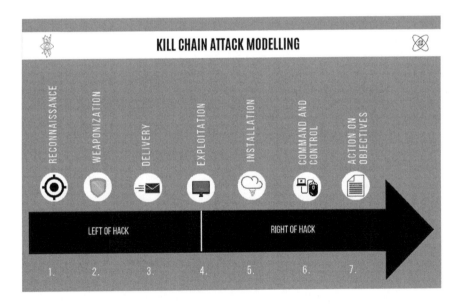

FIGURE 7.4   Kill chain attack modelling.

Phase 1 Probing | 'Reconnaissance': The perpetrator collects data prior to instance of breach. The data might be accumulated from the net, which is accessible to the public. This stage includes selection of target, organization details identification, understanding the host's vertical legislative needs, information over technological alternative, social network activity assessment, or extraction of mail recipient list.

Phase 2 Armament | 'Weaponization': Malicious payloads are created by the perpetrator for sending it to the victim system. This shall come in a myriad of configurations: Internet service exploitation, malware that's custom designed, embed vulnerabilities in a document (for example, through a PDF or through a different format), or watering hole attacks (an attack paradigm that involves analyses of the websites the target uses and use them as ancillary means to infect the target). Primarily, the weaponization is done while being cognisant of the intelligence of the host and in an opportunistic manner.

Phase 3 Distribution | 'Delivery': The package that is infected is then delivered to the host by the attacker through employment of some means of communication. The payload might be sent by the perpetrator via email as an attachment or a link that will download the payload.

Phase 4 Abuse | 'Exploitation': This particular phase encompasses the period when the exploitation/misuse happens. When the host has transferred the payload into their system, then phase 4 is initiated. The payload will compromise the assets and establish a foothold in the environment. In this stage, the aggressor wants the help of the victim. This might be one of the only phases where the process chain can be terminated, by disallowing the download of the malicious package that is sent.

Phase 5 Establishing | 'Installation': This step involves installing payload on the victim's system. To exploit the host's system, the package may run automatically or shall be required to be run by the victim.

Usually, the malicious package's malware is discrete in its execution, accumulating tenacity at points that it is able to access. The attacker shall then be able to command this application without alerting the organization.

Phase 6 Directive | 'Command and control': Here, in the command and control phase, perpetrators gain command of the resources of the victim or exploit the target via different methods of control, to name a few: DNS, ICMP, various networks, and websites. Via the installed payload, the perpetrator establishes a control and command channel so as to access the victim's internal assets. During this part, the perpetrator has, with success, gained control of the victim's system. The data gathering tools used under the perpetrator's command include keystroke logging, password decoding, screen captures, network inspection for login credentials, collection of privileged information, and documents.

Phase 7 'Action on objectives': The action on objectives phase encompasses the methodology the attacker uses to exfiltrate data and/or damages IT assets while concurrently dwell time in the target. The perpetrators complete their goal through the host's system that's infected by the malicious payload. The perpetrator may gain valuable information from the database by using an online server.

7.3.3 ATTACK GRAPH TECHNIQUE

Attack graphs are abstract flowcharts employed in order to assess and map the attack process when a target is attacked. This is often vital to investigate cyber threats on a system or network. This modelling paradigm is derived from tree structured graphs, which through a single root bears children on many levels. The attack graph technique is among the conventional means of discovering system vulnerabilities (security vulnerability assessment of the system), that is inherited by many individuals as well as to develop a tool for efficacious security by examining the network. The given graph primarily comprises of nodes and may be complicated in structure when interacting with a particular case of an attack. Complex attack graphs are computationally difficult to model and realize, they may contain thousands of nodes and a myriad of paths. This computational impediment makes it tedious for analysts to use attack graphs for modelling attacks that are complex in nature.

There exists a plethora of tools and paradigms that generate the attack graphs, these include techniques like Topological Analysis of Network Attack Vulnerability (TVA), Network Security Planning Architecture (NETSPA), and Multihost, Multistage, Vulnerability Analysis (MULVAL) [17]. The techniques given above aids us to draw coherent attack graphs to ascertain the rationale behind an attack instead of how the attack happens. The central idea of an attack graph is understanding the path attackers take to the host's network. Attack graph techniques facilitate to identify intrusions and also the vulnerabilities of the system. An example of an attack graph is demonstrated in Figure 7.5 to depict a sample case. They are often helpful in several areas of computer network security as well as intrusion detection, forensic analysis, risk analysis, and cyber defence. A network admin employs an attack graph to recognize:

- Vulnerability of the system or a network
- Process of an action and the rationale
- Group of set actions or steps that are likely to impair the perpetrator from realizing their goal

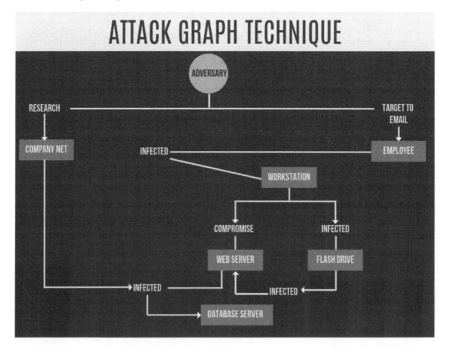

FIGURE 7.5 Attack graph modelling.

7.4 ASSESSING IOT-ENABLED CYBER ATTACKS

7.4.1 TAXONOMIC ASSESSMENT AND INTRUSION DETECTION

Intrusion detection (ID) techniques and paradigms have ameliorated drastically through the course of time and the major development materializing in the past few years. Primarily invented to automate complex and time enduring log file parsing activity, intrusion detection systems (IDSs) have now evolved to composite, "real-time applications" which possess the potential to monitor and analyse the network traffic and to identify any hostile activity. IDSs have the ability to deal with networks with massive speeds and composite traffic, and provide comprehensive information, which earlier was not available, pertaining to active cyber threats to web-based information services that are critical. IDSs have proved to be an essential and pivotal element of all holistic computer security programmes, as they supplement conventional mechanisms for security. Dennings developed the first known model based [56] on a intrusion detection paradigm, through which a plethora of IDSs have been proposed in the research literature and commercial fields. Despite the distinctness in features related to data gathering and examination of these systems, many depend on a common architecture as given in Figure 7.6, which entails the following given elements:

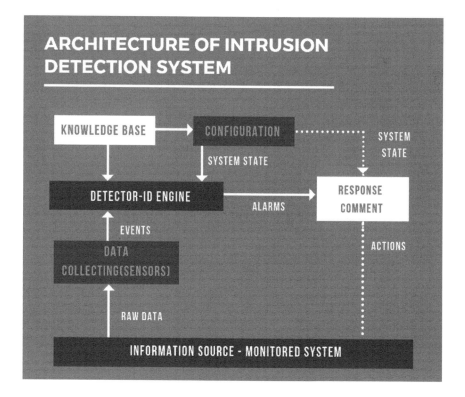

FIGURE 7.6 Basic architecture of intrusion detection system (IDS).

- A *data collecting device* is tasked to collect information from the system that is being monitored.
- *The detector* (*intrusion detection examining engine*) processes the information acquired from the data collecting device (sensor) to detect intrusive activities.
- The *knowledge base* (*database*) comprises of data gathered through the data collecting device, in processed format (such as the database of intrusions along with the signatures, data profiles, processed information, and so on). The information for these databases are commonly made accessible by the network/cybersecurity experts.
- The *configuration device* supplies data pertaining to the existing state of the IDS.
- *Response component* starts the response when the 'intrusion' is detected, these responses can be either automated or through a manual interaction.

There exists two elementary methods for assessing activities to find attacks. They are *misuse detection* (MD) and *anomaly detection* (AD). Misuse detection relies on in-depth information of detected attacks and weak points of the system supplied by experts in a similar manner to a knowledge system. Parallelly, MD rummages around for attackers that decide to execute these attacks or gain advantage of system vulnerabilities. Though MD is often terribly correct in sleuthing well-known attacks, these techniques cannot identify cyber threats that are not known to the system knowledge base. Anomaly detection depends on the assessment of profiles that exhibit conventional behaviour of connections in the network, users of the system, and the host. Anomaly detectors (ADs) identify conventional authorized cyber activity by employing a plethora of methods and then employs a range of quantitative and qualitative indicators to identify aberrations from the outlined conventional activity as a prospective anomaly. Here the advantage is that ADs can detect unknown attacks with the drawback of having a high false notification rate. It may be noted that the aberrations identified by AD algorithms may not be the instance of aberrations and is in fact cases of legitimate but unconventional system behaviour. MD-based approaches can be divided into the following sections:

1. Data-mining techniques
2. Rule-based approach
3. Algorithms based on state-transition analysis
4. Signature methods

AD algorithms can be classified into the following five sets:

1. Statistics based
2. Rule-based approach
3. Distance-based technique
4. Profiling methods
5. Model-enabled approaches

The reaction of the intrusion detection system to detected attacks can be either active or passive in nature. Conventionally, these systems have passive response and essentially notify the chief security officer or network administrator in an organization of the incident without taking any action to counter the attack. The standard methodology for these alerts is via pop-up messages or on-display alerts or cataloguing results in a single location/repository for perusal. The alerts in question can be quite diverse: some may contain just elementary details and some may contain details pertaining to the IP address of the source, target details, port of interest, tools employed, and the damage it has caused. Few ID systems are designed to trigger remote notification via alarms/alerts to cellular devices that include phones and pagers which are often carried by the network security team. Although, the alerts sent through emails may be unreliable, as they are susceptible to ancillary attacks from the perpetrator. Specific intrusion detection systems such as the one used by Cisco, employ Simple Network Management Protocol (SNMP) to trap and alert messages to report generated alarms to a network management system, where network operations personnel will investigate them. A system response that is passive in nature is used for offline analysis. The detection systems may give responses that are active in nature to critical incidents, these included responses like 'patching' vulnerability of a system, forcing the user to log off, resetting router and firewall reconfiguration, and port disconnection. Being cognisant of the attack speed and frequency, a perfect system reacts to these attacks at machine speed automatically without any external aid such as intervention by the operator. However, this expectation is impractical in nature, mostly because of the difficulty in eliminating false alarms. Nonetheless, IDSs may still deliver a plethora of response methods that are active in nature which are employed at the discretion of the designated personnel for network security.

7.4.2 COMPUTER NETWORK ATTACKS (CNAs) ANALYSIS USING TECHNIQUES BASED ON DATA MINING

Data-mining paradigms comprise of tools and algos that are used in order to consolidate data from disconnected data sources in one single repository that is useful for analysis under the ambit of data warehousing. Scientific and engineering development in recent phases have catalyzed and led to accumulation of a Brobdingnagian amount of information. The data we encounter today accumulates quickly and is of tremendous quantity that has transcended the ability for humans to understand it without using appropriate tools. The current estimation of database size for a tech giant such as Google is 10 exabytes, which is 10 million terabytes. Ergo, in recent years, the digital geographic information sets and multidimensional information has augmented quickly in recent years in terms of scope and coverage. The information sets in question encompass all types of digital information developed and spread by governments as well as personal agencies pertaining to climate data, land use, and immense volume of knowledge non-inheritable via remote sensing devices and other monitoring systems. Presently, there exists a significant engrossment in applying data-mining techniques to IDSs. An ID-based problem can be reduced to a specific case of data mining that aims to classify data. An example can be an algorithm

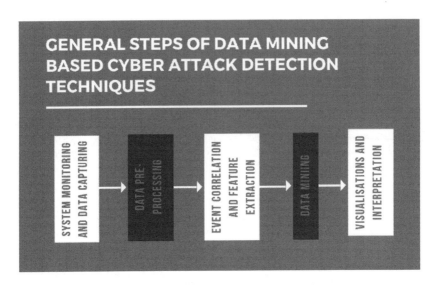

FIGURE 7.7   General steps of data-mining-based cyber attack detection techniques.

where a group of data points pertaining to separate categories (normal activities, contrasting intrusion) are to be separated through modelling.

Data-mining-based cyber attack detection involves five broad phases, as illustrated in Figure 7.7: (1) system observation and data capturing through sensors, network and sniffing agents, and security devices; (2) information pre-processing (e.g., cleansing, filtering, normalization, etc.) at local information stores; (3) event correlation and feature extraction (e.g., through Hadoop Distributed File System [HDFS], and processing big data), (4) data mining (dimensionality reduction, classification, clustering) to observe misuse or anomaly; and (5) visualizations and interpretation of mining results.

In conclusion, these stages may be put into three phases, namely processing, analysis, and visualization, with processing covering the first two stages and analysis the middle two stages. The techniques and methodologies in data mining have been perpetually transforming. A variety of ingenious and instinctive methods are there that have surfaced which fine-tune data processing ideas so as to enlighten companies with detailed insights into their own data and technological trends in the future for the same. Data-mining experts employ a plethora of steps. A few of them are:

Identifying fragmentary data: This step is dependent on the data that is already existing; therefore in case the data is inadequate results, in this case will not be up to the mark. Therefore, it becomes essential to possess the ability to detect fragmentary or incomplete data. One such method is self-organizing maps, or SOMs, that facilitate to identify missing data through multidimensional modelling and visualization of complex data. "Multitask learning" is used for omitted inputs where single valid information that exists is analyzed with its procedures are parallel with a similar data set (compatible), that is incomplete. Preceptors that are multidimensional in nature employing intelligent algorithms to formulate imputation tools can be used to solve the issue of incomplete features of information.

Database analysis: Databases possess pivotal information in a structured composition, therefore algorithms designed employing SQL macros, language which is attributed to the databases, to search out implicit patterns in an ordered information is quite advantageous. The algorithms of this type are embedded in the data flow, such as functions that are user defined and results of the report that are *ready to refer* along with relevant analysis. Possessing a snapshot of data from a huge repository in a cache file is a beneficial technique as it enables examination at a later point. Correspondingly, these algorithms must have the ability to fetch data from a multiple and diverse sources and calculate the trends

Dealing with relational and complex data with efficacy: Query-based algorithms that are interactive, support data warehousing for all classes of functions including but not limited to categorization, association, clustering, and estimating trends. Alternative ideas which catalyze data processing that is interactive in nature are examining graphs, meta-rule-based mining, aggregated query, digital image processing, swap randomization, and applied mathematics analysis.

Some data-mining tools:
1. Orange
2. Rapid Miner
3. Sisense
4. SSDT
5. Apache Mahout

7.5   SECTOR MAPPING FOR CYBER ATTACKS

Attacks that focus on IoT devices that have resource constraints have increased significantly in the past few years. The vulnerabilities in the security sector of IoT technologies are being incessantly identified. These technologies are used in both industrial and home environments such as sensors, industrial actuators home appliances, medical devices, etc. The current state of affairs is exacerbated by defects in application, hardware chips that are faulty, and tamperable devices along with misconfigurations.

This section aims to use a risk-like approach to examine cyber attacks with respect to IoT-enabled devices, so as to highlight the existing threat landscape and isolate hidden and covert attack paths taken against critical infrastructure. Figure 7.8 illustrates critical attack vectors enabled by IoT devices where an attacker exploits any vulnerability along with his/her skill to compromise the device. Post this exploitation of all connections, the attacker will eventually attack the actual target that is the critical system.

In IoT-enabled cyber attacks, the device is the amplifier or the enabler of an attack, the perpetrator identifies and takes advantage of inherent vulnerabilities related to one or multiple layers of the device so as to achieve his/her goal. We

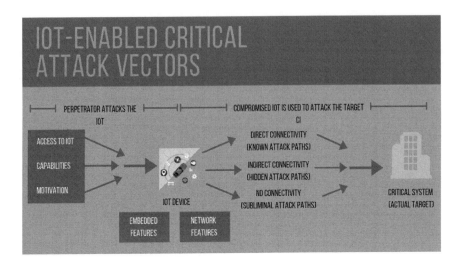

FIGURE 7.8   IoT-enabled critical attack vectors.

classify IoT vulnerabilities in two primary classes: embedded vulnerabilities and network vulnerabilities.

Through harnessing capabilities of the network, the IoT devices gains the ability to connect with different systems, through means which can not be simply percept thus facilitating management of control systems remotely. The attacker might exploit these connections to attack critical infrastructure and associated systems. Table 7.1 maps the category of target with the type of vulnerability the attacker can potentially use.

Direct connections (with CI): Here, there exists a direct connection between the IoT device and the continuous integration (CI) system. The connection may be

TABLE 7.1
Assailability in IoT Device

Assailability in IoT Device	Type of Vulnerability
Hardware layer	• Lack of tamper resistance • Weak embedded crypto algorithms • Weak hardware implementations
Software layer	• Firmware layer • Operating system • Application layer
Communication protocols	• Link and network layer protocol threats • Application layer protocol threat • Network design flaws
Key management	• Absence of support for public key exchange • Easily extractable communication keys • Employing of common or no key

physical or via a logic gate. Essentially, this direct connection produces attack vectors that are easily found and the potential impact can be assessed.

Indirect connection (with CI): There have been a myriad of cases where the device has not been linked to a CI system directly but in an implicit/indirect manner and the attack has been orchestrated. Attacks of this nature mostly misuse device's communication protocols and can be equally impactful as direct connectivity-based attacks, largely owing to the fact that they are not noticed and further underestimated thus impeding the system's preparedness to counter them.

Absence of connectivity (with CI): Intelligent IoT devices which don't seem to be connected, even indirectly with CI systems, are susceptible to cyber attacks. Even proximity in the physical sense could enable attacks against these systems. Further, in other scenarios, the primary issue is the availability and volume of vulnerable IoT devices that have access to the web and thus can be exploited by the attackers.

7.5.1 INDUSTRIAL SYSTEMS AND SCADA SYSTEM

Control systems used in manufacturing processes and industries are extremely process critical in nature and they require high availability. These systems may be automated either partially or fully. These industrial control systems (ICS) collect data pertaining to the status of a process through a myriad of devices which nowadays are enabled by IoT for efficacy.

SCADA, or supervisory control and data acquisition systems, are used to cater to the monitoring and control operations systems that are distributed in nature over huge geographical regions. These include electrical power distribution from grids; manufacturing processes; and the oil and gas sector to monitor and control pipelines, supply of water, and so on. The challenge of augmenting the security of these systems has been daunting especially when this looked at in reference to conventional IT infrastructures. There have been a myriad of choke points and vulnerabilities reported for these systems. The attacks in these cases have been classified based on the surface of attacks and the SCADA field instruments the attack is going to impact. The direct targets can be the control instruments that are connected to the Internet such as programmable logic controllers (PLCs), improvised explosive devices (IEDs), and remote terminal units (RTUs). Further, this may be achieved by breaching the workspace of higher levels of SCADA such as the control centre and the info-tech network so as to use them as ancillary means of entry into the SCADA network. Considering alternative scenarios for ICS that employ IoT-enabled instruments, the perpetrator may attempt to breach the end devices directly. The various attack vectors in this type have been illustrated in Figure 7.9.

Therefore, the attacks on SCADA systems can be trifurcated into three categories:

1. Attack on corporate/IT network, control centre
2. Attack on IoT-enabled the PLC, RTU, IED
3. Attack on IoT field instruments and devices

One of the most significant examples of a cyber attack on industrial systems was STUXNET [19] in June 2010. In 2013, Trend Micro, a network security firm,

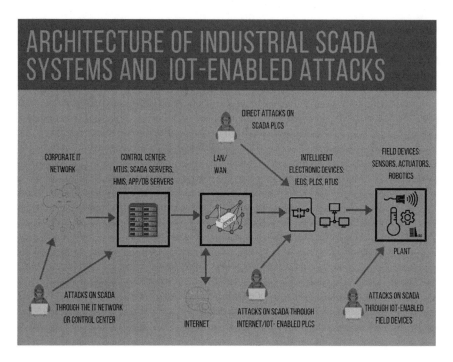

FIGURE 7.9   Architecture for Industrial SCADA systems and IoT-enabled attacks.

employed a control system based on network of Honeypots. These systems were deployed in eight states to collect information about cyber attacks on ICS [20]. The system in a span of 3 months observed 74 intrusions, sourced to 16 states with nearly 60% of them originating from the Russian Federation and with 11 of the instances termed as critical.

7.5.2   TRANSPORTATION SYSTEM

Intelligent transportation systems (ITS) [21] aim to provide innovative and advanced application in transportation services. This encompasses intelligent automobiles and infrastructures, management systems for railways, intelligent naval vessels, and management systems for aircraft and air traffic control. The inception of ITS catalyzed development in this field and led to improving efficacy in terms of transportation, safety, traffic management, etc. But this development has been accompanied by increased susceptibility of the transportation sector to cyber attacks, which may have severe consequences to both people and the infrastructure. European Union Agency for Cybersecurity in its report highlighted the severe lack of policy and regulatory instruments to safeguard cybersecurity of intelligent transportation. Further, the lack of interest in terms of budget allocation and awareness of certain operators exacerbate the threat index in this sector.

In this section we examine attacks catalyzed by IoT technologies in three subsectors of transportation:

Attacks on smart cars and traffic control infrastructures: Elementary discussions [22] highlight the surfeit of intrusions in order to compromise the Control Area Network (CAN) bus. Specifically designed messages can be transferred to the bus, and the system can be easily manipulated to control the display of speedometer, shut down the engine, trigger emergency brakes, etc. These susceptibilities of the CAN bus were extensively highlighted by Valasek [23]. Ergo, an intrusion of this nature requires exploiting the target vehicle by means of physical tampering and hence making it dissimilar from other IoT-enabled attacks.

The attacks on intelligent vehicles enabled by IoT are trifurcated as follows:

1. Using communication protocol
2. Using vehicle's in-built information/entertainment system
3. Intrusion based on augmenting IoT sensors of the vehicle

Attacks on railway control systems: On the proximate level, attacks on railway infrastructure may seem unlikely, however, there exists a myriad of cases [24–26] that highlights the destructive capability of cyber attacks on infrastructure on this nature. The attacks in this subsector use a similar mechanism of attacks as on smart automobiles (as discussed earlier).

The types in this category are:

1. Explicit/active attack on ICS of railway and associated infrastructure
2. Implicit/passive attacks to exploit passenger data

IoT-enabled attacks on aircraft: Airplanes and traffic management systems are advanced, refined, and extremely linked systems that are susceptible to numerous security issues. Instances of attacks in this sector encompasses breaching and impeding the operations of the passport management network, inflicting denial of service attacks to systems employed to direct flight plans, etc. However, the attacks do not conform with the usual IoT-enabled instances per se like we have seen in other cases. But IoT applications in aircraft are delegated to systems such as communications and navigations which may empower the attacker to cause severe damage as they can gain remote access to more critical instruments through them and further exacerbate the severity.

Instances of attack in this sector can be:

1. Exploiting weak points of wireless surveillance network
2. Creating frailty in the in-flight entertainment system

7.5.3 Medical Systems and IoT Health Devices

Internet of Things has played a crucial role in the biomedical and health sector, and has been employed in this sector for various purposes such as accurate monitoring of clinical activities, optimizing the process of patient follow-up, and so on. This has provided efficacy in operation and ameliorated quality of services to the patients. Instances of attacks on medical instrumentation enabled by IoT include: change

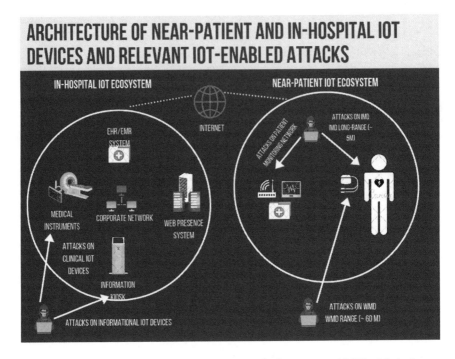

FIGURE 7.10   Architecture of attack vectors in medical systems and IoT health devices.

or denying a treatment, abusing the functionality of the instrument (for example increasing the electricity levels in a CAT scan, changing the settings in a defibrillator), tampering the surgery schedule, or altering an inventory list.

The attacks on medical systems can be classified into the following two subsections as illustrated in Figure 7.10:

1. Attacks on near-patient IoT devices: using the susceptibilities of internal medical devices and wireless medical devices, breaching home monitoring network, stealing health data from personal health devices
2. Attacks on in-hospital IoT devices: harnessing the vulnerabilities of medical instruments, changing system configurations and settings to cause damage, stealing medical records

7.6   FURTHER ADVANCEMENT AND CONCLUSION

The contemporary world has become more reliant on systems, with essential utilities that now are dependent on the Internet. With every new technology comes a possibility of new weapons and consequently with that comes a soldier aiming to wield it for their advantage. This digital revolution has spread in both the civilian and military domain. Now, the weapons are guided by satellites, UAVs are remotely piloted from a command centre, and modern warships are transformed into massive data processing centres. In the current era, bits and bytes are more damaging than bombs and bullets.

There exists a litany of challenges on legal and technical fronts with respect to cyber attacks. There exists a lack of a universally agreed definition of cyber attack and other terms. Further, the Shanghai Cooperation Organisation states and NATO states find each other at diametrically opposite ends on issues ranging from state responsibility, attribution, and limits to use of force. There has been an active call for a safeguarding system from intellectual property threats which have cost trillions of dollars to the global economy. Adequate steps among states are required to converge so as to initiate plans and policies for ensuring perpetrators of cyber attacks are deprived of safe havens to conceal both their bases of operation as well as the source of their attacks. Participation in cyberspace has been increasing drastically ever since its inception and but this has brought in new threats and has exacerbated the existing threats. This has changed the capacity to commit crime, which has asymmetrically changed the capacity to cause-real world terror. A plethora of sector-oriented standardisation steps have been taken, such as industry 4.0, with the aim to include this perpetually changing landscape. We believe that with the collective resolve of the global community and a desire to achieve peace, we can find an answer to this predicament in regulatory reform in modern warfare.

REFERENCES

1. Ezez. 2019. "Blog: 10 mind-boggling figures that describe the Internet of Things (IoT) | Cleo." *Cleo.* https://www.cleo.com/blog/Internet-of-things-by-the-numbers.
2. Razzaq, Mirza Abdur, Sajid Habib Gill, Muhammad Asif Qureshi and Saleem Ullah. 2017. "Security issues in the Internet of Things (IoT): A comprehensive study."
3. Review, NATO. 2019. "The history of cyber attacks - a timeline." *NATO Review.* https://www.nato.int/docu/review/2013/Cyber/timeline/EN/index.htm.
4. International Law Commission. 2001 November. "Draft articles on responsibility of states for internationally wrongful acts." Supplement No. 10 (A/56/10), chp.IV.E.1. https://www.refworld.org/docid/3ddb8f804.html [accessed 4 July 2019].
5. Schmitt, Michael N. 2013. "Tallinn manual on the international law applicable to cyber warfare". 1st edition, Cambridge, UK: Cambridge University Press, ISBN-13: 978-1107613775.
6. Mishra, Bimal Kumar and Hemraj Saini. 2009. "Cyber attack classification using game theoretic weighted metrics approach." *World Applied Sciences Journal*, vol. 7, pp. 206–215.
7. Lupu, T. G.. 2009. "Main types of attacks in wireless sensor networks." *Recent Advances in Signals and Systems*, pp. 180–185.
8. Al-Shurman, Mohammad, Seong-Moo Yoo and Seungjin Park. 2004. "Black hole attack in mobile ad hoc networks." *Proceedings of the 42nd Annual Southeast Regional Conference on - ACM-SE* 42, pp. 96–97. ACM Press. doi:10.1145/986537.986560.
9. Saad Al Shahrani, Abdullah. 2011. "Rushing attack in mobile ad hoc networks." *Proceedings - 3rd IEEE International Conference on Intelligent Networking and Collaborative Systems,* INCoS 2011, pp. 752–758. doi:10.1109/INCoS.2011.145.
10. Khalip, Andrei. 2018 February 19. "U.N. chief urges global rules for cyber warfare." https://www.reuters.com/article/us-un-guterres-cyber/u-n-chief-urges-global-rules-for-cyber-warfare-idUSKCN1G31Q4.

11. Phillips, C. and L. P. Swiler. 1998. "A graph-based system for network-vulnerability analysis." *Proceedings of the 1998 Workshop on New Security Paradigms, ser. NSPW '98*. New York, NY: ACM, pp. 71–79. [Online]. doi:10.1145/310889.310919.

12. Mulazzani, M., S. Schrittwieser, M. Leithner, M. Huber and E. Weippl. 2011. "Dark clouds on the horizon: Using cloud storage as attack vector and online slack space." *USENIX Security Symposium*, San Francisco, CA.

13. Caltagirone, S., A. Pendergast and C. Betz. 2013. "The diamond model of intrusion analysis." *DTIC Document, Tech. Rep.*

14. Lin, Xiaoli et al. 2009. "Threat modeling for CSRF attacks." *2009 International Conference on Computational Science and Engineering*, IEEE. doi:10.1109/cse.2009.372.

15. U. S. J. C. of Staff. 2000. "Joint tactics, techniques, and procedures for joint intelligence preparation of the battlespace." *Joint Chiefs of Staff.*

16. Barik, Mridul Sankar, Anirban Sengupta and Chandan Mazumdar. 2016. "Attack graph generation and analysis techniques." *Defence Science Journal* 66(6), p. 559. Defence Scientific Information and Documentation Centre. doi:10.14429/dsj.66.10795.

17. Ezez. 2019. *Cisco.com*. https://www.cisco.com/c/dam/global/tr_tr/assets/docs/SAFE_Code-Red.pdf.

18. Falliere, N, L. O. Murchu and E. Chien. 2011. "W32.stuxnetdossier." *White Paper, Symantec Corp., Security Response, vol. 5.*

19. Wilhoit, Kyle. 2017. "The SCADA that didn't cry wolf. Ebook." *Trend Micro*. https://www.trendmicro.de/cloud-content/us/pdfs/security-intelligence/white-papers/wp-the-scada-that-didnt-cry-wolf.pdf.

20. Skorput, Pero, Hrvoje Vojvodic and Sadko Mandzuka. 2017. "Cyber security in cooperative intelligent transportation systems". *2017 International Symposium ELMAR*. IEEE. doi:10.23919/elmar.2017.8124429.

21. Koscher, Karl et al. 2010. "Experimental security analysis of a modern automobile." *2010 IEEE Symposium on Security and Privacy*. IEEE. doi:10.1109/sp.2010.34.

22. Miller, C. and C. Valasek. 2014. "A survey of remote automotive attack surfaces." *Black Hat USA*.

23. Leyden, J. 2008. "Polish teen derails tram after hacking train network." *The Register*, 11.

24. Zetter, Kim et al. 2019. "Hackers breached railway network, disrupted service." *WIRED*. https://www.wired.com/2012/01/railyway-hack/.

25. Ezez. 2019. "Hackers are holding San Francisco's light-rail system for ransom". *The Verge*. https://www.theverge.com/2016/11/27/13758412/hackers- san-francisco- light- rail- system- ransomware- cybersecurity- muni.

26. Ezez. 2019. "Medical devices hit by ransomware for the first time in US hospitals." *Forbes.com*. https://www.forbes.com/sites/thomasbrewster/2017/05/17/wannacry-ransomware-hit-real-medical-devices/.

27. Manadhata, Pratyusa K. and Jeannette M. Wing. 2011. "An attack surface metric." *IEEE Transactions on Software Engineering* 37(3), pp. 371–386. Institute of Electrical and Electronics Engineers (IEEE). doi:10.1109/tse.2010.60.

8 A Review of Cyber Attack Analysis and Security Aspect of IoT-Enabled Technologies

Joy Chatterjee, Atanu Das, Sayon Ghosh,
Manab Kumar Das, and Rajib Bag

CONTENTS

8.1   INTRODUCTION

Nowadays, from almost everywhere, different devices, sensors, embedded software, etc. are either connected to communicate with each other or they are responsible for accessing each other via some gateway and for providing the facility to access these devices through a smart phone or computer. This is known as the Internet of Things (IoT). These devices are accessed from a distance, and that is why the concept of IoT is an important research topic. IoT furnishes intelligent services to create a new advanced world for human beings[1].The terminology 'things' in IoT integrates different types of devices, i.e. sensors, which monitor and collect heterogeneous data on systems from human activities[2].The leading aim of the IoT is to serve the facility to create a network infrastructure by maintaining some communication protocols along with some embedded software to allow physical sensors, personal computers, or smart gadgets such as tablets and smart phones. The importance of IoT is to connect various sensors (i.e. IoT devices) and gadgets for interaction among themselves in the absence of human interventions in a wide variety of networks. These devices are not only connected to each other, they can also communicate among themselves to exchange information and to correlate decisions [3]. In the concept of Internet of things, different types of IoT devices sense the data from different sources as an input and transfer these input data over a network without any human interaction. IoT defines the embedded systems, which are capable to interact with different types of IoT devices with the users in different smart automation system [4].Our traditional devices which were used regularly are changed to common household objects by the quick expansion of IoT [5]. The concept of IoT can be applied in different fields of smart automation systems like home automation systems, health monitoring, smart transportation systems, surveillance, and weather forecasting. Classification of security issues, threats, and different types of attacks are a few of the major challenges in IoT (Figure 8.1)

8.1.1 IoT Devices

Different types of hardware components allow creation of the digital world enabling the IoT system. Home appliances, smart vehicles and buildings, healthcare gadgets, etc. can be termed as smart things [6]. A variety of sensors (e.g. humidity, temperature, moisture, weight, pressure, radio frequency identification [RFID], etc.) plays a

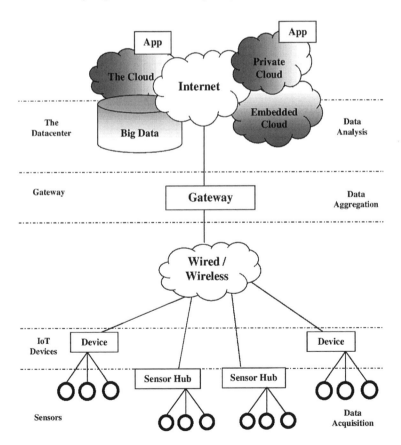

FIGURE 8.1   A structural diagram of data flow in IoT network.

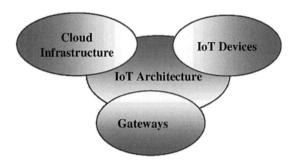

FIGURE 8.2   Integration of different components in IoT.

major role in the IoT system by collecting information from the physical environment. These sensors do not work as individuals; they communicate through some gateway (e.g. Arduino, Raspberry Pi, etc.) by creating a network. One IoT device communicates with another IoT device with the help of other components by maintaining some protocol (Figure 8.2). The classification of IoT devices depends on size and whether they are connected. Security mechanism, memory and power management, and computational performances are major challenges in concept of IoT implementation.

8.1.2   CLOUD INFRASTRUCTURE

The infrastructure of cloud computing plays an important role for the IoT revolution. IoT produces big data, and all these data, which are sensed by any IoT sensors (devices) as an input from physical environment. This is stored in the cloud server for further processing and the desired output is provided through a smartphone or PC. Good cloud infrastructure facilitates data transportation through dedicated links or the Internet.

8.1.3   GATEWAYS

The important role of IoT gateways is to move the information from an IoT device to the cloud or vice versa (Figure 8.3). It may be hardware or software which acts as a connecting point between controllers and cloud. It also integrates IoT protocols for networking and provides facilities for data flow securely between edge devices and the cloud. It provides additional security for the IoT network and data transmission in both directions. The concepts of different types of attacks are associated with the concept of gateways, so different types of security aspect are introduced here.

8.1.4   ARCHITECTURE OF IoT

Architecture of IoT supports connection between heterogeneous devices through the Internet [8].This architecture is based on the concept of three layers: application, network, and perception.

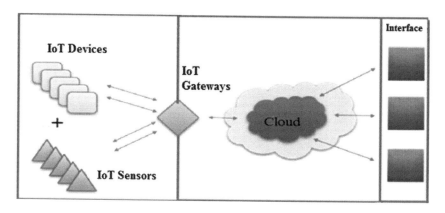

FIGURE 8.3   Association of components to build IoT architecture.

8.1.4.1   Application Layer

The application layer in the specified architecture provides different types of solutions in various fields. But there are some security issues with IoT layers, which create major problems in the IoT architecture. This layer is responsible for solving the security problem or privacy protection along with authentication. Data security is very challenging task for IoT research domain.

8.1.4.2   Network Layer

The network layer provides security during transmission of data over the IoT network. In this layer, data can be routed through any unguided media. A major issue of security pertains in this layer, while the sensing data transmit through the networks. Various types of attacks such as man-in-the-middle and denial of service (DoS) can hamper genuine data when they transmit/route over the network.

8.1.4.3   Perception Layer

The perception layer plays an important role in collecting information from different sensing devices. These sensing devices are attached with microchips for wireless data transmission. Also some common attacks are associated with this layer of IoT architecture such as denial of service attack and node attack (Table 8.1).

8.2   OVERVIEW OF IOT-ENABLED TECHNOLOGIES AND SERVICES

The activity of different type of sensors (devices) in IoT is responsible to collect or gather information from the physical environment. Industrial application, smart home application system, healthcare etc. are implemented by using variety of sensors. Sometimes this sensor does not work individually instead they work with the help of some other IoT devices (e.g. actuators) such as relay and servomotor.

8.2.1   Sensors Used in Various IoT Applications

See Table 8.2.

TABLE 8.1

Security Description and Protocols Used in Different Layers of IoT Architecture

Name of the Layer in IoT	Security Description	Protocols Used
Application layer	Security of web browsing such as TLS/SSL and security to provide user authentication for communication devices.	Constrained Application Protocol (CoAP), XMPP, MQTT, AMQP, etc.
Network layer	Different types of security for information over the Internet and Wi-Fi security.	Different types of routing protocols such as Channel-Aware Routing Protocol (CARP), RPL, CORPL, etc.
Perception layer	Security on RFID and GPS as well as security on wireless sensor networks for different applications such as smart cities, smart homes, and factory automation systems.	Basically responsible to acquire or collect the data from the physical world through different types of sensors.

8.2.2 APPLICATIONS OF IoT

IoT plays an important role to live and work smartly. IoT provides businesses a new revolution on implementing automated systems. It reduces human effort and hazards. On the basis of various IoT devices, numerous data are collected from different sources. These data are processed to make unique decisions without and human interference. IoT has become the most important technology among every industry including manufacturing, retails, finance, healthcare, and inventory management. IoT was even introduced in our daily life to monitor our homes, cities, vehicles, etc. (Figure 8.4). It reduces waste and energy consumption also. This technology was also included in farming to increase the level of production.

8.2.2.1 Smart Home Automation System

With smart home automation, one can switch on the air-conditioning or lights before reaching home or switch off lights after leaving home. Sometimes this concept of IoT is also applicable to locking and unlocking home doors. The concept of smart home automation system can make the lifestyle of humans easier like the smart phone.

8.2.2.2 Smart Car

IoT is also applicable on optimizing vehicles' internal functions. It is able to optimize the internal operation of a car and be responsible for maintenance along with passenger comfort using different types of sensors. Some of major car brands, like BMW and Tesla, are working with this technology for further revolution in the automobile sector.

8.2.2.3 IoT Industrial Application

IoT also creates a vast area in the industrial sector. It creates an intelligent machine in industrial engineering by providing sensors, software, data analytics, etc. It is not

TABLE 8.2

Different Sensors and Their Application in IoT

Name of Sensor	Description	Application
Temperature sensor	Used to detect any changes in temperature from any source and measures the energy change to transform into digital data.	Manufacturing processes in industry, environmental control, etc.
Proximity sensor	It detects the presence or absence of any nearby element and generates a signal which can be easy to understand.	Smart car parking, retail industry.
Pressure sensor	It detects change in pressure and generates an electric signal.	Manufacturing process, water management system.
Water quality sensor	It detects different types of particles in water distribution systems to measure quality of water.	To check the water quality for residence or industry supply.
Chemical sensor	Indicates change in composition of liquid or chemical with respect to air.	Industrial environmental monitoring and harmful chemical detection for paints, plastics, and rubber industries.
Gas sensor	Used to check the quality of air and identify the presence of different types of gases.	Oil and gas industry, laboratory research, pharmaceuticals and petrochemical industry.
Smoke sensor	It senses smoke, e.g. airborne particulates, and its level.	Manufacturing industry and multi-storage buildings.
Infrared sensor	Used to detect certain features of its nearby objects by infrared radiation.	Different type of IoT projects, also for the project on healthcare monitoring system.
Image sensor	Used to transform images into digital signals for further processing.	Image sensors are connected with a wide range IoT devices such as the smart car manufacturing industry.
Motion detection sensor	Applied to observe any physical movement of any object in a particular region and responsible to convert it into electric signal.	Automatic door control for home automation, toll plazas, automatic parking systems, energy management systems, etc.
Humidity sensor	It senses quantity of atmospheric water vapour and converts it into an electric signal.	Weather forecast reports or different IoT projects in agriculture and industrial domains.
Optical sensor	It measures the physical quantity of light rays.	Healthcare, environment monitoring, aerospace, and other industrial applications.

possible to manually monitor all the things of a big industry, so to overcome this IoT is helpful for industrial engineering.

8.2.2.4   IoT in Smart Cities

Surveillance, traffic congestion control, water distribution, etc. are popular applications in smart cities. The concept of IoT can be applies in different areas to make

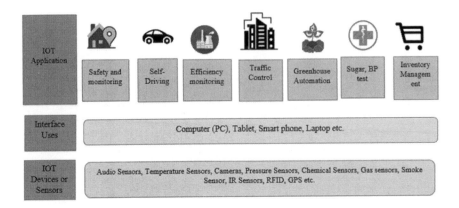

FIGURE 8.4   Application and different devices used for IoT.

cities smart thus providing better lives for residents. As an example of this type of application, sensors can be installed in different areas in a city to collect the data about parking slot availability.

8.2.2.5   IoT in Agriculture

IoT has become more popular in the field of agriculture with the continuous increase in population which leads to a needed increase in food supply. Governments try to help farmers by providing the facilities of advanced technologies, most of which are enabled with the concept of IoT. The most popular growing field in IoT is smart farming. Through this technology farmers can easily get reports of soil moisture, weather forecasts, etc. to yield better return on investment.

8.2.2.6   IoT in Healthcare Monitoring

The concept of IoT can also apply for healthcare monitoring and smart medical devices. In this concept various sensors devices connected with the human body sense/read data like pulse rate, metabolic rate, etc. Collection and analysis of personal information generated from healthcare gadgets play an important role in monitoring health conditions.

8.2.2.7   IoT in Smart Retail

IoT is also popular in the retail stores. Retailers are able to get valuable feedback from customers and get connected with them as well as wholesalers under one umbrella. Smartphones with IoT enabled is the way of communication between retailers and consumers. Retailers can also track the behaviours of consumers and can provide better service and favourable products.

8.3   VULNERABILITIES, ATTACKS, AND SECURITY THREATS ON IOT DEVICES

Threats and different types of attacks are not new for any kind of networking along with IoT. Threats and attacks are the major issues for IoT security when sensitive information routes over a network. Sometimes the privacy and the secrecy of a person can be lost due these types of threats and attacks where intruders may intercept into someone's personal information.

8.3.1   IoT THREATS

Most threats use the advantages of loopholes in a system [9]. These can be categorized into two types: the first being natural threats, viz. earthquakes, hurricanes, floods, etc. Some preventive measures (backup and contingency plans) can be introduced against natural threats, but nobody can prevent them from happening. The second category is threats from human, which we can try to prevent [10]. Different types of protocols are introduced in different layers in IoT architecture to prevent such threats, which are directly related with IoT security.

8.3.2   CYBER SECURITY CHALLENGES

Cyberspace includes IT networks, computer mobile devices, and various intelligent devices. As it is limitless, networks are not bound to geographic boundaries. Security problems also arise rapidly proportional to an increase of the network. More usage of Internet makes cyber space vulnerable. Protection of ICT infrastructure and cyber infrastructure is a major challenge. Due to digitization, people access more and more e-services such as e-commerce and e-governance, which creates concerns about fraudulent personal data.

8.3.2.1   Rapid Changes of Smart Devices

A variety of new product launches day by day rapidly develops businesses and create a vast market place. These products adopt new technology for market expansion and expand the business as well. Sometimes these changes make an organization's cyber security complex, which can be very harmful.

8.3.2.1.1   A Network of Networks

In IoT architecture, different devices are associated with a large network via small networks. The integration of mobile computing with the large network increases the possibility of data hacking. In the IoT system one vulnerable device can control or lead other devices which are connected in a single network at any time from everywhere. It can be a major issue for cyber attacks. IoT devices are expected to provide better service for users, but attacks can interrupt and degrade performance and data will not be secure. Networks provide the opportunity for cyber criminals to access the valuable information when the data is routed through billions of devices.

8.3.2.1.2   Malicious Mobile Applications

Nowadays different types of apps are the highlights of our smart mobile devices. These are the integrated parts of the smart phone revolution. Personal data can be hacked through malicious apps or malware when users access their smart devices. It is very challenging to protect devices from unauthorized access.

8.3.2.2   Data Privacy Issues

Most smart devices (e.g. laptops, smart phones, tablets) contain the personal information of users. These devices also hold different types of sensitive information such as bank details and details of the user's daily life. Most users are not comfortable sharing such information by using third-party applications for data privacy. This concept is also applicable for IoT systems. IoT devices collect a variety of information from the physical environment to provide service to the users, which will be valuable and sensitive. In IoT, different types of techniques (e.g. data encryption, stenography) for data privacy are introduced to secure this information.

8.3.2.3   Utilization of Bandwidth

Millions of sensing devices communicate with each other via a single server for data transmission. Data can be flooded due to traffic congestion in the single server, so the possibility of missing data will increase and also be a major problem for data security. Most of the devices (e.g. sensors, actuators) utilize unencrypted dedicated links for data communication to overcome this problem.

8.3.2.4   Cloud Security Issues

The IoT concept is based on cloud computing because it provides the platform for different devices to communicate with one another. Different types of challenges can arise in cloud security. Security in the cloud is vital in IoT because security of all sensitive information relates with it. Here the risk can be increased in spam of cloud servers.

8.3.3   Attacks in IoT

In IoT, millions of physical devices are communicating with other devices such as software, sensors, and actuators through a network. Security is one of the major problems for communication, so the possibility of different types of attacks increases day by day. There are different types of attacks introduced in IoT systems. Some of them are very common for everyone like man-in-middle attack, sniffer, IP spoofing, DoS attack, malware, etc. [11]. A few other attacks are present in network topology such as sinkhole attacks, wormhole attacks, and black hole attacks [12].Most of them cause users to lose their privacy and mental suffering [13].

8.3.3.1   Impact of Different IoT Attacks
See Table 8.3.

8.3.3.2   Different Attacks over Different Layers in IoT Architecture
See Figure 8.5

8.4   COMPARATIVE STUDY OF DIFFERENT TECHNIQUES IMPLEMENTED TO RESOLVE CYBER SECURITY AND IOT ATTACKS

Many algorithms and techniques on IoT framework were developed to give importance over minimization of security issues. Major security problems on IoT devices

TABLE 8.3
Various Attacks in IoT along with Description

Name of Attack	Description
Man-in-middle	In this attack any malicious or unknown person inserts him-/herself between the conversation of sender and receiver and tries to access the valuable information. Through this attack, data can be hacked during the conversation.
Sniffer attack	This is an application that basically captures the data packet during the movement of a network. It is used to hack any specific network to get sensitive information such as account details and passwords. If the packet data is not encrypted in a proper way, the attackers first capture the packet and easily hack the packet and decrypt it.
IP spoofing	This technique is based on Internet Protocol (IP), where devices are interconnected and identified by a unique number. It is used to hijack the user browsers, which are connected over a network by modifying the packet header with a spoofed source IP address.
DoS attack	In this attack an unauthorized person prevents legitimate users from accessing any specific devices. The concept of this attack is based on incoming traffic flooding when a large number of packets is received from different sources.
Malware attack	Malware is nothing but malicious software which performs different activities on the user's (victim's) system without any knowledge of the user. This type of attack is very harmful and can damage any system. Some popular malware are ransomware and spyware.
Sinkhole attack	In sinkhole attacks any malicious node (system) can attack any neighbour node during data transmission, so the throughput of the attacked node will gradually decrease. This attack is more harmful in combination with another attack.
Wormhole attack	This type of attack can occur when at least two systems directly communicate with each other at different frequencies. In this concept the nodes do not follow the proper path for data routing.
Blackhole attack	Here one malicious node receives packets from other node for routing, but it drops the packet itself. It is very harmful with the integration of any other attack.

were taken up as a vital research topic. Privacy and security were combined together as real issues in data transmission and communication. We have illustrated some of the few research work done in the last few years (also see Table 8.4).

The development of IoT can be classified into information perception, development of artificial intelligence, and intellectual interpretation. Several issues of the perception layer were discussed and the technologies involved in IoT were also elaborated by Xu Xiaohui [14]. Counterfeit attacks and malicious programme attacks are some of the security problems related to sensor networks. Digital certification is the key technology used to ensure privacy and security between different objects during communication.

Arbia Riahi Sfar et al.[15] discussed the classification of security and privacy challenges and the optimal solution in the IoT. Vital solutions about data access control were discussed.

Design issues, viz. heterogeneity, mobility of devices, network connectivity, identification of object, and rules of resource and information exchange in connection

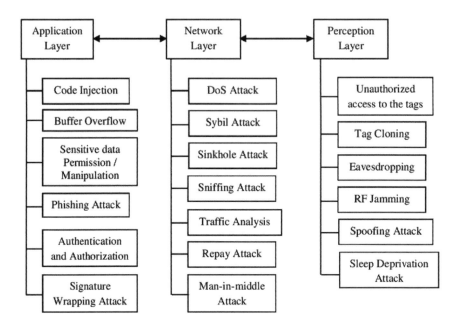

FIGURE 8.5 Pictorial representation of different attacks related with IoT architecture.

to IoT application and its security issues are discussed by Subho Shankar Basu et al. [16]. Common threats like tampering, DDoS, information integrity, and privacy of users were the main areas discussed. To get rid of the aforementioned issues a framework was proposed.

Several security issues such as wireless transmission security, RFID tag security, and information security were discussed by Chen Qiang et al. [17].They investigated existing work on network security. The problem in processing large amount of data collect from IoT devices and providing security and reliability is a main concern. A new security measure for IoT was implemented.

Different protocols were explained to enforce security issues of IoT by Jorge Granjal et al. [18]. The protocols were incorporated to increase security issues in different network layers.

Surapon Kraijak et al. [19] focused on the IoT protocols, architecture, and security issues. The architecture of IoT was divided into five layers on the basis of application. The difference of IoT security and authentication policies and future trends of IoT were very clearly depicted.

Various ways of addressing security and privacy in IoT were explored by Raja Benabdessalem et al. [20]. Various kinds of threats like DDoS attacks and eavesdropping were discussed. Cryptographic algorithms were used to resolve the problem of security during data communication among IoT devices. Security aspects like authorization, authentication, confidentiality, data security, and protection must prevail in IoT.

IoT security challenges are well explored by Mahmud Hossain et al. [21]. An analysis of security challenges on different applications was performed. The critical

TABLE 8.4

Issues and Challenges in IoT

Reference	Focused or Discussed Area	Issues and Challenges
Xu Xiaohui [14]	WSN security issues and data perception and security in processing.	a) Counterfeit attacks b) Malicious programme attacks
Arbia Riahi Sfar et al.[15]	Discussed classification of security challenges and optimal solutions in the IoT. Also discussed cognitive and systematic approaches for the IoT.	a) Intelligent objects in hostile zones may face risks of physical attacks. b) Adequate effort is provided in maximum protocols and algorithms
Subho Shankar Basu et al. [16]	Security challenges like network connection, mobility of devices, addressing and authorization of object, heterogeneity, spatio-temporal services, and rules of resource and information exchange.	a) DoS b) Spoofing c) Tampering d) Repudiation e) Privacy of user f) Replay attack, etc.
Chen Qiang et al. [17]	Wireless security, transmission security, security of RFID tag, privacy protection, and information security.	a) RFID identification b) RFID security issues c) Radio signals attack d) Information security and privacy
Jorge Granjal et al. [18]	Discussed physical, network, application, and MAC layer secured communications.	a) IEEE Standards Communications a) Security in RPL, 6LoWPAN, and CoAP
Surapon Kraijak et al. [19]	Focus on the IoT protocols, architecture, and security issues.	a) Eavesdropping and DDoS attack b) Privacy in IoT and storage devices
Raja Benabdessalem et al. [20]	Various ways of address security and privacy in IoT.	a) Includes authenticity, authorization, integrity, and privacy of personal data b) Attackers transmits malicious programme to tamper their resources
Mahmud Hossain et al. [21]	Discussed security constraints related to hardware, software, and networks.	a) Computational and energy conservation, memory, and embedded software constraint b) Mobility and scalability if the number of devices increases
Glenn A. Fink et al. [22]	Societal impact of IoT in maintaining standards, security and privacy, and different types of vulnerabilities.	a) Crime b) Data privacy c) Protocols d) Cyber warfare
Qi Jing et al. [23]	Discussion on security issues and the cross-layer heterogeneous integration.	a) Uniform coding, trust management, and conflict collision b) Low storage space and low processing capability

(Continued)

TABLE 8.4 (CONTINUED)
Issues and Challenges in IoT

Reference	Focused or Discussed Area	Issues and Challenges
Ahmad W. Atamli et al. [24]	Discussed three major things –malicious programmes, external adversaries, and bad manufacturers –that can create risks to the communication and transmission in IoT devices.	a) Device tampering b) Information disclosure c) DDoS attack d) Spoofing
Kai Zhao et al. [25]	Mentioned several IoT security aspects that occur in the three-layer architecture and specified the solutions of the problems in the perception layer, network layer, and application layer.	a) Node capture, malicious programme and false node, DDoS, routing issues b) Compatibility problems c) Clustering problems d) Identification, authentication authorization, and privacy
Rwan Mahmoud et al. [26]	Discussed technological and security issues of IoT at each layer.	a) Wireless communication, energy conservation, and distributed nature b) Data integrity, authentication, privacy, etc.
Emmanouil Vasilomanolaki et al. [27]	Discussed different architectures of IoT.	a) Privacy, integrity, authentication b) Accountability and revocation c) Pseudonymity and unlinkability d) Data and entity trust
Hui Suo et al. [28]	Focused on various security issues and compared the advancement of research on IoT.	a) Power consumption and storage mechanism b) Authentication and privacy protection
Gupreet Singh Matharu et al. [29]	Described problems in IoT like interoperability, robust connection, privacy and security, identity management, and selection of encryption techniques.	a) Technological diversity and services b) Identification of each object connected in IoT c) Hacking may cause data altercation d) Secure data transmission
Zhi-Kai Zhang et al. [30]	Highlighted the major issues related to IoT such as object identification, information integrity, and authorization.	a) Man-in-the-middle attack creates problem in naming system b) Object certification in IoT c) Anonymization needs access control and cryptographic method
Eleonora Borgia [31]	Explored security aspects in machine-to-machine communication, end-to-end trust, and device and IoT data management.	a) Communication IoT devices b) Security and privacy of data during transmission and storage
Sye Loong Keoh et al. [32]	Discussed standardization of system and data communication security.	a) Interoperable and Datagram Layer Security b) Maintaining privacy during transmission

(Continued)

TABLE 8.4 (CONTINUED)
Issues and Challenges in IoT

Reference	Focused or Discussed Area	Issues and Challenges
Omar Said et al. [33]	Explained security and privacy issues on information gathering and problems in transmission of data.	a) Data collected by RFID b) Data anomaly on wireless transmission c) Addressing devices in IoT and problems in RFID reading, writing, and transmission of data
Jiang Du and ShiWei Chao [34]	Discussed machine-to-machine information security.	a) Frontend actuator and sensors b) Network support c) IT systems support
Vaishnavi J. Deshpande and Dr. Rajeshkumar Sambhe [35]	Focused on the ccyber security and cyber crime aspect and its recent trends.	a) Advanced persistent threat b) Cloud computing and its services Ic) ntrusion attempts and cyber harassment
Ravi Sharma [36]	Focused on national cyber security in terms of saving infrastructure and information.	a) Personal data are the main target area in cyber attack mainly in social media b) Data stored in government sites c) Digital transactions

privacy problems and mitigation processes in IoT were explained. IoT architecture, protocols, and interoperability between connected nodes were presented. Different IoT devices, controllers, and sensor bridges were analyzed to find out the security issues.

Societal impact of IoT in maintaining standards, security and privacy, and different types of vulnerabilities were depicted by Glenn A. Fink et al. [22]. Cyber crime, cyber welfare, social and regulatory issues, etc. were discussed. Different types of vulnerabilities and solutions to mitigate them were elaborated.

Qi Jing et al. [23] analyzed each layer of IoT in terms of security issues and cross-layer heterogeneous integration. Security solutions in terms of RFID, RFID sensor network (RSN), and wireless sensor networks (WSN) were offered. The technology involved in solving the features of the given solutions was analyzed. An overall IoT security architecture was finally driven.

The three major things – malicious programmes, external adversaries, and bad manufacturers – can create risks to the communication and transmission in IoT devices. The requirement to build secure IoT was discussed by Ahmad W. Atamli et al. [24]. Various security issues concerning IoT devices, RFID, and sensors were analyzed and discussed, with emphasis on authentication to ensure privacy and integrity about user information exchange.

Kai Zhao et al. [25] mentioned several IoT security aspects and specified the answer to the privacy aspect in all layers. Some the major problems in the application layer such as identity authentication, software vulnerabilities, and data privacy was elaborated.

Security issues of IoT at each layer were analyzed by Rwan Mahmoud et al. [26]. More stress was given on technological and security issues. Low energy consumption and scalability in wireless communication were analyzed. Various security challenges in the perception layer like node capture attacks and replay attacks were discussed. Some of the attacks like man-in-the-middle, related to network layer, were also discussed.

Different architectures of IoT were discussed and analyzed by Emmanouil Vasilomanolaki et al. [27]. Distinguished features like heterogeneity, controlled resources, and scalability were examined. The privacy and security aspect such as trust, integrity, and identity, and network security related to different IoT architectures were compared and strengths and weaknesses described.

Hui Suo et al. [28] focused on various security issues and compared the advancement of research on IoT. At each level of IoT, security architecture and features were discussed. Security requirements in each level were also mentioned in detail. A summary of research directions on several key challenges in IoT were explained.

The layered architecture and challenges in IoT such as interoperability, robust connection, privacy and security, standardization, identity management, and selection of encryption techniques were described by Gupreet Singh Matharu et al. [29]. Security aspects concerning different layers of the IoT architecture was discussed. Also suggested were different strategies for reducing security issues and vulnerabilities.

Authentication, authorization, malware detection, information restoration, and security are the key topics for ongoing research on IoT. The major issues related to IoT such as object identification, information integrity, and authorization were highlighted by Zhi-Kai Zhang et al. [30]. They also explored malware and a lightweight cryptosystem. Data anonymization was used to reveal the privacy of gathered data.

The rapid improvement of technologies lead to the main challenges faced by researchers in maintaining the privacy and security aspect in any IoT application was explored by Eleonora Borgia [31]. They elaborated on major technologies such as sensing, identification, and communication in IoT applications. They identified scenarios of IoT application in terms of security and privacy. The fundamental architecture and characteristics of IoT were mentioned and they presented a direction for research over different challenges.

IoT protocols and communication security was the main focus area of Sye Loong Keoh et al. [32].The configuration of IoT based on four modes, i.e. Certificate, NoSec, Raw Public Key, and PreShared Key, were discussed. To prepare security on the IoT devices, deployment of Datagram Transport Layer Security (DTLS) was considered as main secured system.

Omar Said et al. [33] suggested a concept of the IoT database and its architecture. The future vision of IoT was also discussed. They presented three- and five-layer architecture and other special purpose architecture that support the IoT database concept. The future research objective on various challenges and serious problems in IoT were analyzed.

To analyze the existing security system of IoT, three methods was offered by Jiang Du and ShiWei Chao [34]. Network, frontend sensors, and IT systems were the security issues in M2M's structure. Security standards in term of data integrity and privacy, user authentication, and communication security were described. The privacy and confidentiality issues of the IoT system were analyzed.

Vaishnavi J. Deshpande and Dr. Rajeshkumar Sambhe [35] focused on the cyber security and cyber crime aspect and recent trends. Different threats related to data communication were discussed. Cyber security solutions, its ethics, and revolutionize the overview of cyber security were focused upon.

Ravi Sharma [36] focused on national cyber security in terms of saving infrastructure and information.

8.5 DIFFERENT TECHNIQUES APPLIED TO RESOLVE THE ISSUES OF IOT DATA AND ACCESS PRIVACY

See Figure 8.6.

8.5.1 DATA PRIVACY

Important techniques of data privacy can be categorized into anonymization-based privacy, conventional cryptography, and lightweight public key primitives. Anonymization-based privacy includes k-anonymity, t-closeness, and l-diversity to ensure the data privacy and preservation. Lightweight primitives include asymmetric and symmetric algorithms, hash function, and pseudo random generator. These techniques have revolutionized in data privacy introducing block ciphers and stream ciphers.

8.5.1.1 Anonymization-Based Privacy

The development of vast networking technologies and data storage mechanisms has made information storage and sharing easier. To access information from a centralized or distributed environment for knowledge discovery we use data mining techniques. So privacy preservation is the main focus area for protecting someone's personal data stored in a public sector or government sector database and to convey correct data instead of diluting privacy.

The anonymization method actually identifies the data indistinguishably from a large set of data. In a large dataset, the attributes can be categorized as unique identifiers which identify the individual, sensitive attributes, including personal information, which must be protected. Quasi identifiers are the attributes connected to an external dataset to identify the individual data [37].

8.5.1.1.1 k-Anonymity

Sweeney [38] proposed group anonymization using the k-anonymity model to solve the problem of privacy preservation of the individual. This model furnishes k-anonymity protection, i.e. any data present in the collected dataset must be identical with a minimum of $k-1$ individuals. So each collection of data of the quasi identifiers in the k-anonymous dataset must be matched with a minimum k record. Therefore the k-anonymous model ensures that private data cannot be recognized by external linking attacks. Generalization and suppression techniques are used in k-anonymity.

Generalization is a technique which converts the actual dataset to a perturbed dataset converting the actual data into a generalized data. In order to minimize

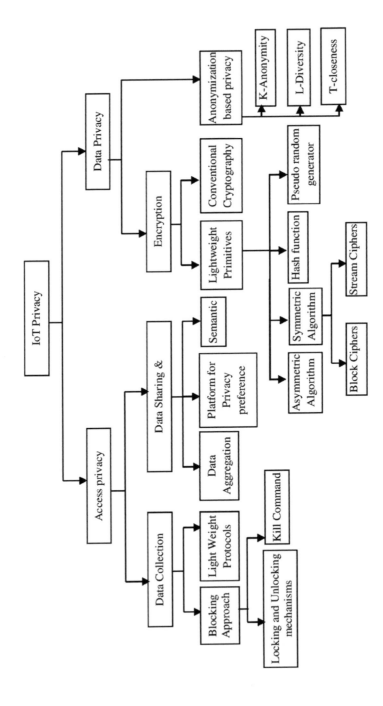

FIGURE 8.6   Pictorial representation of data and access privacy.

the granularity of representation, the records are generalized to a certain range. Generalization is represented by a taxonomy tree. The value of a node can be converted by a node existing in the path between itself or the parent node. The current technique applied on numeric attributes can also be differentiated into fragments of interval values.

Suppression technique removes the value of the attribute fully before the dataset is utilized for analysis. These techniques can be applied anywhere to the dataset. In case of global application, the same type of conversion is done for all objects of the dataset. In case of local application, the conversion is done for a specific object of the dataset. This converted information may in some extent result in information anomalies.

8.5.1.1.2 l-Diversity

Machanavajjhala et al. [39] discussed that the homogeneity attack and background attack can be vulnerable using k-anonymity. So to solve this issue, the l-diversity method was proposed. This method mainly focused on keeping the diversity of very valuable attributes. A congruent class contains this diversity if there is at the minimum well-illustrated information for the valuable attributes. Distinct, entropy, and recursive are the different applications of this technique. In distinct l-diversity there must be l distinct information in each congruent class. In entropy l-diversity there must be l distinct information and values are also distributed equally. Recursive l-diversity ensures the most frequent information does not appear too often. Some of the disadvantages of this technique are pointed by N. Li et al. [40]. This defined technique is vulnerable to skewness and similarity attacks. In some cases this technique is unable to preserve attribute disclosure.

8.5.1.1.3 t-Closeness

In order to enhance the concept of l-diversity, the t-closeness technique was proposed. The key concept of the proposed technique is that the congruent class must have t-closeness. The space between the dispersal of a valuable attribute in the class must be adjacent to their dispersal of the attribute in the whole dataset and must be less than a threshold t.

8.5.1.2   Block Ciphers

In this era a large amount of data are generated from diverse sources from different intelligent objects to minimize energy utilization and to increase performance and efficiency. Various cryptographic primitives are inefficient in preserving privacy. In cryptography, there is a most prominent algorithm operating on definite number of bits, known as a block, which is transformed into cipher text of the identical length applying a symmetric key. This algorithm works as basic components in the building of multiple cryptographic protocols and is very commonly used to apply encryption of bulk data. To make communication more secure, lightweight block ciphers were implemented towards the end of the 1990s. This method consists of two algorithms of encryption and decryption represented as E and D respectively. These algorithms need two inputs, a block of length n bits and a symmetric key of length k bits, and both create n bits of resultant block. The decryption algorithm is treated as a reverse process of encryption. An encryption function is specified in a block cipher, i.e.

$$E_K(P) := E(K,P) : \{0,1\}^k \times \{0,1\}^n \rightarrow \{0,1\}^n$$

$$D_K(C) := D(K,C) : \{0,1\}^k \times \{0,1\}^n \rightarrow \{0,1\}^n$$

Here, inputted is a key K of range k bits, and a string P of range n bits, which produce a string C of range n bits. P and C are termed as the original text and cipher text, respectively. For individual K values, the $E_K(P)$ function needs to be a reverse mapping on $\{0,1\}^n$. The decryption function can be interpreted as

Now, consider K and C to get the original value P, i.e.

$$\forall K : D_K(E_K(P)) = P$$

Previous research on block ciphers led to various primitives for IoT, including TEA/XTEA, PRESENT, LBlock, Speck, mCRYPTON, CLEFIA, KTANTAN, LED, KATAN, PRINT Cipher, KLEIN, SEA, CLEFIA, Simon, and DESXL.

8.5.1.3   Stream Cipher

The stream cipher method is commonly known as symmetric key cipher. In this method the original text digits are merged with a pseudo random cipher digit stream. Each primary text digit is encrypted once to the cipher digit stream to create a digit of cipher text stream. It is also known as state cipher on the basis that in the current state of the cipher all digits are encrypted. A digit is represented as a bit. The bit merging operation is done by exclusive-OR (XOR). Some of the pseudo random number generator algorithms applied in stream cipher are WG-8 Espresso, A2U2, and Enocoro v.2.

8.5.1.4   Public-Key-Based Authentication

Public-key-based authentication can resolve authentication problems including identifying users, applications, and devices in IoT. It also restricts the user from unauthorized access and manipulation of devices. In this method the username and password are encrypted by cryptographic schemes to implement the security in communication as well as device access. This mechanism provides some benefits on the application of IoT, i.e. secure communication among remote devices, implementation of innovative services, avoidance of data stealing, and reduced risk of tampering data using third-party services.

This mechanism is widely used on the Internet, but is very vulnerable for a constrained environment like IoT for excessive use of cryptographic schemes. It involves private and public keys, which ensure privacy and security aspects. Traditional algorithms of public key cryptography like RSA are widely used. The public key is applied in encrypting the message in order to generate the cipher text. The cipher text is decrypted by using the corresponding private key. Digital signatures, where cipher text is produced using the private key, is decrypted by the public key.

8.5.2 ACCESS PRIVACY

Research Mechanism in term of access privacy contains data sharing and management, blocking approaches and lightweight protocols. Again data sharing and management technique further enhanced to data aggregation, platform for privacy preference and semantic method. Blocking approach involves locking and unlocking mechanism and kills command mechanism.

8.5.2.1 Blocking Approaches

Blocking techniques are applied during the data collection phase to solve privacy issues. During the transmission of information, the intelligent object that represents the unique identifier of a person who is transmitting may be attacked by an unauthorized user. So to reduce privacy issues, the kill command of RFID tags can be used to stop the operation of an external programme. Locking and unlocking mechanisms are also used for blocking.

8.5.2.2 Lightweight Protocols

Lightweight protocols are used to enforce authentication and identification. They also consist of some other properties such as proof of existence, delegation and restriction, and distance bounding. The code of these protocols performs more efficiently than standard protocols. Some examples of lightweight protocols are the Lightweight Extensible Authentication Protocol (LEAP), Lightweight Directory Access Protocol (LDAP), and Skinny Call Control Protocol (SCCP).

8.5.3 MACHINE LEARNING (ML) APPROACH ON IoT SECURITY

Since the rapid development of IoT, millions of sensors are connected to each other for communication and generate a large dataset. These sensors collect a variety of data from different sensor nodes and the quantity of the data will be very huge (can be termed as big data). Hence, data analysis plays a key role in this field. The concept of machine learning (ML) can be used to solve different types of problems, which can arise during the periods of data analysis and security in IoT (Figure 8.7). Also the concept of classification and the clustering technique in ML can be used to overcome the different problems in IoT applications like industry automation, smart vehicles, weather forecasting for agriculture, etc. Basically, classification and clustering techniques have been implemented in the cloud server for training and testing purposes. Here different machine learning algorithms are adopted to build a model from the training dataset and to produce a predictive output on the basis of current inputs in the testing phase [41].

8.5.3.1 ML Techniques

ML is a technique based on the concept of artificial intelligence (AI) that is capable of learning from the previous dataset and participates as a predictor for decision making (see Table 8.5). It works as an expert system to build a model that is responsible for learning and predicting new things.

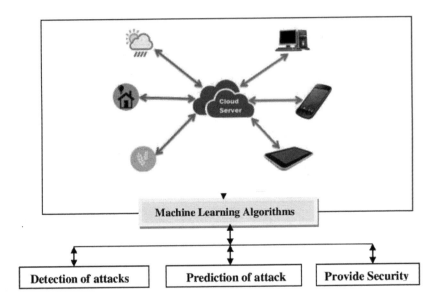

FIGURE 8.7   IoT security using machine learning.

8.5.3.1.1   Classification

Classification is technique in ML which is responsible to predict the classes of a given dataset. This model separate different data set which belonging in the same class. A classifier analyzes the behaviour of a training dataset to understand how new input variables relate to the class. Classification is a part of supervised learning because here the target output depends on the features of the previous input dataset. Some popular classification algorithms which can be applied in different applications of IoT are as follows:

- *k*-Nearest neighbour (KNN)
- Decision tree
- Logistic regression
- Support vector machine(SVM)

These classification algorithms learn the verity of data in the cloud server which are gathered from different sensor devices. To archive the specific goal in data analysis, systems are trained using various algorithms and statistical models [42]. These algorithms are responsible to extract the measurable features from the source dataset and find a correlation between the characteristics to predict a result for new input. Sensor node localization problems and determining the node's geographical position can be resolved by applying the KNN or SVM algorithm [43].

8.5.3.1.2   Clustering

Clustering is another approach in ML where the information about the classes of the data is unknown. There is no idea whether this data can be classified, so this is

TABLE 8.5
Machine Learning Techniques Applied to Solve IoT Network Security Issues

Security Issue	Machine Learning Techniques Applied
Malware analysis	• Recurrent Neural Network (RNN) [44] • PCA [45] • SVM [46] • CNN [47] [48] • VM and PCA [49] • Linear SVM [50]
Anomaly/ intrusion detection	• k-means clustering • Naive Bayes [51] [52] • Decision tree [53] [51] • Artificial neural network (ANN) [54]
Authentication	• Recurrent neural network (RNN) [55] • Dyna-Q [56] • Q-learning [56] • Deep Learning [57, 58, 59]
DDoS attack	• KNN [60] • Q-learning [61] • Decision tree [60] • SVM • Neural network[60] • Random forest [60] • Multivariate correlation analysis (MCA) [62]
Attack detection and mitigation	• Fuzzy C-means (ESFCM) • Deep learning [57, 58, 59] • KNN [63] • SVM [60]

known as unsupervised learning. It is basically used for statistical data analysis in various fields of IoT. Some popular clustering algorithms are as follows:

• k-Means
• Expectation–maximization

The k-means algorithm segregates different data in the cloud server that belong in same class on the basis of their characteristics. The expectation–maximization algorithm can be used to identify and predict attacks based on probability distribution.

8.5.3.1.3 Dimensionality Reduction

Dimensionality reduction is a technique in machine learning where unnecessary features are removed to get the actual features. Some of the algorithms used in this technique are

- Principal component analysis (PCA)
- Singular-value decomposition (SVD)
- Independent component analysis (ICA)
- Non-negative matrix factorization (NMF)
- Linear discriminant analysis (LDA)
- Factor analysis (FA)

8.5.3.1.4   Generative Model

The generative model is very effective in building scanning tools for scanning vulnerabilities in web applications. It is also used in testing unauthorized access. Some of the generative models used in machine learning and deep learning are

- Genetic algorithms
- Markov chains
- Boltzmann machines
- Variational auto encoders
- Generative adversarial networks (GANs)

ML is a perfect tool to be applied in cyber security issues to solve the problem of detection by predictive analyses of tasks and performance of different intelligent devices. It can also monitor the networks for any unauthorized attacks. The primary goal is to predict the attacks and identify threats using different predictive models in machine learning. Network traffic analytics (NTA) in machine learning is a new solution of network security which analyzes the traffic at each network layer to detect threats and anomalies. Prediction, prevention, detection, response, and monitoring are the security mechanisms proposed by Gartner's PPDR model. The ML approach can be used in monitoring endpoints to predict malware and fraud. Anomalies in databases and cloud storage are also predicted. Network protection includes protection of different protocols including Ethernet, wireless, virtual networks, or supervisory control and data acquisition (SCADA) using intrusion detection systems (IDS), which are mostly based on signature identification approaches.

8.6   DATA ENCRYPTION AND DECRYPTION TECHNIQUES TO COMBAT IOT SECURITY-RELATED ISSUES: A CASE STUDY

We implemented an IoT system where one microcontroller takes one temperature value using a DHT 11sensor,generates the cipher text using theBase64 algorithm encryption technique, then sends the cipher text through a wireless network to another microcontroller. After receiving the cipher text, it decrypts the cipher text and shows the result in an output device (Figure 8.8). The microcontroller that takes and encrypts the sensor data is known as the client and the microcontroller that receives and decrypts the message is known as the server. In this system server and client talks to each other in a secret way.

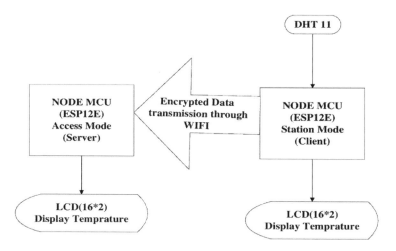

FIGURE 8.8 Pictorial representation of data communication between NODE MCU using cryptographic technique.

8.6.1 Base64 Algorithm

The mechanism of the Base64 algorithm is given next (also see Figure 8.9).

8.6.1.1 Encryption Algorithm

Step 1: Take the plaintext input and divide it into blocks of three characters.

Step 2: Convert each character to ASCII.

Step 3: Convert the ASCII value to binary and each character size is 8 bits, so we will have (8 * 3) 24 bits with us.

Step 4: Group them in a block of 6 bits each (24/6 = 4 blocks). Because 2^6 = 64 characters, with 6 bits we can represent each character in a character set table. (If we have less than six binary values in the last output block, zeros will be append to make it a six binary value.)

Step 5: Convert each block of 6 bits to its corresponding decimal value. The decimal value obtained is the index of the resultant encoded character in the character set table.

Step 6: So for each three characters from input we will receive four characters in output.

Step 7: If the last 3-byte block has only 2 bytes of input data, pad 1 byte of zero (\x00). After encoding it as a normal block, override the last one character with one equal sign (=), so the decoding process knows 1 byte of zero was padded.

Step 8: If the last 3-byte block has only 1 byte of input data, pad 2 bytes of zero (\x0000). After encoding it as a normal block, override the last two characters with two equal signs (==), so the decoding process knows 2 bytes of zero were padded.

8.6.1.2 Decryption Algorithm

Step 1: Calculate the how many '=' are present in cipher text and store within x, and delete all'='.

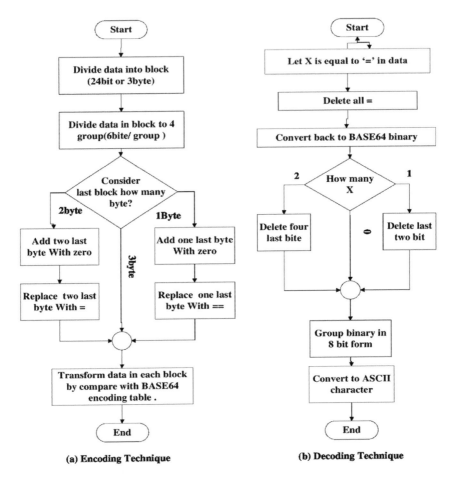

FIGURE 8.9 (a) Flow diagram of encoding technique. (b) Flow diagram of decoding technique.

Step 2: Convert each character in cipher text in base 64 binary.
Step 3: If x = 1, delete the last two bits; if x = 2, delete the last 4 bits.
Step 4: Now make a group of 8 bits.
Step 5: Convert 8 bits data to a decimal number.
Step 6: Convert the decimal to character as per the ASCII (Table 8.6).

8.6.2 Implementation

In this implementation we use the ESP8266 WIFI microcontroller (NODE MCU 12E) as a client and server, and for the display we use LCD (16*2). We can use the ESP8266 microcontroller in two different modes:

- Access Point (AP): In this mode of ESP8266 microcontroller is allowed it to create its own wireless network and have other devices connect to it.

TABLE 8.6
The Base64 Index

Index	Char	Index	Char	Index	Char	Index	Char	Index	Char	Index	Char
0	A	11	L	22	W	33	h	44	s	55	3
1	B	12	M	23	X	34	i	45	t	56	4
2	C	13	N	24	Y	35	j	46	u	57	5
3	D	14	O	25	Z	36	k	47	v	58	6
4	E	15	P	26	a	37	l	48	w	59	7
5	F	16	Q	27	b	38	m	49	x	60	8
6	G	17	R	28	c	39	n	50	y	61	9
7	H	18	S	29	d	40	o	51	z	62	+
8	I	19	T	30	e	41	p	52	0	63	/
9	J	20	U	31	f	42	q	53	1	padding	=
10	K	21	V	32	g	43	r	54	2		

- Station (STA): This mode allows the ESP8266 to connect to a wireless network.

In our implementation the client microcontroller behaves like the station (STA), and the server microcontroller behave like the access mode. So the server creates its own wireless network and client is used to get the ESP module connected to a wireless network established by a server. So the station mode takes the temperature data from DHT11 and generates cipher text using theBase64encryption algorithm, then sends the cipher text through the wireless network to the server. The access mode receives the cipher text and generates the plain text using theBase64 algorithm (Figure 8.10).

8.6.3 Circuit Diagram

See Figure 8.11 and Figure 8.12.

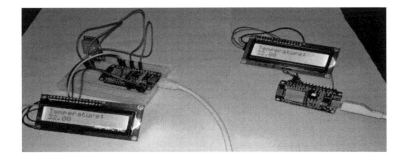

FIGURE 8.10 Communication of temperature data between two devices on encryption mode.

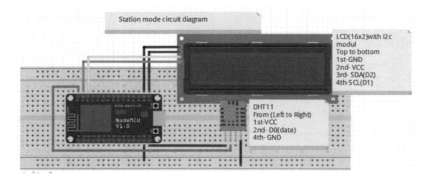

FIGURE 8.11   Circuit diagram of station mode.

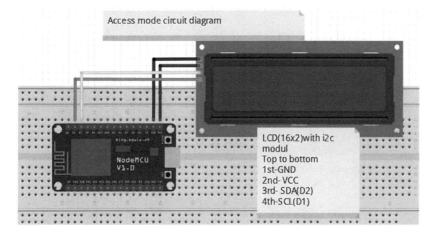

FIGURE 8.12   Circuit diagram of access mode.

8.6.4   WORKFLOW DIAGRAM

See Figure 8.13.

8.6.5   ANALYSIS OF SECURED DATA TRANSMISSION BETWEEN IOT DEVICES

See Figure 8.14, Figure 8.15, and Figure 8.16.

8.7   CONCLUSION AND FUTURE SCOPE

This chapter provides a perception of the detailed concept of IoT and covers the main issue of security in the data generated through IoT devices. Initially, a list of all the required sensors for collecting data through IoT devices was stated. The layers present in an IoT architecture is mentioned followed by various threats and attacks that hinder functions in IoT. This chapter contains survey ofabout25research papers dealing with the reduction of the breach of security in these types of devices. Along with

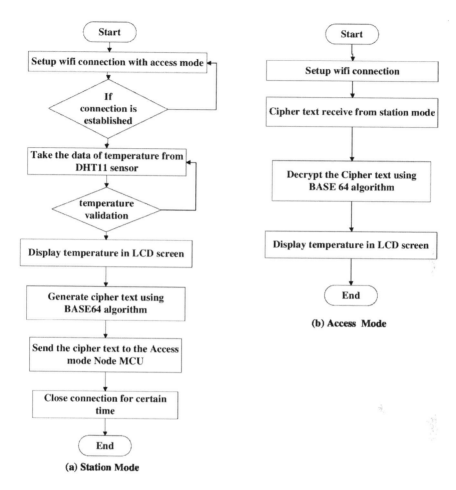

FIGURE 8.13 (a) Flow diagram of station mode. (b) Flow diagram of access mode.

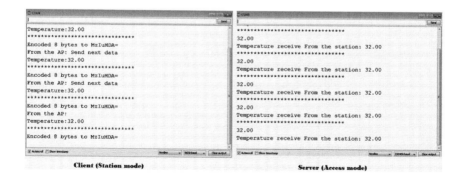

FIGURE 8.14 Client data is encrypted using Base64 algorithm.

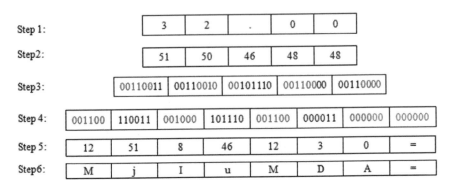

Step 1:	3	2	.	0	0

Step2:	51	50	46	48	48

Step3:	00110011	00110010	00101110	00110000	00110000

Step 4:	001100	110011	001000	101110	001100	000011	000000	000000

Step 5:	12	51	8	46	12	3	0	=

Step6:	M	j	I	u	M	D	A	=

FIGURE 8.15 Encoding process at client end taking temperature equals to 32 and got encrypted text (MjIuMDA=).

Step 1:	M	j	I	u	M	D	A	=

Step 2:	12	51	8	46	12	3	0	=

Step 3:	001100	110011	001000	101110	001100	000011	000000	000000

Step 4:	00110011	00110010	00101110	00110000	00110000

Step 5:	51	50	46	48	48

Step 6:	3	2	.	0	0

FIGURE 8.16 Decoding process at server end to get temperature equals to 32°C from cipher text (MjIuMDA=).

these, a tabular approach containing the threats and challenges mentioned in these papers is highlighted. This chapter also showcases case study in which temperature is being transferred from one device to the other. Two microcontroller devices have been considered: one as the server and the others the client. The temperature is encrypted when sent to the receiver and decrypted back when reached and read at the receiver. The security issue has been implemented through the existing Base64 algorithm. Future research in this field should include incorporation of novel algorithms to ensure maximum security while dealing with huge amounts of data as is generally generated by these types of devices.

As already mentioned, ensuring data privacy is a prime issue in IoT devices and further research in this field should be deployed as almost all devices are being transferred to being "smart" and it will take no time when we humans have to imbibe this "smart" technology into our daily lives. The main vulnerable case that shall arise as a result of this is that as all of our data is being stored and also shared and sold to different organizations; unconstrained access shall lead to a high breach of security leading to fatal consequences. Space limitations, authorization and authentication problems, regular updating of devices, secured message transfer, and combating vulnerabilities to maintain the integrity of data should be the main areas of concern that should be prioritized while handling IoT data. It is anticipated that adopting a multi-layered precautionary design could lead to easy dealing with security problems. As IoT technology plays a key role in commercialization of various applications, security and privacy is of utmost importance. Machine learning techniques area modern approach to solving issues related to IoT security. IoT devices collect a variety of data which may not dependent on each other. Different algorithms in the machine learning approach may be used for IoT data and security solution to ensure public data privacy and to build consumer confidence. Therefore different hybrid algorithms and efficient data representation techniques may be introduced in future IoT security research domains.

REFERENCES

1. Yan, Zheng, Peng Zhang Athanasios and V. Vasilakos. (2014). A survey on trust management for Internet of things. *Journal of Network and Computer Applications*, pp. 120–134.
2. Alaba, Fadele Ayotunde, Mazliza Othman, Ibrahim Abaker Targio Hashem and Faiz Alotaibi. Internet of things security: A survey.
3. Al-Fuqaha, Ala, Mohsen Guizani, Mehdi Mohammadi Mohammed Aledhari and Moussa Ayyash. (2015). Internet of things: A survey on enabling technologies protocols and applications. *IEEE Communication Surveys & Tutorials, Fourth Quarter*, vol. 17, no. 4, pp. 23–47.
4. Mendez Mena, Diego, Ioannis Papapanagiotou and Baijian Yang. *Internet of Things: Survey on Security*, Taylor & Francis, pp. 1939–3547. ISSN: 1939–3555.
5. Jing, Qi, Athanasios V. Vasilakos, Jiafu Wan Jingwei Lu and Dechao Qiu. *Security of the Internet of Things: Perspectives and Challenges*. New York: Springer Science Business Media. doi:10.1007/s11276-014-0761-7.
6. Abomhara, Mohamed and Geir M. Køien. 2015. Cyber security and the Internet of things: Vulnerabilities, threats, intruders and attacks.
7. Køien, G. M. .2011. Reflections on trust in devices: An informal survey of human trust in an Internet-of-things context. *Wireless Personal Communications*, vol. 61, no. 3, pp. 495–510.
8. Vithya Vijayalakshmi, A. and L. Arockiam. 2016. A study on security issues and challenges in IoT. *International Journal of Engineering Sciences & Management Research*, ISSN 2349-6193.
9. Brauch, H. G.2011. Concepts of security threats, challenges, vulnerabilities and risks. In *Coping with Global Environmental Change, Disasters and Security*, Springer, pp. 61–106.
10. Abomhara, Mohamed and Geir M. Køie. 2015. Cyber security and the Internet of things: Vulnerabilities, threats, intruders and attacks. *Journal of Cyber Security*, vol. 4, pp. 65–88.
11. Asim, Makkad. 2017. Security in application layer protocols for IOT: A focus on COAP. *International Journal of Advanced Research in Computer Science*, vol. 8, no. 5, pp. 2653–2656.
12. Sharma, Rahul, Nitin Pandey and Sunil Kumar Khatri. 2017. Analysis of IoT security at network layer. 6th International Conference on Reliability, Infocom Technologies and Optimization (ICRITO) (Trends and Future Directions), pp. 585–590.
13. Abdul-Ghani, Hezam Akram, Dimitri Konstantas and Mohammed Mahyoub. 2018. A comprehensive IoT attacks survey based on a building-blocked reference model. *International Journal of Advanced Computer Science and Applications*, vol. 9, no. 3, pp. 335–373.

14. Xiaohui, Xu. 2013. Study on security problems and key technologies of the Internet of things. International Conference on computational and Information Sciences, pp. 407–410.

15. Riahi Sfar, Arbia, Enrico Natalizi, Yacine Challal and Zied Chtourou. 2018. A roadmap for security challenges in the Internet of Thing, Elsevier. *Digital Communications and Networks*, pp. 118–137.

16. Shankar Basu, Subho, Somanath Tripathy and Atanu Roy Chowdhury. 2015. Design challenges and security issues in the Internet of things, IEEE Region 10 Symposium, pp. 90–93.

17. Qiang, Chen, Guang-ri Quan, Bai Yu and Liu Yang. 2013. Research on security issues on the Internet of things. *International Journal of Future Generation Communication and Networking*, pp. 1–9.

18. Granjal, Jorge, Edmundo Monteiro and Jorge Sa Silva. 2015. Security for the Internet of things: A survey of existing protocols and open research issues. *IEEE Communication Surveys and Tutorials*, vol. 17, no. 3, pp. 1294–1312.

19. Kraijak, Surapon and Panwit Tuwanut. 2015. A survey on Internet of things architecture, protocols, possible applications, security, privacy, real-world implementation and future trends. Proceedings of ICCT, pp. 26–31.

20. Benabdessalem, Raja, Mohamed Hamdi and Tai-Hoon Kim. 2014. A survey on security models, techniques, and tools for the Internet of things. International Conference on Advanced Software Engineering & Its Applications, pp. 44–48.

21. Hossain, Md. Mahmud, Maziar Fotouhi and Ragib Hasan. 2015. Towards an analysis of security issues, challenges, and open problems in the Internet of things. IEEE World Congress on Services, pp. 21–28.

22. Fink, Glenn A., Dimitri V. Zarhitsky, Thomas E. Carroll and Ethan D. Farquhar. 2015. Security and privacy grand challenges for the Internet of things. IEEE, pp. 27–34.

23. Jing, Qi, AthanasiosV, Vasilakos, Jiafu Wan, Jingwei Lu and Dechao Qui. Security of the Internet of things: perspectives and challenges, Springer. *Wireless Networks*, vol. 20, pp. 2481–2501.

24. Atamli, Ahmad W. and Andrew Martin. 2014. Threat-based security analysis for the Internet of things. International Workshop on Secure Internet of Things, pp. 35–43.

25. Zhao, Kai and Lina Ge. 2013. A survey on the Internet of things security. IEEE, International Conference on Computational Intelligence and Security, pp. 663–667.

26. Mahmoud, Rwan, Tasneem Yousuf, Fadi Aloul and Imran Zualkernan. 2015. Internet of Things (IoT) security: Current status, challenges and prospective measures. International Conference for Internet Technology and Secured Transactions (ICITST), pp. 336–341.

27. Vasilomanolakis, Emmanouil, Jorg Daubert, Manisha Luthra, Vangelis Gazis, AlexWiesmaier and PanayotisKikiras. 2015. On the security and privacy of Internet of things architectures and systems. International Workshop on Secure Internet of Things, pp. 49–57.

28. Suoa, Hui, Jiafu Wana, Caifeng Zoua and Jianqi Liua. 2012. Security in the Internet of things: A review. International Conference on Computer Science and Electronics Engineering, pp. 649–651.

29. Matharu, Gurpreet Singh, Priyanka Upadhyay and Lalita Chaudhary. 2014. The Internet of things: Challenges & security issues. IEEE, International Conference on Emerging Technologies (ICET), pp. 54–59.

30. Zhang, Zhi-Kai, Michael Cheng Yi Cho, Chia-Wei Wang, Chia-Wei Hsu, Chong-Kuan Chen and Shiuhpyng Shieh. 2014. IoT security: Ongoing challenges and research opportunities. IEEE International Conference on Service-Oriented Computing and Applications, pp. 230–234.

31. Borgia, Eleonora. 2014. The Internet of things vision: Key features, applications and open issues, Elsevier. *Computer Communications*, pp. 1–31.
32. Loong Keoh, Sye, Sandeep S. Kumar and Hannes Tschofenig. 2014. Securing the Internet of things: A standardization perspective. *IEEE Internet of Things Journal*, pp. 265–275.
33. Said, Omar and Mehedi Masud. 2013. Towards Internet of things: Survey and future vision. *International Journal of Computer Networks (IJCN)*, vol. 1, no. 1, pp. 1–17.
34. Jiang, Du and Shi Wei Chao. 2010. A study of information security for M2M of IOT. IEEE International Conference on Advanced Computer Theory and Engineering (ICACTE), pp. 576–579.
35. Sharma, Ravi. 2012. Study of latest emerging trends on cyber security and its challenges to society. *International Journal of Scientific & Engineering Research*, vol. 3, no. 6.
36. Deshpande, Vaishnavi J. and Dr. Rajeshkumar Sambhe. 2014. Cyber security: strategy to security challenges- a review. *International Journal of Engineering and Innovative Technology (IJEIT)*, vol. 3, no. 9.
37. Dhanalakshmi, S. and P. S. Ahammed Shahz Khamar. Data preservation using anonymization based privacy preserving techniques – a review. *IOSR Journal of Computer Engineering (IOSR-JCE)*, pp. 18–21. e-ISSN: 2278–0661,p-ISSN: 2278-8727.
38. Sweeney, L. 2002. k-anonymity: A model for protecting privacy. *International Journal of Uncertainty, Fuzziness and Knowledge-Based Systems*, pp. 557–570.
39. Machanavajjhala, A., D. Kifer, J. Gehrke and M. Venkitasubramaniam. 2007. L-diversity: Privacy beyond k-anonymity. *ACM Transactions on Knowledge Discovery from Data (TKDD)*, vol. 1, no. 1, p. 3.
40. Li, N., T. Li and S. Venkatasubramanian. 2007. t-closeness: Privacy beyond k-anonymity and l-diversity. Data Engineering, 200, ICDE 200, IEEE 23rd International Conference, pp. 106–115.
41. Ruta, Michele, Floriano Scioscia, GiuseppeLoseto, Agnese Pinto and Eugenio Di Sciascio. 2017. *Machine Learning in the Internet of Things: A Semantic-Enhanced Approach*. IOS Press, pp. 1–17.
42. Zantalis, Fotios, Grigorios Koulouras, Sotiris Karabetsos and Dionisis Kandris. 2019. A review of machine learning and IoT in smart transportation. pp. 1–23.
43. Malik, Rashid Ashraf, Asif Iqbal Kawoosa and Ovais Shafi Zargar. 2018. Machine learning in The Internet of things – standardizing iot for better learning. *International Journal of Advance Research in Science and Engineering*, vol. 07, no. 04, pp. 1676–1683. ISSN:2319-8354.
44. Haddad Pajouh, H., A. Dehghantanha, R. Khayami and K.-K. R. Choo. 2018. A deep recurrent neural network based approach for Internet of things malware threat hunting. *Future Generation Computer Systems*, vol. 85, pp. 88–96.
45. An, N., A. Duff, G. Naik, M. Faloutsos, S. Weber and S. Mancoridis. 2017. Behavioral anomaly detection of malware on home routers. International Conference on Malicious and Unwanted Software (MALWARE), pp. 47–54.
46. Zhou, W. and B. Yu. 2018. A cloud-assisted malware detection and suppression framework for wireless multimedia system in IoT based on dynamic differential game. *China Communications*, vol. 15, pp. 209–223.
47. Su, J., D. V. Vargas, S. Prasad, D. Sgandurra, Y. Feng and K. Sakurai. 2018. Lightweight classification of iot malware based on image recognition. vol. abs/1802.03714.
48. Azmoodeh, A., A. Dehghantanha and K. R. Choo. 2018. Robust malware detection for Internet of (battlefield) things devices using deep eigenspace learning. IEEE Transactions on Sustainable Computing, pp. 1–1.
49. Esmalifalak, M., N. T. Nguyen, R. Zheng and Z. Han. 2013. Detecting stealthy false data injection using machine learning in smart grid. IEEE Global Communications Conference (GLOBECOM), pp. 808–813.

50. Ham, H. S., H. H. Kim, M. S. Kim and M. J. Choi. 2014. Linear SVM-based android malware detection for reliable IoT services. *Journal of Applied Mathematics*, p. 10.
51. Viegas, E., A.Santin, L.Oliveira, A.Frana, R.Jasinski and V.Pedroni. 2018. A reliable and energy-efficient classifier combination scheme for intrusion detection in embedded systems. *Computers & Security*, vol. 78, pp. 16–32.
52. Pajouh, H. H., R. Javidan, R. Khayami, D. Ali and K. R. Choo. 2018. A two-layer dimension reduction and two-tier classification model for anomaly-based intrusion detection in IoT backbone networks. IEEE Transactions on Emerging Topics in Computing, pp. 1–1.
53. Shukla, P.2017. Ml-ids: A machine learning approach to detect wormhole attacks in Internet of things. Intelligent Systems Conference (IntelliSys), pp. 234–240.
54. Caedo, J. and A. Skjellum. 2016. Using machine learning to secure IoT systems. 14th Annual Conference on Privacy Security and Trust (PST), pp. 219–222.
55. Chauhan, J., S. Seneviratne, Y.Hu, A. Misra, A. Seneviratne and Y. Lee. 2018. Breathing-based authentication on resource-constrained IoT devices using recurrent neural networks, vol. 51, pp. 60–67.
56. Xiao, L., Y. Li, G. Han, G. Liu and W. Zhuang. 2016. Phy-layer spoofing detection with reinforcement learning in wireless networks. *IEEE Transactions on Vehicular Technology*, vol. 65, pp. 10037–10047.
57. Diro, A. A. and N. Chilamkurti. 2018. Distributed attack detection scheme using deep learning approach for Internet of things. *Future Generation Computer Systems*, vol. 82, pp. 761–768.
58. Abeshu, A. and N. Chilamkurti. 2018. Deep learning: The frontier for distributed attack detection in fog-to-things computing. *IEEE Communications Magazine*, vol. 56, pp. 169–175.
59. Rathore, S. and J. H. Park. 2018. Semi-supervised learning based distributed attack detection framework for IoT. *Applied Soft Computing*, vol. 72, pp. 79–89.
60. Doshi, R., N. Apthorpe and N. Feamster. 2018. Machine learning ddos detection for consumer Internet of things devices. IEEE Security and Privacy Workshops (SPW), pp. 29–35.
61. Li, Y., D. E. Quevedo, S. Dey and L. Shi. 2017. Sinr-based dos attack on remotestate estimation: A game-theoretic approach. *IEEE Transactions on Control of Network Systems*, vol. 4, pp. 632–642.
62. Tan, Z., A. Jamdagni, X. He, P. Nanda and R. P. Liu. 2014. A system for denial-of-service attack detection based on multivariate correlation analysis. *IEEE Transactions on Parallel and Distributed Systems*, vol. 25, pp. 447–456.
63. Ozay, M., I. Esnaola, F. T. Y. Vural, S. R. Kulkarni and H. V. Poor. 2016. Machine learning methods for attack detection in the smart grid. *IEEE Transactions on Neural Networks and Learning Systems*, vol. 27, pp. 1773–1786.

9 Authentication of Devices in IoT

Daneshwari I. Hatti and Ashok V. Sutagundar

CONTENTS

9.1 INTRODUCTION

IoT (Shah and Yaqoob 2016; Paul and Saraswathi 2017) is the interconnection of devices constituting non-living and living things, communicating over Internet. By 2020, 50 billion devices are predicted to be connected and communicating through the Internet (Kim et al. 2016). Communication and interaction among several heterogeneous devices are to be secured through some mechanisms for privacy preservation. Every device or some devices in the network are supposed to communicate confidential information hence it has to be secured. In the security aspect, privacy, authentication, and authorization of devices are important in networks for secure communication and efficient utilization of resources. Several issues in IoT are data management, resource management, interoperability, and security. Authentication plays a vital role in the field of IoT, as devices in the environment have varied resources capability. The resources required for devices is changing with respect to applications and for authentication. The resources are misused if proper authentication mechanisms are not employed. The mechanisms consume more resources for detection of attacks and securing the information. Authentication is done by performing in multistep or multifactor authentication mechanisms. Prior to authentication, the trust level of a device is determined by the fuzzy approach. Fuzzy logic (Singla 2015) considers the fuzzy values instead of crisp values for decision-making similar to human reasoning and decision-making. Humans decide based on factors affecting the system and act accordingly. In correlation to the human perspective, fuzzy makes decisions by framing the set of rules using antecedents and consequents (Keshwani et al. 2008). The individual rules are converted to crisp values and aggregated to obtain a single crisp value by a defuzzifier, or individual rules are aggregated and the aggregated rule is converted to a crisp value by a defuzzifier. The output of a defuzzifier decides the trust level of the device, and mutual authentication is performed for identifying the identity for communication. The authentication score is used in this work for finding the degree of access to resources to the devices. The authentication architecture, namely centralized, distributed approach, and combinational approach, is considered in the proposed network scenario. Fog computing (Bonomi et al. 2012; Yi et al. 2015) is used in this work for processing the request by the devices, as it provides services towards the edge of the devices with reduced latency and within the response time of the devices. In Section 9.2 the importance of authorization and authentication is addressed and a few works related to authentication in cloud computing, grid computing, fog computing, and IoT are discussed. It also highlights a few challenges and research directions in authentication of devices in IoT. In Section 9.3 various mechanisms for the static and dynamic environment employed locally and distributed is discussed. In Section 9.4 the fog computing (Vaquero and Rodero-Merino 2014; Stojmenovic 2014) paradigm for authentication with reduced computational resource usage is addressed. Due to heterogeneity of devices in the network environment, the agent technology used for interaction among devices is discussed. The importance of authentication and various mechanisms employed for authenticating devices for efficient accessing of resources in a static and dynamic environment is concluded in Section 9.5.

9.2 AUTHENTICATION AND AUTHORIZATION IN IOT

Authorization and authentication helps in authenticating the devices and authorizing for accessing all or any specific type of services based on the mechanisms employed. Authentication is the first step for revealing the identity of the device, then the next authorization is employed for accessing specific resources. If the device is not authenticated, there is no provision for communication in the network, and if authenticated the device does not have full authority for accessing or modifying the parameters in the network. It is controlled by authorization mechanisms (Shen et al. 2017), and based on these mechanisms the devices can fully or partially access various resources. After performing both mechanisms the device may or may not be secured, hence the algorithms are efficiently designed to overcome the attacks occurring in the network with less usage of computational resources with reduced communication cost. The fog computing paradigm is employed for the authentication mechanism, as it contributes to less latency and provides services to the edge

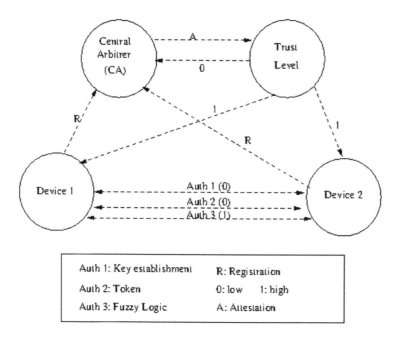

FIGURE 9.1 Flow diagram of the proposed work.

of devices by reducing traffic to the cloud. The edge devices in the perception layer are called the mist layer. The mist layer comprising heterogeneity among devices becomes tedious for processing the entire request at that layer, so fog computing at the top of the mist layer provides services to the end devices. Figure 9.1 shows the process of authentication and flow to access the network resources by the devices. It consists of two devices communicating through the Central Arbiter (CA) and can be extended to multiple devices in the network.

In Figure 9.1 'R' represents registration, 'A' represents attestation, '0' indicates the trust level is low, '1' indicates the trust level is high, Auth 1 represents authentication level 1 using the key establishment phase using elliptic curve cryptography (ECC), Auth 2 represents authentication level 2 using the token approach to authenticate the devices, and Auth 3 indicates the third level using the fuzzy approach to decide the authenticity of devices. Auth 1(0) shows the first level failed in authentication and proceeds with the second level Auth 2. It is also 0, then enters for the third level which is 1 indicating authentication is successful, hence communication to access the resources of the devices. If Auth 3 (0), then authentication fails, and hence the device is locked and does not allow for participation in the network. Few works that are regarding authentication in grid computing, wireless sensor networks, mobile ad hoc networks (MANET), and IoT are discussed in the literature survey.

9.2.1 LITERATURE SURVEY

Several works regarding authentication and authorization in IoT have been carried out and some are listed in this chapter. In Kim et al. (2016) authentication and authorization are automated with the local authorization entity 'Auth'. Authorization is done for registered entities by Auth with session key distribution. An entity uses only the session keys provided by Auth for authentication and authorization. Ferrag, Maglaras, and Derhab 2019) proposed biometrics-based authentication schemes for mobile IoT devices. The features considered were human physiological and behavioural, for example voice and signature. In this work various authentication protocols for different network environments are surveyed. Work related to privacy and authentication in the IoT environment is addressed (Ferrag et al. 2017). Ye et al. (2014) concentrated particularly on efficient mutual authentication and secure key establishment using ECC. It reduces storage and communication overheads in the perception layer of IoT. Hongmei Deng, Mukherjee, and Agrawal 2004) proposed efficient key management and authentication mechanisms for securing the ad hoc network. An identity-based cryptography provides authentication and confidentiality, and does not comprise the centralized certification authority for distributing the public keys and the certificates. It improvises the tolerance of the network by saving of network bandwidth. Shivraj et al. (2015) focused on lightweight-identity-based ECC and Lamport's one-time password (OTP) algorithm for authentication between devices in IoT. Alhothaily et al.'s (2018) one-time authentication system leveraging user computing devices does not create nor remember any identities, mitigating the risks of static username and password. For a resource constraint device other researchers proposed a 'lightweight anonymous credential construction' with 'Nguyen's dynamic accumulator' under the 'witness update outsourcing' paradigm (Yang et al. 2016). Porambage et al. (2014) implicitly designed a two-phase authentication protocol using certificates for WSNs in distributed IoT applications. The Physical Unclonable Function (PUF)-based Authentication Scheme (PAS) with session key for ensuring the secure interaction between smart devices in IoT is addressed by Muhal et al. (2018). The work of Ourad, Belgacem, and Salah (n.d.) proposed a concept of blockchain solution to allow users to connect and authenticate securely to their IoT devices. The Reviving-under-Denial of Service (RUND) protocol is proposed by Yao et al. (2019). In Alcaide et al. (2013), the complete decentralized anonymous authentication protocol for privacy-preserving IoT target-driven applications is proposed. Since devices in IoT are resource constraint, lightweight token-based authentication is proposed by Dammak et al. (2019). Fuzzy logic (Magdalena 2015) has been applied for checking the behaviour of users to ensure continuous authentication (Mondal and Bours 2014). Ye et al. (2014) described authentication in multiple phases, namely the initialization phase, mutual authentication phase, and key establishment phase. Trusting of devices using certificate authority, authentication using ECC is carried out. In the proposed work multiple steps including trusting, authentication between devices in IoT is addressed using fuzzy logic and the mutual authentication mechanism by reducing communication overhead of a device through locally centralized globally distributed authentication architecture is addressed. Challenges are discussed in the next section.

9.2.2 Challenges and Research Issues

Security and privacy are very crucial aspects in IoT for preserving the identity and information of devices. Various works related to authentication, security, and privacy

(Karthiban and Smys 2018) are discussed in the literature survey. Some of the challenges and issues raised in this context are mentioned in this section. It is most important before designing an algorithm to consider the resource constraint device's capability, computation cost, and communication cost. The devices in IoT are scalable and heterogeneous, so the algorithm to be designed has to ensure scalability, interoperable, and adaptable to the context. The energy consumption, computational delay, bandwidth requirement, and memory usage has to be considered for designing algorithm. The need of authentication is addressed in previous sections, but the mechanisms involved in authentication of devices have to ensure the efficiency and validate whether the device or user is real or fake. The devices in IoT are heterogeneous and constitute communicating between many diversities including human to human, human to device, and device to device (static or mobile). The devices after manufacturing are assigned with a MAC address and connected to the network that issues an IP address, but human beings communicating to a device have to be trusted initially for avoiding the severity of network failure. Several spoofing techniques (Xiao et al. 2018) for MAC address and IP address are studied similarly attacks caused by human has to be studied for preserving the privacy and confidentiality among the networks. Another challenge is interoperability among standards for authentication between devices is significant. The algorithm for trusting, authenticating, privacy of data (Kulkarni, Durg, and Iyer 2016), confidentiality, and interoperability (Mohamad Noor and Hassan 2019) for all types of networks with less energy consumption, bandwidth, and tolerable to several attacks of all the layers is essential. In this work the idea is proposed to verify and validate the device one step before authentication by considering MAC and IP addresses of the device followed by the fuzzy approach (Jain, Wadhwa, and Deshmukh 2007) for deciding the degree of authenticity and trust.

9.3   AUTHENTICATION MECHANISMS IN IOT

Due to growth of devices in IoT, authentication is given first priority compared to security and privacy mechanisms. Authentication (Wu et al. 2018) ensures the identity of devices, and further privacy operations are employed for preserving the data and achieving secure communication. Authentication is performed by considering the authentication procedure, factors, and architecture. In IoT the devices enter and leave the network as they wish, but it is difficult to manage fickle devices. The devices in static are also fickle as they request the resource and leaves without the reason. In a dynamic environment, fickle devices are defined as devices existing in the network that may leave and new devices enter into the network with or without completing the task. Such devices have to be tracked and authenticated. In the next section both environments are discussed. In such environments handling of devices is done by central authority that is a centralized approach or it can be distributed. In this work locally centralized and globally distributed approach are applied for authenticating.

9.3.1   STATIC ENVIRONMENT AND DYNAMIC ENVIRONMENT

In a static environment the devices are considered to be present in the environment for some duration τ with mobility of factor α. The devices are authenticated by considering different authentication factors. The devices randomly enter and leave the network and are handled by the proposed algorithm. In network environment both static and dynamic devices are considered, and the algorithm handles the static and dynamic (Matos et al. 2018) devices without disturbing other device operations and maintains network connectivity. The dynamic devices are locked for certain duration to complete the evaluation of finding the trust level of devices, so that if these devices leave the network and enter later, the first step is ignored and saves the computations required. Devices not verified are not allowed to enter and access the resources of other devices. The proposed algorithm addresses this problem by different authentication architecture.

9.3.2 CENTRALIZED APPROACH

In a centralized approach (Shan Yin, Yueming Lu, and Yonghua Li 2015), the devices communicate through the Central Arbiter (CA). Initially the devices are clustered based on availability of resources in the devices, and then these devices register to the nearest CA. The CA is present in the fog layer having resources required for managing the device registrations, attestation, and authentication. The CA is authenticated and encrypted with existing security mechanisms. Figure 9.2 shows the communication between devices through CA. The devices register to the CA, then the CA attests the devices and allocates task-id (Tid) to the devices, then the devices mutually authenticate to communicate. If any device enters or leaves the network, the network is not affected compared to the distributed approach. In contrast if centralized authority CA fails or discharges energy, then it is difficult to manage the network, hence distributed approach is discussed to overcome the problem.

9.3.3 DISTRIBUTED APPROACH

In a centralized approach the CA was capable of doing registration, attestation, and assigning Tid, but in the distributed approach, the registration and attestation followed by mutual authentication is performed at all devices. This approach consumes

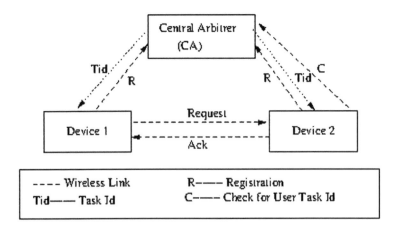

FIGURE 9.2 Centralized approach.

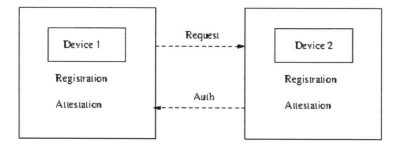

FIGURE 9.3 Distributed approach.

more energy at the device level. The devices involved in communication are heterogeneous, hence computation cost is different for all the devices, it is tedious for managing resource utilization at the device level. Figure 9.3 shows distributed or decentralized (Skarmeta, Hernández-Ramos, and Moreno 2014) approach wherein the devices communicate without CA involved in process.

9.3.4 Locally Centralized Globally Distributed

By this approach drawbacks of centralized and decentralized approach discussed earlier can be minimized. The devices with scarce resources are grouped constituting one network and devices with resources comparatively more than scarce resources are grouped into another network. The resource constraint devices follow a centralized approach and other devices employ a decentralized approach. This approach is illustrated with a network scenario in Section 9.4.

9.4 FOG-BASED IOT DEVICES AUTHENTICATION

Fog computing (Yannuzzi et al. 2014; Luan et al. 2015) provides services to the devices at the edge of the network. Heterogeneous devices require services from the cloud. As it is time consuming and to avoid congestion, fog computing provides services from the cloud to edge devices. As devices are involved in communication, their information is to be secured to ensure privacy for the devices. Hence authentication of devices has to be done beforehand for privacy and security of the information transmitted by the device. In this section the network environment is illustrated to address the proposed work followed by three phases of authentication comprising registration, attestation, and authentication followed by authorization. Agent-based authentication is addressed for distributing the task of the CA to agents for parallel computation and reducing the computation cost and mitigating the effect of failure of individual CA. Table 9.1 represents the notations used in the proposed algorithm and network environment.

9.4.1 Network Environment

As edge devices have scarce resources, the lightweight algorithms for authentication have been surveyed in Section 9.2. In the proposed scenario the edge devices are of resource rich and scarce clustered in different networks – N1, N2, and N3.

TABLE 9.1

Notations

Symbols	Specifications
Di	Devices
	di = {d1, d2, ..., dn}
CAi	Central Arbitrer
	CAi = {CA1, CA2, ..., Can)
Tidi	Task Id
	Tidi = {Tid1, Tid2, ..., Tidn}
NWidi	Network Id
	NWidi = {NWid1, NWid2, ..., NWidn}
MACidi	MAC address of device
	MACidi = {MACid1, MACid2, ..., MACidn}
IPi	IP address
	IPi = {Ip1, Ip2,..., Ipn}
AS	Authentication Score
h()	One-way hash function
Yi	Private key
Zi	Public key
C	Code

The perception layer comprising of various devices are prone to several attacks, namely device failure, denial of service, a malicious device changing the identity of a device, spoofing identities of devices, etc. Hence in the proposed work the trust of the devices is ensured through human decisions based on fuzzy approach. The trusting of devices is dealt in the fog layer as resources required for computation cannot be afforded by all heterogeneous devices. Edge computing performs the maximum resource utilization proposed by Sutagundar, Attar, and Hatti 2019), hence services are given to devices at the fog layer. Figure 9.4 shows the network scenario in which three layers are considered, namely perception, fog, and cloud.

In the perception layer devices are present, and the fog layer provides services to the devices through the cloud. A few services are extended from the cloud to the fog layer for device operation. A locally centralized and globally distributed approach is used for authenticating devices. In the Figure 9.4 {CA1, CA2, ..., Can} are central arbiters acting as gateways for the edge devices existing in the fog layer. It monitors the entering and leaving devices, activity of the devices, and mainly checks and verifies whether the devices are trusted or fake. CAi clusters the devices with fewer resources and then registers the nearby devices. The devices are attested and task ids (Tidi) are issued to devices and stores in database of the CA. If d1 present in N1 wishes to communicate with d5 in N2, CA1 and CA2 communicates, and provide Tidi accordingly. D1 of N1 will be allocated a network id (NW_{idi}), and then it has to register with CA2. It verifies the id given by CA1, allowing for access into N2 and the devices. In this work three phases are carried out in authentication of devices.

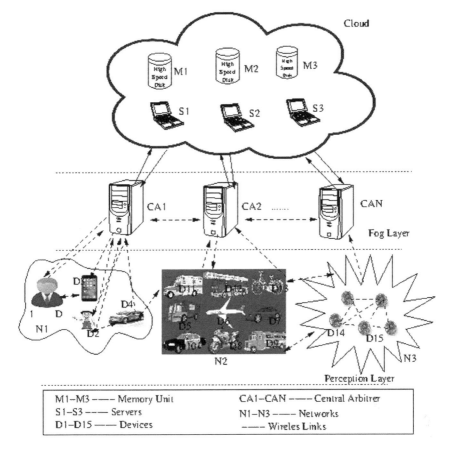

FIGURE 9.4 Network scenario.

1. Registration phase

In this phase the devices register using their MAC id and IP address to the CA. Consider register devices at CAi = {d1, d2 ,..., d10}. Each device di has D = {di, MACidi, IPi, Nwidi, Ti}. Nwidi, Ti is the network id and time stamp issued by CAi. Figure 9.5 shows the CA block. CA comprises of registering, attesting, and storing in the database. Once registered the output 'D' is sent to the attestation phase.

2. Attestation phase

The output of the registration phase is a crisp value fed to the CA. Figure 9.6 shows the functions of the attestation phase. It has a fuzzifier, fuzzy interference system (FIS) of Mamdani-type (Magdalena 2015; Uppalapati and Kaur 2009), defuzzifier, and allocator for sending Tid to devices and storing in database. This phase is required to ensure the trust level of a device. The device is trusted through FIS; once trusted it can communicate by entering the network. The FIS system gives a

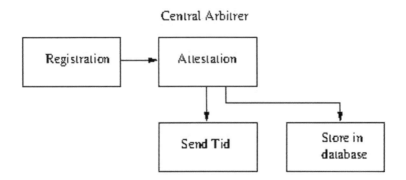

FIGURE 9.5 Central Arbiter (CA).

trust id to devices registered; one copy is stored in the database and forwarded to devices. Fuzzy logic is used for decision-making (Javanmardi 2012; Jiang et al. 2013) to measure the level of trust. The FIS system comprises a set of fuzzy rules to decide the trust level of a device. If the level is below threshold, the device is sent for the registration phase, and if greater than threshold medium trusted and it is sent for the next phase. Simultaneously its activity is observed periodically by the CAi. If trust level is equal to threshold, then it is assured and carried to the authentication phase.

Table 9.2 illustrates a few fuzzy rules for design of FIS to evaluate the trust level of a device. Fuzzifier converts all input to linguistic variables for formation of rules. If–then rules are used for framing rules, and to decide the trust level, the threshold value is chosen based on membership values.

The set of fuzzy rules are framed by considering fuzzification of input variables:

$$D = \{di, MACidi, IPi, Nwidi, Ti\}.$$

The IP address membership value is $\mu_A(x)$, the MAC address of the device membership value is $\mu_B(x)$, and the Network Id membership value is $\mu_C(x)$. The membership values are chosen based on checking spoofing of the IP address and MAC address. The membership values ranges from 0 to 1. The values decide the levels of spoofing

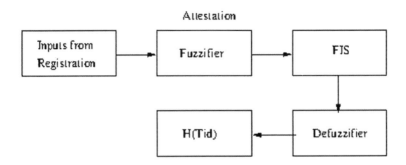

FIGURE 9.6 Attestation block.

TABLE 9.2

Fuzzy Membership Values

Parameters	Membership Values		
	X1	X2	X3
T1 (very low)	0.4	0.2	0.4
T2 (low)	0.5	0.3	0.5
T3 (medium)	0.7	0.5	0.45
T4 (high)	0.65	0.7	0.6
T5 (very high)	0.8	0.9	0.75

TABLE 9.3

Sample Set of Fuzzy Rules

Rule 1: If X1 is 0.4 AND X2 is 0.2 AND X3 is 0.4 the output is T1 (very low level of trust).
Rule 2: If X1 is 0.5 AND X2 is 0.3 AND X3 is 0.5 the output is T2 (low level of trust).
Rule 3: If X1 is 0.7 AND X2 is 0.5 AND X3 is 0.45 the output is T3 (medium level of trust).
Rule 4: If X1 is 0.65 AND X2 is 0.7 AND X3 is 0.6 the output is T4 (high level of trust).
Rule 5: If X1 is 0.8 AND X2 is 0.9 AND X3 is 0.75 the output is T5 (very high level of trust).

in an address and help in finding the identity and trust level of a device as tabulated in Table 9.2. Table 9.2 is reference for framing the fuzzy rules tabulated in Table 9.3.

A few combinations of membership values are chosen and framed the rules. As the number of rules increases, the high degree of trust. Hence it ensures the device is true, not fake or malicious. The rules framed are aggregated and then defuzzified using the Centre of area method (COA). FIS aggregates by combining all individual outputs as

$$\text{Fag} = G\left\{ \mu_{T1}(x), \mu_{T2}(x), \mu_{T3}(x), \mu_{T4}(x), \mu_{T5}(x)\right\},$$

where G is the function of aggregation, and defuzzifies the Fag as Tid = D(Fag), where D is the defuzzification method. The COA method is represented as Equation 9.1 (Ross, Booker, and Parkinson 2002):

$$Tid = \frac{\sum_{i=1}^{5} \mu_x(T_i).(T_i)}{\sum_{i=1}^{n} \mu_x(T_i)} \tag{9.1}$$

The defuzzified output is a crisp value for the trusting of the device and the value Tid is used for the device. It is stored in the CAi. The device request another device in the network for communication by using Tid along with IP, MAC, network address, and time of being assigned the id. The next step involves a mutually authentication process for ensuring authenticity and allows for communication discussed in the authentication phase.

3. Authentication phase

Authentication is two way, and multistep authentication is chosen for the highest authentication. The trust ids are issued from the CA, and then devices are given authority to communicate with devices by taking the consent of device that is involved in communicating. The mutual authentication using a low level, which involves exchange of private key and public key, is proposed in algorithm 1. In mid-level token-based authentication is performed in algorithm 2. At a high level, the fuzzy approach is used to decide the authentication of devices illustrated through a set of fuzzy rules. The three levels are used in multistep and it is selected based on the activities of users or devices involved in communication. If the activity is varied, then a second level or medium level of authentication is applied. If device fails in authenticating at the second level, then a fuzzy approach considering multifactors is applied for verifying the device at the third level (A. Roy and Dasgupta 2018). Based on fuzzy decision, the device is reregistered or locked and not allowed for further interaction in the network.

```
Algorithm 1: Mutual Authentication between device 1 (d1) and
device 2 (d2) (Low level)
Nomenclature: h( )-one way hash function; Y1-private key of
d1; Y2- private key of d2; Z1- public key of d1;Z2- public key
of d2;
Begin
Step 1: d1 task id is Tid1 is hashed P1=h(Tid1),
        Compute private key Y1=s P1; Public key Z1= aY1;
        d1 sends (req,Tid1,MAC1,IP1,Z1,T1) to d2; T1 is current
        time of d1
Step 2: d2 receives the request from d1, d2 sends P1 to CA for
        checking whether it is trusted;
        If P1= h(Tid1)|| T1 checks in database, if equal then
        its trusted or else rejects the request
        and forward to registration phase.
Step 3: d2 performs P2=h(Tid2);
        Compute private key Y2=s P2; Public key Z2= bY2;
        d2 sends (Tid2, MAC2,IP2, Z2, T2) to d1
Step 4: d1 checks whether T2 is valid, if valid finds public
        key Z1'=a P and
        sends (Tid1||MAC1||IP1||Tid2||MAC2||IP2||Z1'||Z1'') to
        d2
Step 5: d2 verifies h(Tid1||MAC1||IP1||Tid2||MAC2||IP2||Z1'|
        |Z1'')
        If Z1''=s⁻¹Z1 then its authenticated or else
        authentication fails;
```

    d2 performs Z2'=bP then sends (Tid2||MAC2||IP2||Ti
    d1||MAC1||IP1||Z2'||Z2'') to d1
Step 6: d1 receives from d2 finds Z2''= s⁻¹Z2;
    Verifies (Tid2||MAC2||IP2||Tid1||MAC1||IP1||Z2'||Z2'')
    =
    (Tid1||MAC1||IP1||Tid2||MAC2||IP2||Z1'||Z1'') then d2
    is authenticated to d1 hence
    mutually authenticated.
Step 7: If malicious devices is suspected apply medium level
    mutual authentication that is token
    based discussed in algorithm 2.
Step 8: If still threat exists employ third level
    authentication mechanism using fuzzy approach
    for deciding whether to reregister or lock the device.
End

    In algorithm 2, the medium level of authentication is proposed. Tokens are of several types. In this work a software token (ST) is used. It is two-factor authenticate security device which authenticates the use of computer services. The ST is to be stored on a device for authentication of a device. The key including the token is comprised of a MAC id, IP address, Tid of device, and token.

    $K = \{MAC_{idi}||IP_i||T_{idi}||Nw_{idi}||Token_i||T_i\}$ is the key used by a device for authenticating with another device. If malicious devices identify one of the key parameters, but the complete key cannot be found and fails to be authenticated. All the combinations have to be same as that of the original device, then only it will be authenticated. The software token has a disadvantage because the programme sending or receiving the code can be hacked. So in this work, the token is constructed by use of some arithmetic logic in finding the matching between code sent to the user and in the database. The token construction is discussed in algorithm 2.

Algorithm 2: Mutual authentication d1 to d2 (Medium level)
Nomenclature: Di-Devices; CAi- Central Arbiter; Code- C;
OC- obtained code
Begin
Step 1: Devices Di registers to CAi,
Step 2: Attestation phase is completed
Step 3: The trust id is sent to Di along with some random
numbers
Step 4: The random numbers are computed to form equivalent
code and it is stored in database of CAi.
Step 5: The random numbers forming a equivalent key, if hacked
and perform some operations on it and outputs random code.
Step 6: Random code is checked in the database of CAi to
verify whether the respective Di random number is having same
code.
Step 7: If random key = code (C) stored in database matches
random key= obtained code (OC) then Di is authenticated and
allowed to communicate.

Step 8: If in three attempts OC is obtained then it is allowed, in ten attempts it obtains OC then forwarded to third level of authentication.
Step 9: If failed in ten attempts then it is locked by the CAi and not allowed in communication in the network.
End

The third level is fuzzy approach in which input parameters are login activities, history of communication in the network, and trust level indicators by the neighbouring devices or the CA having the particular device. The trust level is ensured by considering the opinion of the particular device by the neighbouring device. If three parameters have very low membership values, they are locked by the CA and not allowed in communicating with any network. The authorization or access to the devices is provisioned by evaluating the authentication score. The steps adopted for finding AS are discussed in the following section.

9.4.2 Authorization

Authorization deals in deciding to give access for resources to authenticated devices. It is a process of deciding how much resources are to be permitted for accessing. In this work, the authentication score (AS) is used for determining the percentage of resource to be accessed. ASi is evaluated by considering the basis of authentication phase, the time required for authenticating, attempts made to authenticate, and levels of authentication. AS1, AS2, and AS3 are three scores indicating high, medium, and low scores. The three scores are calculated as

$$\{AS1 = Tr < TT + 1 \text{ attempt for authenticating } + \text{Auth 1 } (\text{first level of authentication})\};$$

$$\{AS2 = Tm = TT + 2 \text{ attempts for authenticating } + (\text{Auth 1} + \text{Auth 2}) (2 \text{ levels})\};$$

$$\{AS3 = Th > TT + 3 \text{ attempts for authenticating } + (\text{Auth 1} + \text{Auth 2} + \text{Auth 3}) (3 \text{ levels})\};$$

where Tr, Tm, and Th are the times taken for the authentication process by devices; TT is the total time taken for the authentication process given as $TT = TY + TZ + Treq + 2TCA + Th + Tt + Tf$. Any devices possessing are given access to resources by paying. If the device possesses AS1 as a score, then the requested resources are completely given access. If AS2 is possessed, then partially requested resources are given access, and if it is AS3, then 20% of resources requested are given access. The CA monitors the login activity of the device if it seems to be genuine. Then part-by-part access is granted and if found to be suspicious, the access is denied and device is locked. AS helps in reducing the wastage of resources and increases lifetime of the network. If any devices leave the network during accessing of the partial resources of required are penalized and then allowed to leave. The method discussed is of the centralized approach and CA has to perform all the operations required for authentication, hence consumes more power and delay is increased. This problem is addressed by the agent-based approach for authentication in a distributed way.

9.4.3 AGENT-BASED APPROACH FOR AUTHENTICATION

Agents (Mahmoud 2000; Sabir et al. 2019) are software programs categorized as static, mobile, reactive, communicative, learning, and or goal-oriented based on the features. Agent technology can be used in the network environment to track the activities of devices that seem to be suspicious. They have the capability of communicating with all types of devices, hence CA power consumption can be reduced by assigning a few task of the CA to agents. In the security aspect, if any malicious device is trying to hack randomly, the agents are deployed which are capable of tracking the device login attempts and sending the alert message to the CA for controlling the device communication in the network. In the adaptive environment the agent plays a vital role in managing device registration and authentication in IoT. An agent incorporates assisted learning to help in making decisions based on the information received by the previous user. It has a knowledge base and tries to solve the problem based on learnt situations in the network. The learning method can be supervised and non-supervised learning. It learns by seeing the network and adapts to the network scenario and increases the efficiency of the system. Figure 9.7 shows the agent-based approach for authentication and authorization of devices. Central Arbitrer Agency (CAA) comprises of static and mobile agents (Manvi and Venkataram 2004).

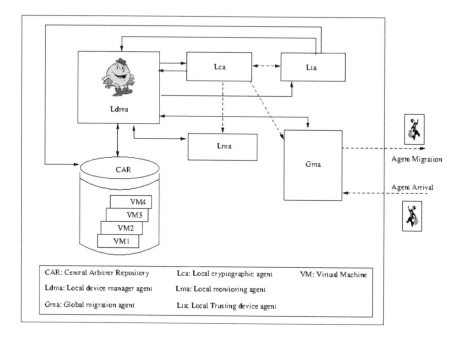

FIGURE 9.7 CA Agency (CAA).

CAA creates CAR, Ldma, LTa, Lca, Lma, and Gma for storing, managing devices, checking the trust level, cryptographic agent for security of agents, monitoring the activities of device, and migrating agents, respectively.

i. CA Repository (CAR) – Stores all the registered devices' ids and trust ids. It maintains the status of devices entering and leaving.
ii. Local device manager agent (Ldma) – It is a static agent that monitors the device entering and leaving the network 1 and migrating to network 2 (shown in Figure 9.4).
iii. Local monitoring agent (Lma) – It is a static agent that monitors the login activity of a device at the time of access to resources in the network. If any fluctuation in the activity is observed, it immediately communicates to Ldma to lock the device or for any action.
iv. Local trusting the device agent (LTa) – It is a mobile agent that computes the trust level of a device and assigns to devices as Tid. Tid is stored in the CAR for further comparing the new devices entering in to the network.
v. Local cryptography agent (Lca) – It is a mobile agent comprised of symmetric and asymmetric key cryptographic techniques. It checks the security of other agents periodically as the agent security may be altered due to environmental factors or network conditions. It is lightweight and embedded with all devices.
vi. Global migrating agent (Gma) – It is a mobile agent responsible for directing the device to another network CA or allowing it to communicate with the nearest CA based on the resources available in the CA to avoid failure of the system.

The CAA helps in performing the tasks in parallel by dividing among static and mobile agents. The communication overhead is bandwidth required for communicating with the CA and other devices instead of individual CA. The computation takes less time as the tasks are divided among various agents for performing operation in parallel compared to the non-agent approach.

9.4.4 Performance Parameters

Some of the performance parameters considered for evaluating the proposed work are computational cost and communication cost.

9.4.4.1 Computational Cost

The total cost of the proposed approach includes cost at the registration phase, attestation and authentication phase, and authorization. The cost incurred at the registration phase is given by $Cr = Tr + Ta$; Tr and Ta represent the registration time and assigning network id respectively.

Cost at attestation phase is evaluated as $Ca = Tf + 2Th + Td + nTs$, where Tf is time required for the FIS system, Th is for one hash function, Td is the time required for storing in the database, and Ts is sending time of Tidi to devices. Cost at the registration phase includes time required for multistep authentication and is calculated as $Cau = TY + TZ + Treq + 2TCA + Th + Tt + Tf$, where TY is the time required for computing the private key and TZ for the public key, Treq is the requesting time from d1 to d2, TCA is time required for sending and checking Tid in the database, Th is for one-way hash function, Tt is for token-based authentication, and Tf is for the fuzzy approach. Cost of the authorization process (Cath) includes evaluating ASi. Th is the time required for estimating ASi = {AS1, AS2, AS3}. The total cost is C = Cr + Ca + Cau + Cath. The cost incurred is more compared to the rest of the work, but the added advantage is trusting of devices at several levels ensures they are for tolerable to a man-in-the-middle attack (T. Roy and Dutta 2011), eavesdropping (Aliyu, Sheltami, and Shakshuki 2018), replay attacks (Pawar and Anuradha 2015), and node attack, and avoids IoT failure.

9.4.4.2 Communication Cost

The cost for communication is given as Cc with communication parameters Cc ={Yi, Zi, Tidi, MACidi, IPi, Nwidi, ASi}. The communication overhead is reduced by the centralized approach and increased in the distributed approach. So as to maintain balanced communication cost among devices in the network, the combinational approach is proposed. The energy consumption of the device is reduced comparatively as CA is performing most of the operations.

9.5 CONCLUSION

The proposed work mainly trusts the device before authenticating as in the process of authentication the devices are trusted but the level of trust among devices is not identified. The device behaves as real by hacking the device identity and degrades the network efficiency by capturing the confidential information. To solve such issues the proposed work's fog-based IoT device authentication ensures trust of the devices as the first priority and mutual authenticity is ensured to communicate, and authorization allowed for accessing the devices in the network. Multistep and multifactor authentication using the agent approach is performed by devices for a high degree of assurance for claiming to be real and with reduced delay. Accessing of resources is provisioned on the basis of AS in balancing the usage of network resources. The proposed work is tolerable to a man-in-the-middle attack, eavesdropping, and a replay attack. Fog computing and agent technology together provide services in time without degrading the efficiency of the network. The future work is to minimize the computation and communication cost by considering other environmental factors affecting the devices and tolerable to few more attacks in different layers.

REFERENCES

Alcaide, Almudena, Esther Palomar, José Montero-Castillo, and Arturo Ribagorda. 2013. "Anonymous Authentication for Privacy-Preserving IoT Target-Driven Applications." *Computers & Security* 37(September): 111–23. doi:10.1016/j.cose.2013.05.007.

Alhothaily, Abdulrahman, Arwa Alrawais, Chunqiang Hu, and Wei Li. 2018. "One-Time-Username: A Threshold-Based Authentication System." *Procedia Computer Science* 129: 426–32. doi:10.1016/j.procs.2018.03.019.

Aliyu, Farouq, Tarek Sheltami, and Elhadi M. Shakshuki. 2018. "A Detection and Prevention Technique for Man in the Middle Attack in Fog Computing." *Procedia Computer Science* 141: 24–31.

Bonomi, Flavio, Rodolfo Milito, Jiang Zhu, and Sateesh Addepalli. 2012. "Fog Computing and Its Role in the Internet of Things." In *Proceedings of the First Edition of the MCC Workshop on Mobile Cloud Computing*, 13–16. MCC '12. New York, NY: ACM. doi:10.1145/2342509.2342513.

Dammak, M., O. R. M. Boudia, M. A. Messous, S. M. Senouci, and C. Gransart. 2019. "Token-Based Lightweight Authentication to Secure IoT Networks." In *2019 16th IEEE Annual Consumer Communications Networking Conference (CCNC)*, 1–4. doi:10.1109/CCNC.2019.8651825.

Ferrag, Mohamed Amine, Leandros A. Maglaras, Helge Janicke, Jianmin Jiang, and Lei Shu. 2017. "Authentication Protocols for Internet of Things: A Comprehensive Survey." Research article. *Security and Communication Networks*. https://www.hindawi.com/journals/scn/2017/6562953/.

Ferrag, Mohamed Amine, Leandros Maglaras, and Abdelouahid Derhab. 2019. "Authentication and Authorization for Mobile IoT Devices Using Biofeatures: Recent Advances and Future Trends." *Security and Communication Networks* 2019(April). doi:10.1155/2019/5452870.

Deng, Hongmei, A. Mukherjee, and D.P. Agrawal. 2004. "Threshold and Identity-Based Key Management and Authentication for Wireless Ad Hoc Networks." In *International Conference on Information Technology: Coding and Computing, 2004. Proceedings. ITCC 2004*, Vol. 1., 107–11, Las Vegas, NV: IEEE. doi:10.1109/ITCC.2004.1286434.

Jain, V., S. Wadhwa, and S. G. Deshmukh. 2007. "Supplier Selection Using Fuzzy Association Rules Mining Approach." *International Journal of Production Research* 45(6): 1323–53. doi:10.1080/00207540600665836.

Javanmardi, Saeed. 2012. "A Novel Approach for Faulty Node Detection with the Aid of Fuzzy Theory and Majority Voting in Wireless Sensor Networks." *International Journal of Advanced Smart Sensor Network Systems* 2(4): 1–10. doi:10.5121/ijassn.2012.2401.

Jiang, Haifeng, Yanjing Sun, Renke Sun, and Hongli Xu. 2013. "Fuzzy-Logic-Based Energy Optimized Routing for Wireless Sensor Networks." *International Journal of Distributed Sensor Networks* 9(8): 216561. doi:10.1155/2013/216561.

Karthiban, K., and S. Smys. 2018. "Privacy Preserving Approaches in Cloud Computing." In *2018 2nd International Conference on Inventive Systems and Control (ICISC)*, 462–7. doi:10.1109/ICISC.2018.8399115.

Keshwani, Deepak R., David D. Jones, George E. Meyer, and Rhonda M. Brand. 2008. "Rule-Based Mamdani-Type Fuzzy Modeling of Skin Permeability." *Applied Soft Computing* 8(1): 285–94. doi:10.1016/j.asoc.2007.01.007.

Kim, Hokeun, Armin Wasicek, Benjamin Mehne, and Edward A. Lee. 2016. "A Secure Network Architecture for the Internet of Things Based on Local Authorization Entities." In *2016 IEEE 4th International Conference on Future Internet of Things and Cloud (FiCloud)*, 114–22. Vienna, Austria: IEEE. doi:10.1109/FiCloud.2016.24.

Kulkarni, S., S. Durg, and N. Iyer. 2016. "Internet of Things (IoT) Security." In *2016 3rd International Conference on Computing for Sustainable Global Development (INDIACom)*, 821–4.

Luan, Tom H., Longxiang Gao, Zhi Li, Yang Xiang, Guiyi Wei, and Limin Sun. 2015. "Fog Computing: Focusing on Mobile Users at the Edge." *arXiv:1502.01815 [Cs]*, February. http://arxiv.org/abs/1502.01815.

Magdalena, Luis. 2015. "Fuzzy Rule-Based Systems." In *Springer Handbook of Computational Intelligence*, edited by Janusz Kacprzyk and Witold Pedrycz, 203–18. Berlin Heidelberg: Springer. http://link.springer.com/chapter/10.1007/978-3-662-43505-2_13.

Mahmoud, Qusay H. 2000. "Software Agents: Characteristics and Classification." *School of Computing Science, Simon Fraser University*, 1–12.

Manvi, S.S, and P Venkataram. 2004. "Applications of Agent Technology in Communications: A Review." *Computer Communications* 27(15): 1493–508. doi:10.1016/j.comcom.2004.05.011.

Matos, E. de, R. T. Tiburski, L. A. Amaral, and F. Hessel. 2018. "Providing Context-Aware Security for IoT Environments Through Context Sharing Feature." In *2018 17th IEEE International Conference On Trust, Security And Privacy In Computing And Communications/ 12th IEEE International Conference On Big Data Science And Engineering (TrustCom/BigDataSE)*, 1711–5. doi:10.1109/TrustCom/BigDataSE.2018.00257.

Mondal, Soumik, and Patrick Bours. 2014. "Continuous Authentication Using Fuzzy Logic." In *Proceedings of the 7th International Conference on Security of Information and Networks - SIN '14*, 231–8, Glasgow, Scotland, UK: ACM Press. doi:10.1145/2659651.2659720.

Muhal, M. A., X. Luo, Z. Mahmood, and A. Ullah. 2018. "Physical Unclonable Function Based Authentication Scheme for Smart Devices in Internet of Things." In *2018 IEEE International Conference on Smart Internet of Things (SmartIoT)*, 160–5. doi:10.1109/SmartIoT.2018.00037.

Noor, Mohamad, Mardiana binti, and Wan Haslina Hassan. 2019. "Current Research on Internet of Things (IoT) Security: A Survey." *Computer Networks* 148(January): 283–94. doi:10.1016/j.comnet.2018.11.025.

Ourad, A. Z., B. Belgacem, and K. Salah. 2018. "IOT Access Control and Authentication Management via Blockchain.", In *Proceedings of Internet of Things - ICIOT 2018 - Third International Conference*, Seattle, WA, June 25–30. Lecture Notes in Computer Science 10972, Springer.

Paul, P. V., and R. Saraswathi. 2017. "The Internet of Things — A Comprehensive Survey." In *2017 International Conference on Computation of Power, Energy Information and Communcation (ICCPEIC)*, 421–6. doi:10.1109/ICCPEIC.2017.8290405.

Pawar, Mohan V., and J. Anuradha. 2015. "Network Security and Types of Attacks in Network." *Procedia Computer Science* 48: 503–6. doi:10.1016/j.procs.2015.04.126.

Porambage, P., C. Schmitt, P. Kumar, A. Gurtov, and M. Ylianttila. 2014. "Two-Phase Authentication Protocol for Wireless Sensor Networks in Distributed IoT Applications." In *2014 IEEE Wireless Communications and Networking Conference (WCNC)*, 2728–33. doi:10.1109/WCNC.2014.6952860.

Ross, Timothy J., Jane M. Booker, and W. J. Parkinson, eds. 2002. *Fuzzy Logic and Probability Applications: Bridging the Gap*. ASA-SIAM Series on Statistics and Applied Probability. Philadelphia, PA/Alexandria, VA: Society for Industrial and Applied Mathematics ; American Statistical Assoxciation.

Roy, Arunava, and Dipankar Dasgupta. 2018. "A Fuzzy Decision Support System for Multifactor Authentication." *Soft Computing* 22(12): 3959–81. doi:10.1007/s00500-017-2607-6.

Roy, Tumpa, and Kamlesh Dutta. 2011. "Mutual Authentication for Mobile Communication Using Symmetric and Asymmetric Key Cryptography." In *Trends in Network and Communications*, 88–99. Springer, Berlin, Heidelberg.

Sabir, Badr Eddine, Mohamed Youssfi, Omar Bouattane, and Hakim Allali. 2019. "Authentication and Load Balancing Scheme Based on JSON Token For Multi-Agent Systems." *Procedia Computer Science* 148: 562–70. doi:10.1016/j.procs.2019.01.029.

Shah, S. H., and I. Yaqoob. 2016. "A Survey: Internet of Things (IOT) Technologies, Applications and Challenges." In *2016 IEEE Smart Energy Grid Engineering (SEGE)*, 381–5. doi:10.1109/SEGE.2016.7589556.

Shen, J., D. Liu, Q. Liu, X. Sun, and Y. Zhang. 2017. "Secure Authentication in Cloud Big Data with Hierarchical Attribute Authorization Structure." *IEEE Transactions on Big Data*, 1–1. doi:10.1109/TBDATA.2017.2705048.

Shivraj, V. L., M. A. Rajan, M. Singh, and P. Balamuralidhar. 2015. "One Time Password Authentication Scheme Based on Elliptic Curves for Internet of Things (IoT)." In *2015 5th National Symposium on Information Technology: Towards New Smart World (NSITNSW)*, 1–6. doi:10.1109/NSITNSW.2015.7176384.

Singla, J. 2015. "Comparative Study of Mamdani-Type and Sugeno-Type Fuzzy Inference Systems for Diagnosis of Diabetes." In *2015 International Conference on Advances in Computer Engineering and Applications*, 517–22. doi:10.1109/ICACEA.2015.7164799.

Skarmeta, A. F., J. L. Hernández-Ramos, and M. V. Moreno. 2014. "A Decentralized Approach for Security and Privacy Challenges in the Internet of Things." In *2014 IEEE World Forum on Internet of Things (WF-IoT)*, 67–72. doi:10.1109/WF-IoT.2014.6803122.

Stojmenovic, Ivan. 2014. "Fog Computing: A Cloud to the Ground Support for Smart Things and Machine-to-Machine Networks." In *2014 Australasian Telecommunication Networks and Applications Conference (ATNAC)*, 117–22, IEEE.

Sutagundar, Ashok V., Ameenabegum H. Attar, and Daneshwari I. Hatti. 2019. "Resource Allocation for Fog Enhanced Vehicular Services." *Wireless Personal Communications* 104(4): 1473–91. doi:10.1007/s11277-018-6094-6.

Uppalapati, S., and D. Kaur. 2009. "Design and Implementation of a Mamdani Fuzzy Inference System on an FPGA." In *NAFIPS 2009 –2009 Annual Meeting of the North American Fuzzy Information Processing Society*, 1–6. doi:10.1109/NAFIPS.2009.5156408.

Vaquero, Luis M., and Luis Rodero-Merino. 2014. "Finding Your Way in the Fog: Towards a Comprehensive Definition of Fog Computing." *ACM SIGCOMM Computer Communication Review* 44(5): 27–32. doi:10.1145/2677046.2677052.

Wu, F., X. Li, L. Xu, A. K. Sangaiah, and J. J. P. C. Rodrigues. 2018. "Authentication Protocol for Distributed Cloud Computing: An Explanation of the Security Situations for Internet-of-Things-Enabled Devices." *IEEE Consumer Electronics Magazine* 7(6): 38–44. doi:10.1109/MCE.2018.2851744.

Xiao, L., X. Wan, X. Lu, Y. Zhang, and D. Wu. 2018. "IoT Security Techniques Based on Machine Learning: How Do IoT Devices Use AI to Enhance Security?" In *IEEE Signal Processing Magazine* 35 (5): 41–49.https://doi.org/10.1109/MSP.2018.2825478.

Yang, Yanjiang, Haibin Cai, Zhuo Wei, Haibing Lu, and Kim-Kwang Raymond Choo. 2016. "Towards Lightweight Anonymous Entity Authentication for IoT Applications." In *Information Security and Privacy*, edited by Joseph K. Liu and Ron Steinfeld, 265–80. Lecture Notes in Computer Science. Springer International Publishing, Switzerland.

Yannuzzi, M., R. Milito, R. Serral-Gracià, D. Montero, and M. Nemirovsky. 2014. "Key Ingredients in an IoT Recipe: Fog Computing, Cloud Computing, and More Fog Computing." In *2014 IEEE 19th International Workshop on Computer Aided Modeling and Design of Communication Links and Networks (CAMAD)*, 325–9. doi:10.1109/CAMAD.2014.7033259.

Yao, Qingsong, Jianfeng Ma, Rui Li, Xinghua Li, Jinku Li, and Jiao Liu. 2019. "Energy-Aware RFID Authentication in Edge Computing." *IEEE Access* 7: 77964–80. doi:10.1109/ACCESS.2019.2922220.

Ye, Ning, Yan Zhu, Ru-chuan Wang, Reza Malekian, and Lin Qiao-min. 2014. "An Efficient Authentication and Access Control Scheme for Perception Layer of Internet of Things." *Applied Mathematics & Information Sciences* 8(4): 1617–24. doi:10.12785/amis/080416.

Yi, S., Z. Hao, Z. Qin, and Q. Li. 2015. "Fog Computing: Platform and Applications." In *2015 Third IEEE Workshop on Hot Topics in Web Systems and Technologies (HotWeb)*, 73–8. doi:10.1109/HotWeb.2015.22.

Yin, Shan, Yueming Lu, and Yonghua Li. 2015. "Design and Implementation of IoT Centralized Management Model with Linkage Policy." In *Third International Conference on Cyberspace Technology (CCT 2015)*, 1–5. doi:10.1049/cp.2015.0859.

10 Software-Defined Networks and Security of IoT

Ahmed Gaber Abu Abd-Allah, Atef Zaki Ghalwash, and Aya Sedky Adly

CONTENTS

10.1 INTRODUCTION

The technology SDN (software-defined networking), despite its disorderly quality compared with the traditional networking, is important among network methodologies and technologies. SDN decouples three plans – application, control, and data – and they are managed by programmed controller(s). SDN will become significant and an enabler for IoT that develops and loads by network growth. SDN will be a very critical aspect to IoT. Professionals forecast that IoT possibly will include a massive 21 billion devices by 2020 (Infosys 2019). To achieve better support for these procedures, the SDN market will multiply to approximately $133 billion by 2022 forecasts an article by Allied Market Research, a universal marketplace research organization (2019). The industry terminology for the knowhow SDN evolves rapidly as a response to data access and transmission emerging requirements. To turn SDN into a standard model, the increasing capacity of moveable and other connected devices must start to develop the data they generate among each other. Moreover, subdivisions concerned with retrieving information from these associated devices will need to generate related security procedures to hand over and stock this information. Specialists define the variations the entire network will undertake to have the ability to keep up with IoT, and it is termed elasticity (Flauzac et al. 2015).

SDN's elasticity becomes dominant and important as we go into an age ruled by huge amounts of data. Outdated networks are not prepared. It should remain premeditated to cope with the flow of huge amount of data and the regular innovativeness. Such data must be analyzed to support it, otherwise it won't help. That's because the SDN is flexible; the elasticity and agility permits organization of the correct type of data in this new era. We will clarify the importance of integrating SDN with IoT by giving an example. In the healthcare field, hospitals likely have a lot of equipment and patient terminal devices. All these are likely to be IoT based, permitted, and linked. Therefore, a patient is assigned to therapeutic scanning equipment that has heavy data flow on the specified network. You need flexible architecture to handle this huge flow being loaded into the network.

Also, hospital networks are filled with critical personal data that must be secured and be associated with point-to-point information safety. If the hospital equipment in this situation is moveable, the equipment must be linked in the new area directly and all network guidelines and rules that were earlier created in the network must be transported to the new spot as well. It is essential that the huge amounts of data are on an SDN that is elastic, and that the network is flexible, adaptable, and quickly replies. There is great opportunity for development in this global network.

By using the SDN architecture's network security, numerous workings have been examined. Also using firewalls, IPS11, NAC7, or IDS modules (Hakiri et al. 2014) instead of the SDN controller, or security rules upon OpenFlow forwarding devices have been attempted. The appearance of the next generation of Internet structures need uniform and advanced security, such as verifying network tools, operators, and items linking to operators making use of mutually wired and wireless information. Moreover, a monitoring process must be assigned to the behaviour of operators and items, conviction limitations set up, as well as exploiting secretarial approaches sideways with programmes substantiation. Nevertheless, current security systems do not present these security measures to encounter the security needs of the next generation of Internet structures.

10.1.1   Restrictions of Traditional Architectures

At the Internet edge, traditional security mechanisms such as avoidance systems, interference discovery, and firewalls (Nanyang Huang 2018) are installed. Such tools are used to protect the network from external attacks. However, these mechanisms are no longer sufficient to secure the next generation of the Internet.

In 2020 around 25 billion devices (Martinez-Julia and Skarmeta 2014) will be connected. The scalability issues were unsolved in the traditional environment, so technology had to seek another path to fulfil the needs of the growing network. All these devices are connected through forwarding devices like switch, hubs, and routers, and all these devices need secure environment to work in (Hande and Akkalakshmi 2015). The main issue in the traditional networks is the security of the large numbers of devices. All of these new devices have led to the emergence of various equipment as stated earlier. Consequently, the number of connected devices and the transmission rates have increased, and online services such as e-commerce, e-banking, VoIP, and e-mail have emerged (Martinez-Julia and Skarmeta 2014).

10.1.2 SOFTWARE-DEFINED NETWORK (SDN)

SDN is a novel way of net programming to lead and manage the behaviour of the network dynamically through a controller via open applications (Badach 2018) that disparity depends on closed devices with defined moderate. SDN delivers a universal visualization for administrators toward managing the system and controlling each flow. Conversely, because of the imperfect size of the flow table, the current SDN designs have numerous routine issues such as great delay in the packet arrival, high storage burden, and inapt scalability. Regarding the network technology embedded, the SDN model allows a centralized controller to control the data flow, in order to attach these campaigns that is produced from different sellers. The unified controller establishes all the information, and monitors and preserves an entire data flow view of the network and related devices. This centralization is the latest attempt to achieve the correct management of network roles (Badach 2018). SDN simply decoupled the data plane from the control plane (see Figure 10.1).

Such an environment will be very effective for sustainable and secure architecture for IoT. The reasons behind the importance of the SDN include:

- Application will be included in the network and as a network administrator you can easily manage these applications. An IT specialist can utilize a remote monitoring mechanism to realize the level of awareness of suitable applications via SDN.
- Overload over the network during a certain time of day, the specialist can ensure he routes applications in ways that will deal with that sensitive traffic level. What makes SDN so attractive is that it allows the administrator to develop a policy to make enhanced and central applications during such times of load and pressure on the network.

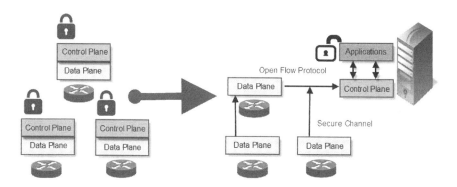

FIGURE 10.1 Traditional vs SDN.

- By using SDN, the bandwidth required by a network becomes much easier for an organization to control because of its adaptable environment. Suppose your organization has both a field office and a switching device as part of the IT scenario. By using SDN, you can deal with these diverse strains in an adaptable manner too, providing the bandwidth to the enterprise or office that needs it most at a certain time.
- Many features of SDN are dynamic, including quality of service (QoS). Speciality switching devices can respond to QoS by making requests and creating special actions of packets within the network. Better still is that the SDN controller can add the extremely beneficial application-awareness feature (Bakhshi 2017). Doing so lets the application become aware of the preferential treatment that is provided.
- With the self-learning option, if there is a heavy usage of the data exchange, such as remote patient consultancy as mentioned in the hospital example earlier, it can accept an adaptable bandwidth based on requirements through an easy-to-use interface.
- Security, and this point especially will be our argument in this work, in IoT is to secure end-to-end data. SDN has enabled connectors to secure the end devices and embed dynamic policies in real-time. These connectors can make the portable or fixed devices connect to the network very fast (real-time) and deploy the policies automatically.

Firewalling, interference discovery, and avoidance systems all these are considered traditional security mechanisms are installed at the Internet edge. These tools are used to guard the network from exterior threats. These appliances are no longer safe enough for the next Internet generation.

10.1.3 OpenFlow Protocol

OpenFlow protocol is an exposed code of behaviour to drive the flow table in diverse types of switches and data-forwarding devices (McKeown 2008; Bakhshi 2017). Traffic could be distinguished and panelled into manufacture in addition to exploration flows. Academics can manage their identifiable flows by picking the paths of their traffic monitor and the handling they obtain (McKeown 2008). By using this processing, routing protocols could be used by researchers, addressing structures, security prototypes, and uniform replacements to the Internet Protocol. On the similar system, the construction stream of traffic is inaccessible and handled alike nowadays. The OpenFlow Switch depends on the Flow Table to route its data or traffic, and the actions submitted for each flow assigned into the table.

The traditional activities maintained by an OpenFlow Switch are comprehensive, which means including vast rules and policies to allow concise management, and this would be supportive for high-performance and low-cost. The information route essential has a prearranged grade of elasticity, which forms the capability to postulate random treatment of each packet and look for an additional degree of limitation, but still value range of actions (McKeown 2008).

Briefly, the OpenFlow Switch comprises at minimum three fragments. A Flow Table, accompanied by actions related to every entrance of flow, in order to contact

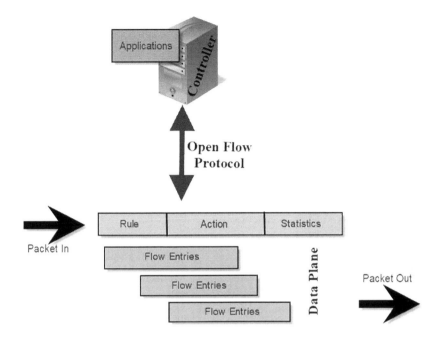

FIGURE 10.2   SDN structure using OpenFlow protocol.

a forwarding device, and the method of proceeding such flow. A Secure Channel which links this switch with the remote control process (controller), permitting orders between a controller as well as the switch. An OpenFlow protocol to direct packets (Lu et al. 2017) that provides a controller with an exposed and typical method in order to interconnect with the forwarding device. Through identifying a standard interface (OpenFlow protocol) (NEC 2011; ONF 2015) between sides which accesses within this Flow Table could be specified superficially, OpenFlow Switch circumvents the requirement for academics to plug in the switch in the network (McKeown 2008).

And as shown in Figure 10.1 and Figure 10.2, the OpenFlow protocol (NEC 2011) is a method of contact among the control plane and data plane as demonstrated earlier. An OpenFlow system eliminates the whole control roles from the switches; so the controller makes each progressing choice (Nanyang Huang 2018).

NOIoT security is determined by the many aspects and security values argued later. Challenges that are faced by IoT security have been the main target of many researchers. In the following section, an analysis of some related work is presented and the contribution of this research is described (Adly, 2019, Adly, 2020).

10.2   RELATED WORKS

10.2.1   Secure SDN Platform for Secured IoT

Flauzac et al. (2015) provide a design to ensure the security of the whole system with the notion of network security implanted in each controller.

The Internet is growing fast and at the end of 2014, 42.3% of people were associated with the Internet. However, the network security breaches increase with Internet development (NOLOT 2015).

Much research is devoted to the security of the Internet of Things, since it will contain every item or device with interacting capabilities, for example sensors in many devices in different disciplines such as medical, aeroplanes, cars, factories, etc.

10.2.2 Architecture Proposed

An architecture of three layers was proposed for ad hoc networks, inspired by the OpenDaylight Controller. The basic framework is the

- Physical layer
- Programmable layer
- Operating system layer

As discussed earlier, an SDN's layers and interfaces are connected to a virtual forwarding device (switch) that is responsible for forwarding the packets between the network objects. The proposed model divides the SDN network into multiple subnetworks called domains. Each domain has a controller responsible for setting the rules on each device and also can accept or decline any connection requests.

As shown in the Figure 10.3, A wants to send to B. A and B are in different domains. The packet of A might include a virus or any thread may be vulnerable on the network. So other parties in other domains must know the identity of senders and receivers, that is senders and receivers must know each other.

When A sends a request to B, the security controller will broadcast a message to explore if the sender and the receiver know each other; if not the connection will

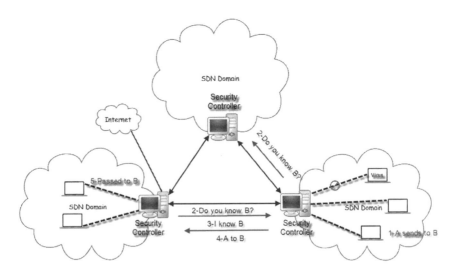

FIGURE 10.3   Grid of security in SDN domain.

be terminated. If the security controller of the receiver accepts the request, then it can receive the packet. After the acceptance, A sends the packet to B and finally B receives the packet.

SDN-based network architectures can include distributed controllers. This solution could be used in the framework of ad hoc networks that IoT depends on.

10.2.3   A Secured SDN Framework for IoT

In 2015 Sahoo et al. offered a secured SDN structure for IoT, since message protection is a foremost critical matter when data is actually transferred to and from different places. Besides, by expending the peak contemporary Internet machinery expansion, a large number of devices are contacting with users exploiting both underwired and wireless frameworks. By way of consequence, handlers or consumers and grid devices too are frequently exposed to possible threats. An exceptional security appliance should be used with the Internet of Things, since it integrates all terminals with grid capabilities. Which means it is an atmosphere where all objects have exclusive identifiers and has aptitude to allocate data without demanding human-to-computer or human-to-human interaction (Sahoo et al. 2015).

In the conventional structure, an interruption detection system (IDS) is prepared at the edge level of the Internet. These structures are exploited to keep the system safe from external hazards, but they are not adequate to handle the security of the ever-changing Internet. IoT based on systems with no borders promotes additional risk to the admission system controller. The main matter for an ad hoc arrangement for IoT is security (Sahoo et al. 2015).

The five basic properties of a secure communications network are integrity, confidentiality, availability, non-repudiation, and authentication (Tayyaba 2017). Spiteful attacks and hazards are considered as grid threads, and network administrators must defend their networks. In parallel, the SDN network framework has to be protected as well as sustain the aforementioned properties. The SANE and Ethane architectures (Valdivieso Caraguay et al. 2014) consider the security features of an inaccessible controller and advancing substructure.

The Ethane and SANE frameworks are similar to each other in that they permit more development tools for both data and control. Ethane is used to address numerous risks as it has several assuring practices that are parallel to SDN architecture. ProtoGENI (Valdivieso Caraguay et al. 2014), a network testbed, has exposed various threats. Any grid system has its policy issues; such countless associated issues are grouped to support in discovering the threats. The current SDN model delivers numerous facilities. The major outlook of application layer is security. The SDN framework is shown in Figure 10.2. Other systems have been roughly composed with dissimilar layers as well as SDN interfaces. Many investigations about SDN security subjects have been recently conducted (Rana et al. 2019).

Sahoo et al. (2015) have proposed a construction which offers a security application and self-motivated conformation to the network, as Figure 10.3 demonstrates. As a result of the network infrastructure, a comprehensive traffic observation process is not possible in ad hoc systems (Waleed Alnumay 2019). Enhanced security will be accomplished via the control plane components to validate the entire network devices. All ports in the switch are directly jammed by the controller each and every time a secure linking is launched among the switch and the controller, which is called data and control plane integration. The client verification process must take place first and then the right flow entries will be pushed into the intended switches. This is similar to IoT settings and features, where the verification procedure contains Internet-permitted activities.

Here there is a suggestion among the grid campaigns with an OpenFlow enabled port, which is associated with the control plane. Each controller's cluster connections have a secure rule with an alternative or extra module. For the purpose of ensuring network security in the domain of SDN, some SDN controllers perform like security protectors of the entire network. On every occasion there remains an essential process that must take place of sending a message amid two nodes in different domains. The original flow entry is moved near the safety control plane. Then the controller requests for each nearby controller, however they can distinguish the target of the demanded flow entry. If from any cluster or domain there is a time-out request, which is a connection failure between two nearby controllers, then a previously chosen controller will become a boundary controller in addition to observing the circulation flow.

10.2.4 IoT–SDN Integration

The extent of the IoT idea has created new challenges to both networking and Internetworking schemes in current and future networks, especially the Internet. To start, as shown in Figure 10.4, networks must support self-defined option and support mixed devices in networking performance and fundamental protocols such as OpenFlow. Each IoT object (device or thing) has been constructed, or even planned, to achieve specific objectives. Moreover, the entire environment where some objects are positioned is usually designed with a specific objective. Finally, IoT suggests the extensive interconnection of several mixed networks (Desai 2016), the objects that

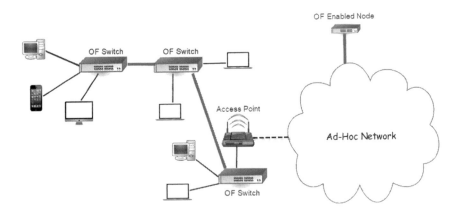

FIGURE 10.4   The extended SDN domain.

combine them, the environments where they are running, the upper and lower layer protocols they are using, and even the unequal objectives they need to achieve.

Martinez-Julia and Skarmeta (2014) have designed an integration framework to achieve the mix between SDN and IoT technologies. However, our point of view is restricted to SDN as the tool to achieve the IoT; it is not a mixture of two technologies.

10.2.4.1 Proposed Framework

An overview of the integration of SDN and IoT, as shown in the Figure 10.5, comprises a minimum set of functional blocks (NOLOT 2015), recognized by the performer and level to which they influence, that is object or network, data, or control plane. Therefore, two objects linked to an SDN-enabled network will be able to cooperate with the IoT controller by deploying their internal IoT agents. The purpose is to provide setting information to the controller for it to make the essential decisions and duplicate them into the necessary network. While the IoT controller is represented as just one practical block, it is an inside segment so new functionality can be added to the IoT overlay without influencing other elements, nor requiring the establishment of new relationships with the SDN controller.

The network generates a path from the requester to the responder. That path may be logical (setting up a determined connection or circuit) or virtual. This is where SDN is in play to allow objects trusting on different (and thus ad hoc) protocols to communicate to each other. Therefore, SDN mechanisms can be exploited to generate a path that links both ends. This is called a forwarding path and it is attained by setting up the essential forwarding rules into all forwarding elements found in such path.

Then, the IoT controller detects its role. It has to receive the communication request from the sender, detect the receiver in the network graph, determine the path using some routing process, compute the forwarding rules depending on the normal state of the protocols used by the objects, and finally share these rules to the SDN controller for it to set them into the forwarders. Communication requests are directed

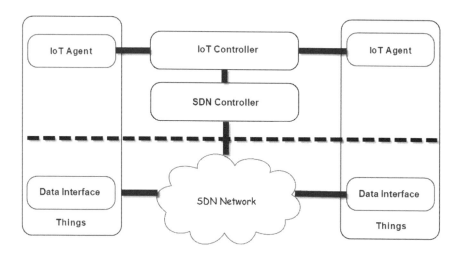

FIGURE 10.5   Integrated framework between IoT and SDN.

to the IoT controller by IoT agents installed into the objects that will be compounded with other mechanisms of the object to find out the necessary facts of the message that is being established, such as the identifier or address of the destination. This data is sent to the IoT controller before the communication is even introduced, so that the network will be prepared for it to progress.

10.3 SDN TECHNOLOGY CHALLENGES

SDN proceeds as technology used in making networks remain adjacent, however, not prompt. In addition, some applications are in positions of scalability, dependability, and security bounded by other structures (Saraswat et al. 2019) that have to be overwhelmed to be measured suitable for stakeholders. Subsequently, such structures are examined. According to what was discussed earlier, a leave-taking between the control as well as data planes permits self-governing expansion and development.

By depending on hardware technology in the data plane, and packet dispensation degree, technologies including Application-Specific Integrated Circuits (ASIC), Application-Specific Standard Products (ASSP), Field Programmable Gate Array (FPGA) (Binlun 2018; Saraswat et al. 2019; Mabel et al. 2019), or multicore CPU/ GPP could be supported. Temporarily, the presentation should also be contingent mainly upon NOS as well as hardware (Beacon, POX, Floodlight), and that is in the control plane. Nevertheless, a bad routine may occur at any of the stages and that can cause major harm, such as package damage or interruption and improper performance of the system of rejection of DDoS. Because of that motive, an equilibrium in routine is essential, along with the capability of expanding SDN components of hardware as well as software.

Furthermore, OpenFlow practices the shared hardware properties of real networks, like flow tables (NEC 2011; ONF 2015, 2019). By physical hardware, SDN could be prolonged outside flow tables and handle extra existing possessions. The combination and exploration of novel structures between data and control plane is a recent hot subject. SD machinery can integrate and professionally use many applications such as flow arrangement, encryption, and analysis, in addition to devices like practice packet processing and middle boxes (Mustafa and Mkpanam 2018). Also, the quantity and the situation of the controllers in a system are negotiable. The decisive influences for selecting the quantity and place are introduced as the terminology and the predictable recital of the system. An essential feature must be considered, which is security. All network applications don't have the same admission rights. The task of outlines, and validation for accessing the network resources are crucial.

Additionally, OpenFlow institutes TLS (Transport Layer Security) (ONF 2015) elective usage like certification instrument amid controllers and switches. Nevertheless, there are still not enough qualifications that offer security for numerous controller schemes that exchange data among them and the forwarding devices. Moreover, OpenFlow creates an unidentified packet (or its packet header) that possibly will be sent totally to the controller; there is a possibility to be exaggerated easily by DDoS threats through transferring numerous unspecified packets to the switch. Evolution amid real system constructions to SDN based on structures is also a known concern. In spite of system policies with OpenFlow support (IBM and NEC) in the market, the total substitution of the system structure is impossible. The conversion period needs machinery, procedures, and borders permitting cohabitation among constructions. Presently, there are significant attempts to realize this aim. The Open Networking Foundation (ONF) issued the IFConfig Protocol (ONF 2015) as a primary phase to be the companion of OpenFlow forwarding devices. Correspondingly, IETF's Forwarding and Control Element Separation Working Group (ForCES) in addition to the European Telecommunications Standards Institute (ETSI) are working on the interfaces standardization for this expertise arrogated expansion.

10.4 REVOLUTION OF IOT WITH SDN

IoT devices are susceptible to security dangers in an assorted system. Insufficient consideration of security features are observed in an SDN-founded IoT system. A proposed protected design for IoT system is founded on SDN (Martinez-Julia and Skarmeta 2014). This security architecture focuses on the verification of any device related to IoT system and its relation with the controller. In a framework like this, the IoT is an ad hoc network when a wireless item establishes an association with the controller which blocks all ports once the linking is proven, and controller switches authenticating that device. Unless the operator is reliable, the controller begins streaming to that operator. In such a network limited controllers assist as security protectors, through interchanging operator authentication and verification. If there is a controller protector disaster, approximately other edge controller is chosen as the controller protector.

As we mentioned before, the scope of this survey concerns the dependability between SDN and IoT as new technologies to achieve and accomplish better security performance. Table 10.1 shows the recent research of frameworks of SDN established depending on IoT security solutions.

IoT will be the main provision in launching highly developed networks and bring forth a revolution in society and industry. It is assessed that about 6.5 billion projects were put into use in 2016, more than in 2015 by 30%. There are about 5.5 million devices associated and connected with the Internet. This quantity is comprehensively growing and it will reach about 50 billion in 2020 (Matt 2018) (Figure 10.6). In line with this growth, companies are confidently spending an enormous amount on IoT; about $656 billion in 2014, and an estimated growth to $1.7 trillion in 2020 (Saleh 2013, 2019). It is estimated that there will be a growth of 90% in the connection of intelligence and smart connectivity in cars in 2020, which was only 2% in 2012. In the broader sense, this swift switch is obliging producers and companies, and consequently, research aims have changed. Around $8 billion will be spawned, matched to only $960 million dollars in 2014; this is a 90% composite growth rate according to International Data Corporation. Software Defined Network application and infrastructure was a top 10 trend during 2015 as per a Gartner report. The yearly data on growth also exceeds restrictions in zettabytes in 2016 and is expected to go beyond the development rate in terms of SDN. There is a rise of 87% in manufacture in the data centre using SDN and $960 million income was spawned in 2015, which is expected to increase to $8 billion by 2018, i.e. a 734% entire growth. The rise

TABLE 10.1
SDN-Based IoT Security Solutions

Approach	Security Parameter	Network	Description	Limitations
Secured SDN Framework (Leontiadis 2012)	Authentication	Ad hoc network	Authentication is activated once the controller prevents all ports from receiving the new flow.	Still not proven execution or reproduction, only a hypothetical basis.
DISFIRE (Luo 2012)	Authentication and authorization	Grid network	Authorization is guaranteed, because several controllers apply a dynamic protection in each sub-hierarchal network cluster.	Assessment of outline is missing. The procedure recycled is OpFlex which is not essentially verified.
Black SDN (Wu 2015)	Location security, confidentiality, integrity, authentication, and privacy	Generic IoT/ M2M communication	By encryption in the connection and custom SDN controller as TTP, this will keep the metadata and the payload safe.	Scalability in indistinct system will generate risk in case of comprehensive retreat.
SDP (Chakrabarty 2016)	Authentication	Ad hoc network/ M2M communication	SDP gathers the IP addresses of all M2M message adept campaigns and stocks them into a rational network. Authenticate founded on info kept.	Scalability drives happenstance routine in case of IoE.
SDIoT (Miyazaki 2014)	Authentication	Generic IoT network	It utilized SD Security appliance leveraging NFV and SDP for ensuring secure access in the network by authentication.	For an only SDSec logical element, it is difficult to handle huge networks and data centres. A tentative evaluation is missing.

Source: Tayyaba, 2017.

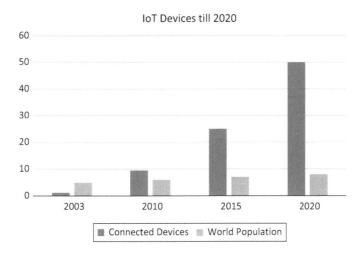

FIGURE 10.6 IoT devices in use worldwide (Tayyaba 2017, 2018).

in both areas obviously foresees a union of machineries and development in SDN-based IoT production.

10.5 CONCLUSION

In this research we determined the recent related works of the combination of a software-defined network and Internet of Things, comparison between traditional applications and SDN, its challenges and how it affects IoT, and why to use SDN. Also a strategic view of the current and the future IoT progress was stated.

The nonlinear development and growth of the entire operation and connected facilities that interchange data and info through the network are the merged perception of IoT. Now the inflexibility of old-style frameworks is incompetent signifying reconsidering novel conducts toward custom the substructure and transportations technology will support in that customization. SDN has appeared as a substitute and solution to the present issues of old-style systems. By allowing network administrators to have an entire view of the whole network, in addition to the occasion to manage and monitor the system confers to the requirements of each business. So, the flexibility offered by SDN could be efficiently used to permit devices connected to heterogeneous (ad hoc) networks to connect with one another. This is independent of the abilities and options of such objects or devices, so it fits perfectly into IoT situations. Partial computations or communication ability is a vital influence that regulates the form of IoT network frameworks and the way they communicate. Moreover, the difficulties and challenges for the employment of SDN in manufacture systems are examined. Remarkably, SDN delivers the tackles to enhance the performance of the management of the network conduct. The usage of such tools is novel and a hot topic area of learning, and also the growth of SDN research is increasing, so one day that pattern will carry advanced ways to monitor and control your network. Also the strategy of explicit protocols for precise determinations which are commonly mismatched and do not permit things to be easily interrelated which affects security. That issue could be resolved by using the machineries obtainable by the software-defined network via constructing a new facility on its top that provides provision to IoT substances, as some of the aforementioned research has attempted. However, this involves other challenges and difficulties that must be determined and solved.Yet, the innovative encounters and exploration purposes are considerably modest than the principles of the dogmatization, so determining the solutions would be a good introduction on the way to have a good strategy of the integrated structural design.

REFERENCES

Adly, A. S. (2019). Technology trade-offs for IIoT systems and applications from a developing country perspective: case of Egypt. *The Internet of Things in the Industrial Sector.* Springer.

Adly, A. S. (2020). Integrating Vehicular Technologies Within the IoT Environment: A Case of Egypt. *Connected Vehicles in the Internet of Things.* Springer.

Badach, Anatol. (2018). "SDN importance for data centers." 10.13140/RG.2.2.27189.78563.

Bakhshi, T. (2017). "State of the art and recent research advances in software defined networking." *Wireless Communications and Mobile Computing,* 2017.

Binlun, J. N., Chin, T. S., Kwang, L. C., Yusoff, Z., & Kaspin, R. (2018, December). "Challenges and Direction of Hybrid SDN Migration in ISP networks." In *2018 IEEE International Conference on Electronics and Communication Engineering (ICECE)* (pp. 60–64). IEEE.

Chakrabarty, S. and D. W. Engels. (2016). "A Secure IoT Architecture for Smart Cities." 13th IEEE Annual Consumer Communications & Networking Conference (CCNC) 812–813.

Desai, A., Nagegowda, K. S., & Ninikrishna, T. (2016, March). "A framework for integrating IoT and SDN using proposed OF-enabled management device." In *2016 International Conference on Circuit, Power and Computing Technologies (ICCPCT)* (pp. 1–4). IEEE.

Flauzac, Olivier, Carlos Gonzalez, Abdelhak Hachani and Florent Nolot. (2015a). "SDN Based Architecture for IoT and Improvement of the Security". 29th International Conference on Advanced Information Networking and Applications Workshops.

Flauzac, O., Gonzalez Santamaría, C. J., & Nolot, F. (2015b). "New security architecture for IoT network." Procedia Computer Science, 52 (2015), pp. 1028–1033.

Flauzac, O., González, C., Hachani, A., & Nolot, F. (2015c, March). "SDN based architecture for IoT and improvement of the security." In *2015 IEEE 29th International Conference on Advanced Information Networking and Applications Workshops* (pp. 688–693). IEEE.

Hakiri, A., Gokhale, A., Berthou, P., Schmidt, D. C., & Gayraud, T. (2014). "Software-defined networking: Challenges and research opportunities for future Internet." *Computer Networks, 75,* 453–471.

Hande, Y. S., & Akkalakshmi, M. (2015). "A Study on Software Defined Networking." *International Journal of Innovative Research in Computer and Communication Engineering,* 3(11).

Huang, N., Li, Q., Lin, D., Lit, X., Shen, G., & Jiang, Y. (2018, June). "Software-defined label switching: Scalable per-flow control in SDN." In *2018 IEEE/ACM 26th International Symposium on Quality of Service (IWQoS)* (pp. 1–10). IEEE.

Infosys. (2019). "Software-Defined-Networking-a-Critical-Enabler-of-Iot." https://medium.com/.

Leontiadis, I., C. Efstratiou, C. Mascolo and J. Crowcroft. (2012). "SenShare: Transforming Sensor Networks into Multi-Application Sensing Infrastructures." European Conference on Wireless Sensor Networks, 65–81.

Lu, Zhaoming, Chunlei Sun, Jinqian Cheng, Yang Li, Yong Li and Xiangming Wen. (2017). "SDN-Enabled Communication Network Framework for Energy Internet." *Hindawi Journal of Computer Networks and Communications* 2017, 13. Article ID 8213854.

Luo, T., H. Tan and T. Quek. (2012). "Sensor OpenFlow: Enabling Software-Defined Wireless Sensor Networks." *IEEE Communications Letters* 16, 11.

Mabel, J. P., Vani, K. A., & Babu, K. R. M. (2019). "SDN Security: Challenges and Solutions." In *Emerging Research in Electronics, Computer Science and Technology* (pp. 837–848). Springer, Singapore.

Martinez-Julia, P., & Skarmeta, A. F. (2014). "Empowering the Internet of things with software defined networking." *White Paper, IoT6-FP7 European research project.*

Matt. (2018). "How Many Billion IoT Devices by 2020?" https://www.vertatique.com/50-billion-connected-devices-2020.

McKeown, N., Anderson, T., Balakrishnan, H., Parulkar, G., Peterson, L., Rexford, J., ... & Turner, J. (2008). "OpenFlow: enabling innovation in campus networks." *ACM SIGCOMM Computer Communication Review*, 38(2), 69–74.

Miyazaki, T., S. Yamaguch, K. Kobayashi, J. Kitamichi, S. Guo, T. Tsukahara and T. Hayashi. (2014). "A Software Defined Wireless Sensor Network." International Conference on Computing, Networking and Communications (ICNC).

Mustafa, A. S., Mkpanam, D., & Abdullahi, A. (2018) "Security in Software Defined Networks (SDN): Challenges and Research Opportunities for Nigeria."

NEC (2010), "Ip8800/s3640 software manual, openflow feature guide (version 11.1 compatible)." NEC, Tech. Rep. NWD-105490-001, May 2010.

Nygren, A., Pfaff, B., Lantz, B., Heller, B., Barker, C. Beckmann, C., ... & McDysan, D. (2015). "Openflow switch specification version 1.5. 1". Open Networking Foundation, Tech. Rep.

Rana, D. S., Dhondiyal, S. A., & Chamoli, S. K. (2019). "Software defined networking (SDN) challenges, issues and solution". *Int. J. Comput. Sci. Eng*, 7, 1–7.

Sahoo, K. S., Sahoo, B., & Panda, A. (2015, December). "A secured SDN framework for IoT." In *2015 International Conference on Man and Machine Interfacing (MAMI)* (pp. 1–4). IEEE.

Saleh, A. A. M. (2013). "Evolution of the Architecture and Technology of Data Centers towards Exascale and Beyond." Optical Society of America, OFC/NFOEC Technical Digest.

Saraswat, S., Agarwal, V., Gupta, H. P., Mishra, R., Gupta, A., & Dutta, T. (2019). "Challenges and solutions in Software Defined Networking: A survey." *Journal of Network and Computer Applications*, 141, 23–58.

Tayyaba, S. K. (2017). "Software Defined Network (SDN) Based Internet of Things (IoT): A Road Ahead." ACM ISBN.

Valdivieso Caraguay, Á. L., Benito Peral, A., Barona Lopez, L. I., & García Villalba, L. J. (2014). "SDN: Evolution and opportunities in the development IoT applications." *International Journal of Distributed Sensor Networks*, 10(5), 735142.

Waleed Alnumay, U. G. and Pushpita Chatterjee. (2019). "A Trust-Based Predictive Model for Mobile Ad Hoc Network in Internet of Things." International Conference on Collaboration Technologies and Systems (CTS), Atlanta, GA, USA.

Wu, D., D. Arkhipov, A. Eskindir, Q. Zhijing and J. McCann. (2015). "UbiFlow: Mobility Management in Urban-Scale Software Defined IoT." IEEE Conference on Computer Communications (INFOCOM).

11 RSA-Based Remote User Authentication Scheme for Telecare Medical Information System

Sumit Pal and Shyamalendu Kandar

CONTENTS

11.1 INTRODUCTION

Progression of the Internet and its related technologies has simplified the domain of communication. Tasks like banking, communicating, ticket booking, payments, etc. that used to be quite tedious due to the amount of physical effort one had to put into those has become quite simple today. However, with such advances come problems as well and one major problem plaguing the world of the Internet today is that of security and privacy. Information is a valuable asset and so even the simplest of security threats to the same can have drastic results. As such, there are four security checks proposed that any system trying to send information through the Internet must satisfy. These are as follows:

- Confidentiality: Confidentiality is the property which ensures that information is not disclosed to any individual, entity, or process of unauthorized nature.
- Authentication: Authentication means that the person sending a piece of information or data over the Internet, is really who he or she claims to be.
- Integrity: Data integrity strives at maintaining and assuring the accuracy and completeness of data throughout its life cycle thus ensuring no unauthorized modification of the data.
- Non-repudiation: In an information exchange between two parties, non-repudiation implies that there is no room for denial for any party regarding being involved in said transaction.

Algorithms deployed on the Internet have been developed integrating one of the security checks or all of them depending on the purpose for which the algorithm is going to be used. For example, the railway reservation system of India, IRCTC, uses user authentication through a username and password before allowing someone to book a ticket. It also ensures confidentiality, as information exchanged is available only between the user and IRCTC and not to any third party. Non-repudiation also is respected as every transaction made is available on the user's personal page and so the user cannot deny making a transaction, and in similar fashion, IRCTC also cannot deny any kind of transaction made or ticket booking done.

In-time medical service is very essential in today's life. It is being found that health centres, hospitals, nursing homes, etc. are situated mainly in urban areas, and people from remote rural areas have to travel a lot to access the service. A telecare medical system uses the advancement of information and communication technology to provide certain healthcare services to people in remote areas, which may be viable in critical situations. This type of service minimizes transportation cost and time, life-threatening situations, etc. This is rising as an emerging field of low-cost but on-time medical service. One of the highly demanding fields where there is a need for various kinds of security protocols is telecare medical information systems (TMIS). In telecare medical information systems, a doctor can remotely access the medical records of a patient and can suggest required remedies without the need for proximity. With the evolution of the Internet of Things, this can be further improved and low-cost communication devices can be used not only for intra-city doctor and patient connections but also to connect rural parts of a country to urban medical facilities, thereby ensuring proper and better medical care. The differences between the traditional and the new TMIS can be found in Figure 11.1.

Even though the prospect of connecting the rural to the urban or providing an in-house medical facility is exciting, one cannot refuse the need for a security system since this field involves the need for exchanging sensitive information, leaking of which can in many ways spell trouble for the patient involved. Attackers specifically target the

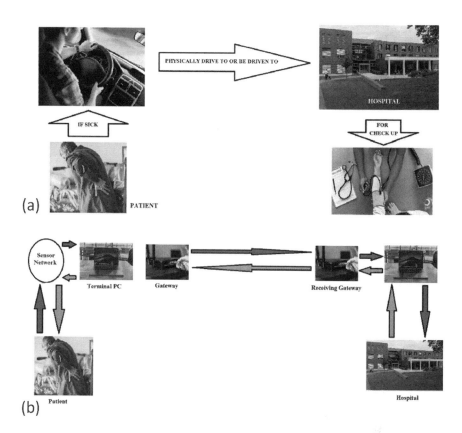

FIGURE 11.1   (a) Traditional medical system. (b) Telecare medical information system.

medical reports of patients, as they are of extremely high value and can be used in a variety of ways. In order to combat situations like these, security protocols, as mentioned earlier, are required. Among all the measures mentioned, research has been made extensively in the field of user authentication, especially in remote user authentication.

Remote user authentication, as the name suggests, denotes the technique in which a remote user willing to access service from a remote server is authenticated first by it before granting access. It involves a series of steps where users are required to provide their credentials, which are then sent to the particular server and when the targeted server authenticates the user as true, access is provided. There are two ways this can be accomplished, either the server maintains a database which contains a record of all the registered users and then a request arrives from the user side; or the server checks the database and then if the credentials stored in the database match with those that the user has provided, then the user is considered as authentic and is provided access. While it is a simplistic process, there are many problems associated with the same. They are:

- Storage overhead: Storing a database in the server side requires additional storage overhead which is incremental in nature. Thus, if a server is popular, then it will end up with millions of users, which will cause storage problems.

- Failure: Storing information in a single or a collection of various sources is always susceptible to failure. Mechanical or electrical faults can occur at any time and due to such a fault, the users will be unable to log in or access resources. As a result, the entire process will come to a standstill. So for a remedy, it is better not to store information in this manner.
- Security: A database stored in a server is susceptible to a wide variety of attacks, both internal and external. An adversary looking to obtain valuable information need only hack a single entity or a group of entities. With the present computing power available, this is quite an easy task.

It is because of these disadvantages that storing the credentials of a user in the database is generally avoided. Instead of this, real-time authentication is preferred where the user is authenticated on the go using various steps. It can be done with or without the aid of various algorithms like the RSA algorithm or elliptic curve cryptography (ECC). This is generally done using one-way hash functions, the exclusive-OR (XOR) operation, and concatenation techniques. In such a system, smart cards have a significant role to play. Smart cards are widely accepted devices used extensively for the purpose of remote user authentication. To exchange, store, and manipulate data, a smart card is used based on connection. Smart cards are of two types: contact and contactless. A smart card contains an embedded IC which can either be a microprocessor with a memory or a simple memory circuit. The memory contains stored information that can be read using a reader in either a contact or contactless fashion. The reader is usually connected to a host computer which then transmits the information from the reader to any desired destination if need be. For biometric-based authentication schemes, smart cards are really useful as one needs a match on the biometric provided. Since the biometric in question requires decomposition into points and matched, a portable processing device like a smart card provides the platform to do so. Figure 11.2 provides the block diagram of a smart card and the different components of the IC in the smart card.

The RSA algorithm is a popular algorithm that can be used for the purpose of intensifying the security of any system. The RSA algorithm is named after its inventor scientists Ron Rivest, Adi Shamir, and Len Adleman. The technique was proposed in 1977. It is extensively used in encryption and decryption processes. It is an asymmetric key cryptosystem, which means that there are two sets of keys generated. One is the public key, which is given to any interested party who wishes to communicate, and the other is known as the private key, which is kept secret for any party. The entire algorithm consists of three phases and each phase is explained next:

- Key generation phase: In the key generation phase, two types of keys namely public and private keys are generated. The key generation process is performed as the following:
  1) Two large prime numbers p and q are taken. The taken prime numbers are approximately of equal size. The product of the prime numbers, $n = pq$, is then computed.
  2) $\emptyset = (p - 1)(q - 1)$ is computed.

3) An integer number e is selected in such a way that the two conditions $1 < e < \emptyset$ and gcd (e, $\emptyset$) $= 1$ are satisfied.

4) Here (n, e) acts as the public key and (n, d) as the private key. The values of d, p, and q are kept undisclosed.

- Encryption phase: This phase takes place on the sender's side and comprises of the following steps:
  1) The sender first obtains the receiver's public key (n, e).
  2) The message being sent by the sender is represented as a positive integer m and the encryption is performed as $c = m^e \bmod n$. Here c is known as the ciphertext.
  3) The received ciphertext transmitted to the receiver through any channel.
- Decryption phase: Once the ciphertext c is received by the receiver, then the process of decryption begins. It is done in the following manner:
  1) $p = c^d \bmod n$ is computed at the receiver side using his private key (n, d).
  2) The message is then extracted from p.

It is extremely difficult to factorize very large prime numbers. The security of an RSA cryptosystem is based on this fact. Indeed, if the size of p and q is in thousands, then n is going to be in 10^6, which becomes quite difficult to find out even for the fastest of factoring algorithms present in today's time. Without n, it becomes quite impossible for anyone to calculate d. Hence, on a whole, the RSA algorithm is

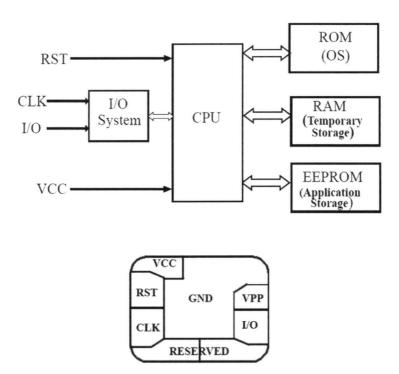

FIGURE 11.2   Various components of a smart card.

quite secure, and thus any algorithm centred on it can be considered secure as well. However, one disadvantage of RSA is its computational complexity. The multiplication and exponential operations contribute to huge computational overheads and that is especially so because we deal with large prime numbers here for the purpose of security. Thus, computational power must be considered when dealing with or implementing RSA in the security system.

In this chapter, the proposed scheme is designed for telecare medical information systems using smart cards. The proposed technique takes the help of the RSA algorithm. In the scheme, the doctor and patient have been considered as separate entities and multivariate in nature, that is, there is more than one patient and doctor present at a time. The scheme has been designed in such a manner that any number of patients can be connected with a single doctor. The proposed technique also supports the provision for mutual authentication as both the doctor and the patient has to be authorized before proceeding with any other type of communication, and once the authentication step is completed, session keys are generated which will encrypt any sort of information being shared. It has to be remembered that session keys are valid for a single session only, and for initiating a second session, a fresh set of session keys has to be generated. For each patient and doctor, the registration process will be a one-time affair unless there is a system problem which requires a re-registration. After the registration of the patient and doctor is done, every time the patient has to avail the medical services, he/she will have to log in and have to be authenticated. Like the patient, the doctor also has to do the same. Once both are authenticated, the session key will be generated which will carry forward further messages. For the purpose of registration, login, and authentication, we have kept a neutral middleman known as the telecare medical server (TMS) which will take care of all the aforementioned processes. The reason for using an extra entity in this process is for two reasons: security and computational complexity. With the addition of a TMS, various computations that either the patient's terminal or the doctor's terminal would have had to perform can now be done using a dedicated system.

The organization of the research chapter has been done in the following manner. Section 11.2 contains the literature survey of various related schemes of this topic. Discussions are done regarding the advantages and the disadvantages of various techniques. Section 11.3 has an in-depth discussion on the scheme including the various steps performed at various instances. Section 11.4 contains an analysis of the scheme including the computational cost analysis and comparative analysis with other state-of-the-art schemes. Finally, some concluding words are presented in Section 11.5.

11.2 LITERATURE SURVEY

The concept of remote user authentication has existed for a long time in the literature. The older proposed schemes in this domain mainly deal with single-server systems or, in other words, systems that are only associated with a single unit with which all kinds of interaction take place. With the addition of smart cards and multi-server architecture, this simple concept has been modified in several manners in order to suit the needs of the architecture. One of the primary works proposed in this domain is by Lamport (1). This one is a password-based authentication scheme which is asserted to be useful for imparting messages in a safe way over an unreliable network condition. Nonetheless, the negative marks of Lamport's strategy is the necessity of a verification table at the server side which may open up conceivable outcomes for different sorts of security dangers. Chang and Wu (2) have presented a similar authentication scheme depending on an identity-based signature scheme (3) introduced by Shamir. However, Hwang and Li (4) have demonstrated some security imperfections in Chang and Wu's scheme. To avoid the use of a verification table, Hwang and Li used El-Gamal's cryptosystem (5). Das et al. came up with a method using dynamic ID (6), which has advantages of user flexibility for password changes and also does not require the presence of verification table at the server side. As indicated by the claim of the authors, the plan is secured against insider attacks, replay attacks, and ID theft. But these have been opposed in Wang et al.'s scheme (7). Password-based authentication methods are easy to utilize and convey, however, they experience numerous bottlenecks. The real inconvenience of password-based authentication is that one can undoubtedly lose passwords or they might be imparted to other individuals (8) and are vulnerable to dictionary and brute force attacks. Other similar schemes of varying strengths and weaknesses have been proposed (9–14).

A biometric-based authentication mechanism uses biometric data like fingerprint, iris, unique mark, palm print, face, etc. for user verification. These are exceptional and can't be copied effectively, with no possibility of getting lost. Due to this, proposed remote user authentication schemes eventually moved on to be more biometric-based rather than password-based. A number of biometric-based RUA schemes can be found (15–18).

With the exception of a few schemes, almost all the schemes are for a single-server system. In the current network scenario, single-server systems are quite rare and most of the support is for a multi-server system. Multi-server architectures present their own set of challenges and they often appear as more complicated than those of single-server architectures. Therefore, schemes designed for environments having more than one server keep those constraints in mind. Liao and Wang proposed a method for operation in such a setup (19) that they claim provides a secure platform for password updating without any third-party involvement. Hsiang and Shih (20), however, proved that Liao and Wang's method was vulnerable to insider and stolen verifier attacks. Lee et al. (21) have proposed a scheme for RUA, which they have claimed to be computationally lighter than other similar schemes. Those claims were brushed aside by Li et al. (22) who proved that Lee et al.'s scheme is vulnerable to forgery attack and server spoofing attack. Chuang and Chen (23) proposed a scheme which was developed based on trust computing. Their scheme is lightweight and they claim that it resists major security threats. A lightweight RUA scheme uses only one-way hash functions, string concatenation, and XOR operation bypassing complex mathematical-computation-dependent key-based cryptographic protocols. Mishra et al. (24), however, have shown that Chuang and Chen's scheme falls prey to denial of service (DoS) attack, server spoofing attack, and impersonation attack. Li et al. (25) proposed an RUA method in a multi-server environment with the assistance of a smart card. The scheme utilizes the idea of a neural network. The drawback of the scheme is the consumption of time for the training of the neural network (26). He et al. (27) proposed a robust remote user authentication scheme employing biometric and elliptic curve cryptography. According to their claim, the three-factor authentication for the multi-server architecture is a first of its kind. However, Odelu et al. (28) demonstrated that He et al.'s scheme falls prey to attacks and also is unable to provide strong user anonymity. Lu et al. (29) presented a three-factor authentication for an RUA scheme in a multi-server environment. Researchers have used a public key cryptography technique in the login and authentication phase, and have claimed the resistance of the proposed technique against most of the attacks. A recent proposal (16) has disapproved the claims and has shown that Lu et al.'s scheme is weak to user's anonymity and server impersonate attack. Truong et al. (30) proposed a multi-server authentication scheme using the concept of elliptic curve cryptosystem and have claimed it to be secure against known kinds of attacks. Later, Zhao et al. (31) have pointed its vulnerability to server impersonation and offline password guessing attacks.

For telecare medical information systems, there has been a plethora of schemes implemented with various established security algorithms like elliptic curve cryptography, RSA algorithm, chaotic maps, etc. The schemes provide a host of features, and various mechanisms are implemented to tighten the security. However, one thing that has to be understood is that every scheme proposed to date has its own sets of strengths and weaknesses. One of the early works in this field was done by Lee et al. (32) who proposed an algorithm to solve the problem of transporting ill and handicapped patients over a long distance. Another similar article was published by Woodward et al. (33) who developed a telemedicine system using a mobile phone. More recent work in this field is done by Wu et al. (34) who enhanced his scheme with the help of a pre-computation process. However, it was proven by He et al. (35) that Wu et al.'s scheme offers no protection against masquerade and insider attacks. He et al. (35) have come up with a new proposal for TMIS which he claimed has better performance and is efficient for low-power mobile devices. But it was disapproved (36), showing that He et al.'s scheme is vulnerable to offline password guessing attack.

Wen et al. (37) proposed an RUA scheme for telecare medical information systems. He also analyzed Wu et al.'s (34) scheme and revealed that it fails in the field of patient anonymity preservation while also being vulnerable to impersonation and server spoofing attack. A scheme proposed by Wen et al. eliminates all the faults of Wu et al.'s scheme. However, it was later proved by Xie et al. (38) that Wen et al.'s scheme is vulnerable to offline password guessing attack while simultaneously not providing user anonymity and forward security. Xie et al. (38) constructed a scheme which followed the three-factor authentication mechanism. Upon construction, they claimed that it was secure, but Xu et al. (39) proved that it is insecure against synchronization attack while simultaneously adding excessive storage load to the server. Xie et al. (40) introduced a three-party anonymous password authenticated key exchange technique in TMIS. The backbone of the scheme is the elliptic curve cryptography.

Mishra et al. (41) proposed a scheme as an improvement to Yan et al.'s (42) scheme. Even though Mishra et al. claimed the proposed scheme to be secure enough, Guo et al. (43) found its weakness to session key attack and impersonation attack.

They included biometrics in TMIS to address the issue of Mishra et al.'s scheme. Giri et al. (44) proposed a technique for TMIS employing the RSA algorithm. Several RSA-based TMIS are available in literature (45, 46).

A chaotic map is extensively used in TMIS. Zhang et al. (47) used Chebyshev chaotic maps to propose a TMIS which they claim to resist various attacks and achieve security. Numerous proposals in TMIS have used Chebyshev chaotic maps (48, 49, 24).

ECC is a cryptographic primitive used for public key cryptography. The beauty of ECC is its smaller key size with equivalent security compared to other non-ECC-based cryptography. Some recent proposals on TMIS have used ECC (50–53).

A blockchain-based telecare medical information system is presented by Ji et al. (54). The beauty of the technique is its reliable and verifiable multi-level location-sharing scheme by which the patient's location can be obtained. Some recent proposals on medical services using blockchain have been addressed (55, 56).

Similar schemes have been proposed by Jiang et al. (18) whose improved scheme followed three-factor authentication for TMIS using the concept of bio-hashing, Tan et al. (57) whose proposal was based around a delegation-based authentication system, Amin et al. (58) whose scheme is anonymity preserving and light weight and as such is useful for IoT environments, Chatterjee et al. (59) who designed their scheme keeping in mind access control and user authentication as such implementing a fine-grained access control, and Xiong et al. (60) whose scheme boasts strong authentication and anonymity.

Along the same lines, various other schemes can be found like El-Gamal's cryptosystem (61), using RFID-based mechanisms (62), using cloud-assisted techniques (63), and using cognitive techniques (64) which have implemented one or more of the existing security protocols and architectures in order to provide a backbone to their methods.

11.2.1 Preliminaries

This section describes the concepts of some of the tools used to develop the proposed method as well as the list of notations used throughout the scheme. While RSA provides the backbone of the scheme, it was already described in detail in the "Introduction" section. Here, the discussion is regarding some of the key concepts used in order to implement the scheme. These concepts include one-way hash function, concatenation operation, and XOR operation.

- One-way hash function: It is widely used as a cryptographic tool and used in various encryption processes because of the fact that it is quite easy to compute but very difficult to reverse it and obtain the original number. It is of the form $h(x) = y$, where $x = \{0, 1\}*$ is the variable whose hash we can calculate and $y = \{0, 1\}^n$ is the calculated hash. It is to be noted that x is a string of arbitrary length and the other string y is of fixed length. A one-way hash function also has to satisfy the following properties:
- Given $m \in x$, it is quite difficult, especially in polynomial time, to find the input m for the given output y.

TABLE 11.1

Notations Used throughout the Scheme

Notation	Definition
P_i	Patient who wishes to use the telecare medical system
$Doctor_j$	Doctor who is a registered member of the medical system
SC	Smart card
ID_i	Identification of P_i
PW_i	Password of P_i
TMS	Telecare medical server
X_s	Secret key stored in the TMS
SK_{ij}	Calculated session key
$\oplus$	Exclusive-OR operator
$h(.)$	One-way hash function
$\|$	Concatenation operator
$S[.]$	A list of all the doctors registered with the TMS

- It is computationally hard to obtain the input $m' \in x$ such that $m' \neq m$ and $h(m) = h(m')$.
- It is similarly computationally difficult finding the pair $(m, m') \in x$, where $m \neq m'$ such that $h(m) = h(m')$.
- Notations: Table 11.1 presents the notations used to explain the proposed method and its related issues.

11.2.2 PRELIMINARY CALCULATIONS

Here, the implementation of the RSA algorithm with respect to the proposed technique is shown. Note that in RSA there is a need for maintaining two keys, one public and one private. In the proposed method, this duty falls into the hands of the TMS. Since every patient and doctor will register at the TMS, it provides all registered parties with a replica of its public key and keeps the private key to itself. Before accepting new registrations. however, the TMS should do the following:

1) The TMS selects two large prime numbers p and q.
2) TMS then calculates a value n equalling to $p \times q$.
3) After the calculation of n, an Euler totient function $\emptyset(n)$ is calculated as $(p - 1) \times (q - 1)$.
4) A number e is then selected such that $1 < e < \emptyset(n)$. The number e is made public.
5) Another number d is calculated such that $(d \times e) \bmod \emptyset(n) = 1$. The number d is private. Thus, the numbers e and d form the public and private key pair of a system.
6) Encryption of a message M is done using $E = M^e \bmod n$ and, simultaneously, decryption is performed using $M = E^d \bmod n$. Here E is the encrypted text.

11.3   DESCRIPTION OF THE PROPOSED METHOD

11.3.1   DOCTOR REGISTRATION PHASE

In this phase, the doctor $Doctor_j$ becomes a registered member of the telecare medical system (TMS). This phase is necessary as unless the doctor becomes a registered member, he or she cannot take part in this algorithm. The doctor registration process is described in detail next and can also be summarized from Figure 11.3.

- $Doctor_j$ enters his license number L_j and password PW_j.
- A registration message $<LPW_j>$ is sent from $Doctor_j$ to TMS, where $LPW = h(L_j \| PW_j)$ via a channel which is secured.
- Upon receiving the message TMS computes:
  1) $D_j = X^e_j \bmod n$
  2) $DK_j = h(X_s \| X_j)$, where X_s is a key which is kept secret by the server and X_j is the random number generated by the server. It is to be noted that X_j is not stored in TMS.
- The telecare medical server sends to $Doctor_j$ via secure channel the values of DK_j and D_j.
- A unique value S_j is stored in the TMS database alongside LPW_j.

11.3.2   PATIENT REGISTRATION PHASE

Similar to the doctor registration, the patient P_i must also register to the TMS in order to get benefited by the telecare medical information system. The process is pictorially represented in Figure 11.4. The detailed description of the process is provided next:

- The P_i enters his ID_i and PW_i.
- A registration message $<IDP_i, A_i>$ is sent from P_i to TMS, where $IDP_i = h(ID_i \| PW_i)$ and $A_i = h(IDP_i)$ via a channel which is secure.

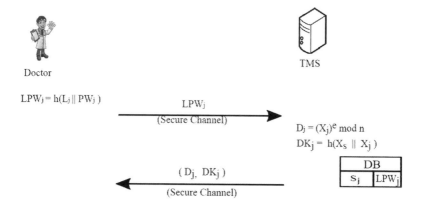

FIGURE 11.3   Doctor registration phase.

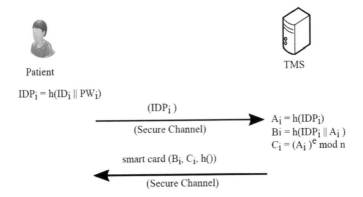

FIGURE 11.4   Patient registration phase.

- Upon receiving the message TMS computes:
  1) $B_i = h(IDP_i \| A_i)$
  2) $C_i = A^e_i \bmod n$
- The server issues a smart card (SC) to P_i via a secure channel by storing B_i, C_i, and $h(.)$ in the memory of SC.

11.3.3   Login and Authentication Phase

This phase is divided into (a) patient login, (b) service selection, (c) doctor authentication, (d) handshaking between patient and doctor, and (e) session key agreement phases. Those are described in the corresponding sections.

11.3.3.1   Patient Login

In the login phase, patient P_i has to log in and authenticate to the medical server TMS. This part of the process is described next:

- The assigned smart card is inserted by P_i into some terminal of the system and it provides its ID_i and PW_i.
- SC computes $IDP_i = h(ID_i \| PW_i)$ and $B^*_i = h((IDP_i \| A_i)$. It verifies whether $B^*_i = B_i$?. If the condition is satisfied, the operation proceeds to next step else P_i is rejected.
- Smart card generates a nonce N_1 and computes:
  1) $M_1 = N_1 \oplus h(A_i)$
  2) $M_2 = h(A_i \| N_1)$

And then sends $<M_1, M_2, C_i>$ to the TMS via any channel. It is noted that this channel is not required to be secured.

This is represented in Figure 11.5.

11.3.3.2   Service Selection

As soon as the login message is received from the patient P_i, the first step of any responsible medical server would be to ensure that the message is from a legitimate

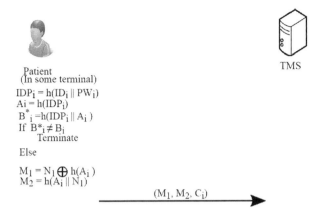

Patient
(In some terminal)
$IDP_i = h(ID_i \| PW_i)$
$Ai = h(IDP_i)$
$B^*_i = h(IDP_i \| A_i)$
If $B^*_i \neq B_i$
    Terminate

Else

$M_1 = N_1 \oplus h(A_i)$
$M_2 = h(A_i \| N_1)$

(M_1, M_2, C_i)

TMS

FIGURE 11.5 Patient login phase.

user and that no foul play is going on. In order to do that, the TMS does the following:

- TMS first calculates $C^d_i \bmod n = A_i$.
- Then from the value of A_i, a hash of A_i is generated and that is XORed with M_1 to calculate the value of N_1.
- The server then calculates $M_2^* = h(A_i \| N_1)$ and checks if the M_2^* that is calculated is equal to the received M_2 or not. If this check satisfies, then the TMS rejects the login message else the TMS considers patient P_i to be a valid and generates a nonce N_2 which is random in nature.
- TMS computes the following messages:
  1) $M_3 = N_2 \oplus h(A_i)$
  2) $M_4 = h(A_i \| N_1 \| N_2)$ and sends those to the P_i with S[.], the list of doctors with unique serial number S_j to prompt the patient to choose the particular doctor.
- Patient can select the particular doctor s/he wants to get an appointment. Before doing that, P_i performs the following:
  1) $N_2 = M_3 \oplus h(A_i)$
  2) $M_4^* = h(A_i \| N_1 \| N_2)$
- And checks if the calculated M_4^* is equal to the received M_4 or not. If this check satisfies, P_i moves towards service selection.
- In service selection process, P_i is provided with the list of doctors with unique serial number S_j. The patient chooses a particular doctor from the list and following steps are performed.
- Two messages M_5 and M_6 are computed as:
  1) $M_5 = h(A_i \| (N_1 \oplus N_2)) \oplus S_j$
  2) $M_6 = h(A_i \| (N_1 \oplus N_2) \| S_j)$

And then it is sent to TMS through any channel.
This is illustrated in Figure 11.6.

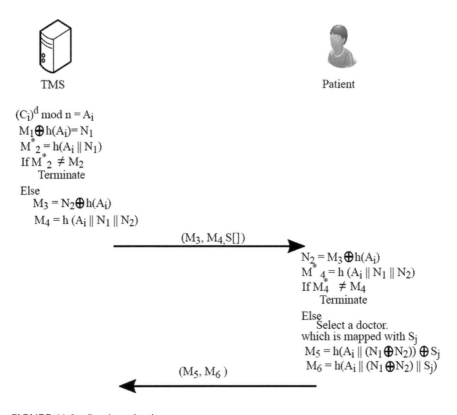

FIGURE 11.6   Service selection.

11.3.3.3 Doctor Authentication Phase

- As we have seen that the patient has selected the particular doctor and sends the serial S_j of the doctor by message passing. This is performed by the following steps:
- Upon receiving the messages, TMS retrieve S_j and checks the authenticity from M_5 and M_6 as
  1) $S_j = M_5 \oplus h(A_i \parallel (N_1 \oplus N_2))$
  2) $M_6^* = h(A_i \parallel (N_1 \oplus N_2) \parallel S_j)$ and checks *if $M_6^* == M_6$*. If satisfied, the process continues, else terminates.
- The TMS searches the database to find out the subsequent LPW_j in the database. From LPW_j, LM is calculated as $h(LPW_j)$ and a message LM is sent to the *Doctor$_j$*
- Upon receiving the message LM, the doctor inputs his L_i and PW_i and calculates the doctor's version of LPW_j^* as $h(L_i \parallel PW_i)$ and then proceeds to calculate LM^* and after that checks if the receiving LM is same as LM^*. If yes, then the next steps are executed.
- *Doctor$_j$* calculates DPW_j as $h(LPW_j) \oplus D_j$. Another variable LDK_j is calculated as $h(LPW_j \parallel DK_j)$, and finally LDK_j and DPW_j is sent to the TMS via any channel.

- Upon receiving these messages, TMS checks to see if the messages are received from an authenticate $Doctor_j$ or not. If the TMS remains unconvinced about it, then the entire process terminates completely. The steps taken by TMS to ensure this are:
  1) $DPW_j \oplus h(LPW_i) = D_j$
  2) X_j is calculated as $(D_j)^d \bmod n$.
- DK_j is calculated as $h(X_s \| X_j)$.

TMS then calculates the value of LDK_j^* as $h(LPW_j \| DK_j)$ and then checks if the calculated value of LDK_j^* is equal to the received value of LDK_j or not. If yes, then TMS is convinced that the $Doctor_j$ who has sent the data is a valid one. Figure 11.7 diagrammatically represents this.

11.3.3.4   Handshaking between Patient and Doctor

From this phase onwards, TMS tries to establish a handshaking between patient and doctor and comes out of the scenario. This is performed by the following steps:

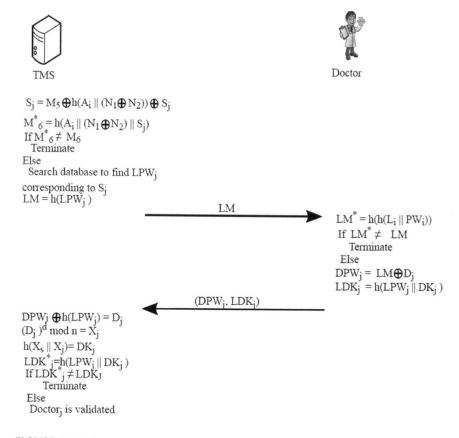

FIGURE 11.7   Doctor authentication phase.

- Once the TMS is convinced about the validity, it generates a nonce N_3 and from the generated nonce, a value V_{ij} is calculated as $V_{ij}=h(DK_j \parallel h(N_1 \parallel N_3))$.
- Once V_{ij} is calculated, TMS calculates the following messages:
  1) $M_7=h(A_i) \oplus N_3$
  2) $M_8=h(A_i \parallel N_1 \parallel N_3) \oplus DK_j$
  3) $M_9=h(A_i \parallel DK_j \parallel V_{ij})$
- The messages M_7, M_8, and M_9 are sent to the patient P_i.
- Simultaneously, another set of messages M_{10} and M_{11} are calculated as:
  1) $M_{10}=h(DK_j \parallel LPW_j) \oplus h(N_1 \parallel N_3)$
  2) $M_{11} = h(DK_j \parallel LPW_j \parallel V_{ij})$. These two messages are sent to the *Doctor_j*.
- On receiving the set of messages M_7, M_8, and M_9, P_i first checks if the source of the messages is a valid one or not. That is done through the following sequence of steps:
  1) $M_7 \oplus h(A_i)=N_3$
  2) $M_8 \oplus h(A_i \parallel N_1 \parallel N_3)=DK_j$
  3) P_i then calculates M_9^* and checks if the value of calculated M_9^* is the same as received M_9 or not. If yes, then P_i is convinced that the TMS is not a fraud.
- On receiving M_{10} and M_{11}, *Doctor_j* performs the following operations.

From M_{10}, *Doctor_j* finds out $h(N_1 \parallel N_3)$ by:

1) $M_{10} \oplus h(DK_j \parallel LPW_j))$.
2) V_{ij} is calculated as $h(DK_j \parallel h(N_1 \parallel N_3))$.
3) M_{11}^* is calculated as a checking to see if received M_{11} is equal to the calculated M_{11}^* or not. If yes, then *Doctor_j* is convinced that the message has been received from a valid source.

The flow of operations in handshaking is represented in Figure 11.8.

11.3.3.5 Session Key Computation

At the patient's side: Once the check is performed, P_i will now move forward towards the calculation of the session key. That will be done in the following few steps:

- SK_{ij}, the session key, is calculated by P_i's terminal as $h(V_{ij} \parallel DK_j \parallel B_i \parallel h(N_1 \parallel N_3)$.
- P_i then calculates two messages M_{12} and M_{13} as:
  1) $M_{12} = h(DK_j \parallel V_{ij}) \oplus B_i$
  2) $M_{13} = h(SK_{ij} \parallel B_i)$
- Once these messages are calculated, they are sent directly to *Doctor_j*. Now that we have established the flow of operations in the patient side, let us focus on the doctor's side.

At the doctor's side: Once *Doctor_j* received the messages M_{12} and M_{13}, session key SK_{ij} is calculated at the *Doctor_j*'s side:

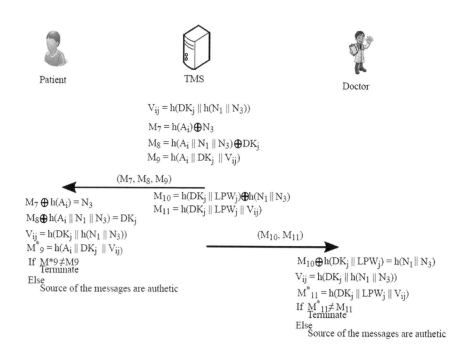

$$V_{ij} = h(DK_j \| h(N_1 \| N_3))$$
$$M_7 = h(A_i) \oplus N_3$$
$$M_8 = h(A_i \| N_1 \| N_3) \oplus DK_j$$
$$M_9 = h(A_i \| DK_j \| V_{ij})$$

Patient — (M_7, M_8, M_9) ←

$$M_7 \oplus h(A_i) = N_3$$
$$M_8 \oplus h(A_i \| N_1 \| N_3) = DK_j$$
$$V_{ij} = h(DK_j \| h(N_1 \| N_3))$$
$$M^*_9 = h(A_i \| DK_j \| V_{ij})$$
If $M^*9 \neq M9$
    Terminate
Else
    Source of the messages are authetic

TMS:
$$M_{10} = h(DK_j \| LPW_j) \oplus h(N_1 \| N_3)$$
$$M_{11} = h(DK_j \| LPW_j \| V_{ij})$$

(M_{10}, M_{11}) →

Doctor:
$$M_{10} \oplus h(DK_j \| LPW_j) = h(N_1 \| N_3)$$
$$V_{ij} = h(DK_j \| h(N_1 \| N_3))$$
$$M^*_{11} = h(DK_j \| LPW_j \| V_{ij})$$
If $M^*_{11} \neq M_{11}$
    Terminate
Else
    Source of the messages are authetic

FIGURE 11.8 Handshaking between patient and doctor.

- From M_{12}, B_i is retrieved as $M_{12} \oplus h(DK_j \| V_{ij})$.
- $Doctor_j$ then calculates the session key SK_{ij} as $h(V_{ij} \| DK_j \| B_i \| h(N_1 \| N_3)$. As we have seen previously, $Doctor_j$ has all the tools required to calculate the session key.
- $M_{13}{}^*$ is then calculated as $h(SK_{ij} \| B_i)$ and then the received M_{13} is checked with the calculated $M_{13}{}^*$. Once this is done, $Doctor_j$ authenticates P_i as being a valid patient.
- Further communication between patient and doctor are encrypted by the computed session key.

The final stages of this phase can be pictorially seen in Figure 11.9.

11.3.4 PASSWORD CHANGE PHASE

The algorithm proposed in this chapter allows P_i to change his or her password. This phase can be conducted in an offline mode, which means that the SC need not be connected to the TMS in order to complete this phase. The various steps required for P_i to change his or her password is given next:

- P_i inserts the SC and enters ID_i and PW_i.
- The SC computes A_i and checks if $B_i = B_i{}^*$. If this check is not satisfied, then the process is terminated.

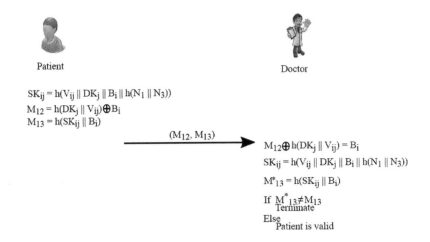

FIGURE 11.9   Session-key-enabled validation.

- SC then prompts P_i to enter a new password PW_i^{new}. Once P_i enters this, SC performs the following calculations:
  1) Calculates A_i^{new} as $h(h(ID_i \parallel PW_i^{new}))$.
  2) Calculates C_i^{new} as $(A_i^{new})^e \bmod n$ and stores it on the SC.
  3) Calculates B_i^{new} as $h((IDi \parallel PW_i^{new}) \parallel A_i^{new}$.
- Thus at the end of the phase the SC contains A_i^{new} and C_i^{new} and P_i successfully updates his password to PW_i^{new}.

11.4   PROPOSED SCHEME ANALYSIS

11.4.1   SECURITY REQUIREMENTS ANALYSIS

This section performs the security examination of the proposed technique. Here the plan is tried against six distinctive assaults and six usefulness necessities identified with RUA schemes using smart cards. In this investigation it can be found that the proposed technique is very vigorous with regard to manage significant sorts of attacks.

1) Attack using stolen smart card: If an adversary A somehow acquires the smart card, it will gain access to B_i and C_i. Any meaningful information from either two variables is impossible to obtain because B_i takes the benefits of cryptographic one-way hash function in order to protect ID_i, PW_i, and A_i, and we have already established inverse operation of a one-way hash function is infeasible. Also, from C_i, unless A has the private keys d and n, it cannot decrypt C_i to obtain A_i. Thus, stealing a smart card will do A no good.

2) Attack using impersonation: When any random user attempts to mask his original identity and masquerades as the legitimate patient P_i to try to log in into a TMS to obtain the services, we say that it is an impersonation attack.

Such an attack on the proposed scheme will end in failure, because in the proposed scheme the ID_i and the PW_i of the P_i is concatenated and then hashed before being sent to the TMS. Thus there is no scope whatsoever to obtain the information as due to the properties of a hash function which makes it one-way only. Thus, we can say that the impersonation attack is resisted in the proposed scheme.

3) Attack using an insider: An insider attack will be successful if any unauthorized person illegally accesses resources of the TMS and uses the values stored there for malicious purposes. No password table is maintained at the TMS and so the IDP_i cannot be obtained directly. Also, the messages sent to the TMS contains the PW_i concatenated with ID_i and then hashed. As a result, in order to guess the correct password the system manager has to guess PW_i at a time which is computationally hard due to inversion of the cryptographic one-way hash function. So, our proposed scheme resists insider attack.

4) Attack using password guessing: Let us take a scenario under consideration that an adversary A attempts to steal the patient P_i's password from the parameters stored on P_i's smart card. The parameters stored in the smart card are B_i and C_i. From B_i, A cannot obtain anything as it is protected using cryptographic one-way hash function. From C_i it is again quite difficult to obtain A_i as A does not know d and n. It might also happen that A tries to obtain P_i's PW_i from the communicating messages. However, in all of the communicating, there is no scope of obtaining PW_i as in all the messages, we have used cryptographic one-way hash function to protect the variable containing the PW_i and as we know, inversion of one-way hash function is computationally hard. Thus the proposed scheme resists password guessing attack.

5) Attack using replay: A successful replay attack is said to have been performed when an attacker intercepts message flows and then using those, replays useful messages to log in the server successfully. Let us consider that an adversary A intercepts the login message M_1, M_2, and C_i, and later resends it to the TMS. After receiving it, the TMS will check the sending login message with the stored message and if it is not equal, then the server will reject the adversary's login request message. The proposed scheme resists the replay attack because every time the login message is changed due to random nonce N_1.

6) Attack by denial of service: To perform a DoS attack on the proposed scheme, an adversary has to submit the correct password of patient P_i to the card reader. But, it has been shown in the aforementioned propositions that there is no chance for an adversary to extract or guess the correct password PW_i of a patient P_i from the smart card's parameters and also from the communicating messages between patient U_i and the TMS. Thus, the proposed scheme is secure against a DoS attack.

7) Attack using stolen verifier: By stolen verifier, it is meant that an inside member can steal or modify the passwords or the user verification tables stored in a server's database. However, the proposed scheme does not need

the presence of a verification table at the TMS. All kinds of verification are done with the help of calculations performed at the TMS, at $Doctor_j$'s terminal, and at P_i's terminal. Even though TMS does maintain a table that is used for information retrial rather than verification and even then, obtaining information from the verification table will in no way provide an attacker with any meaningful information. So, we can say that the proposed scheme resists stolen verifier attack.

8) Forward secrecy: It can be said that flawless forward secrecy has been achieved when in the situation that even though TMS's secret number X_s is compromised, still an adversary cannot derive any previous session keys. The proposed scheme preserves forward secrecy as even if X_s is revealed to an adversary A then it still needs the value of d and n to calculate X_j, which is used for calculating the SK_{ij}. Since A cannot obtain d, he will be unable to obtain any session key. Thus, the proposed scheme resists forward secrecy.

9) Security for session key: A session key is used in the proposed algorithm to protect the communication made between the patient and doctor, keeping in mind the sensitivity of medical information. In the proposed scheme, SK_{ij} is calculated as $h(V_{ij} \parallel DK_j \parallel Bi \parallel h(N_1 \parallel N_3))$. Since the value of nonce N_1 and N_3 are random, it is difficult to guess them correctly. Similarly, DK_j is calculated as $h(X_s \parallel X_j)$. The value X_j is calculated as D^e_j mod n. Since n is difficult to obtain and d is the private key stored in TMS, calculation of X_j and subsequently DK_j are difficult. Also, an attempt may be made to obtain DK_j from M_8 as $M_4 = h(A_i \parallel N_1 \parallel N_3) \oplus DK_j$. But that will be infeasible as the value of N_1 and N_3 will have to be guessed and A_i cannot be obtained easily as it needs n and the secret value. Similarly, B_i and V_{ij} are difficult to obtain for the attacker. Thus, the session key SK_{ij} is secure in the proposed scheme.

10) Mutual authentication: The proposed technique supports the notion of mutual authentication and great steps are taken throughout the technique to uphold the same. As such, it is one of the most vital steps undertaken before the session key is generated. In this process, TMS plays an important role as both the patient P_i and the doctor $Doctor_j$ need to authenticate themselves before further communication can take place. If at any stage the $Doctor_j$ or P_i fails to convince the TMS about its legality then the process is terminated. Thus the proposed scheme has mutual authentication.

11.4.2 COMPUTATIONAL COST ANALYSIS

There are four phases of the proposed method. The first is the doctor registration phase and then patient registration, followed by the login and authentication phase, and concluding with the password change phase. Each phase contains different types of operations like encryption/decryption operation, hash function, concatenation operation and XOR operation. The computational time required for hash, XOR, concatenation, and encryption/decryption operations are represented as T_H, T_X, and T_C respectively. Even though there are encryption and decryption operations running in the proposed scheme and also in the schemes with which the comparison has been made, selection for the parameters has only been done on the basis of which

TABLE 11.2

Computational cost of different phases of the proposed method

Phase	Computational Cost
Registration	$5T_H + 4T_C + 2T_{ED}$
Login and authentication	$40T_H + 19T_X + 44T_C + 2T_{ED}$
Password change	$6T_H + 8T_C + T_{ED}$
Total	$51T_H + 19T_X + 56T_C + 5T_{ED}$

are common among all the three. Since, T_H, T_X, and T_C are common among all three schemes, those have been selected. Another thing to note is that we consider the collective computational cost incurred in each phase for both $Doctor_j$ and P_i. This means that the doctor registration phase and patient registration phase cannot be segregated but are to be considered one collective unit, the registration phase. While there is no significance of doing this operation, we do it for simplicity purposes. Also, it will be useful later during comparison-based analysis of the method. Table 11.2 displays the computational cost involved in the three phases (doctor and patient registration, login and authentication, password change) of the proposed scheme.

11.4.3 COMPARATIVE ANALYSIS

This section provides a comparative analysis among various established schemes in the same domain and the proposed technique. Throughout the literature, many such schemes can be found each having its own set of advantages and disadvantages. The schemes with which the comparative analysis is performed are Mishra et al. (24) who use a chaotic-map-based technique for the key agreement, Chaudhury et al. (65) who use ECC to fit their technique, and Chaturvedi et al. (66) who solve the security problem of a telecare medical information system using a dynamic ID. Upon this analysis we can get an idea as to where the proposed algorithm lies with respect to the other schemes. The computational cost analysis is presented in Table 11.3. and Table 11.4 contains the security analysis. With respect to the security requirements, the attacks and requirements that have been considered can be found in Section 11.4.1. In Table 11.3, stage P1 refers to the combined registration phase, P2 refers to the login and authentication phase, and P3 refers to the password change phase.

From the comparison performed in Table 11.3 we can see that while there is some room for criticism in terms of computational cost in some phases, the reason for that can be traced to the inner dynamics of the scheme. The difference of the proposed scheme than the other schemes lies in the choosing of a doctor by a patient. It presents a choice-based mechanism where the patient P_i can choose a $Doctor_j$ of his or her choice. In addition to that the proposed method performs mutual authentication of both the doctor and the patient through a trusted TMS before the calculation of the session key. Due to these security measures the proposed method consumes more computation cycles. However, we have seen in Section 11.4.1 that due to these

TABLE 11.3

Comparison of Computational Cost

Phase	Mishra et al.	Chaudhury et al.	Chaturvedi et al.	Proposed
P1	$4T_H + 4T_X + 2T_C$	$3T_H + 4T_C + T_X$	$4T_H + 4T_X + 4T_C$	$6T_H + 5T_C$
P2	$16T_H + 5T_X + 22T_C$	$8T_H + 30T_C + 4T_X$	$11T_H + 9T_X + 15T_C$	$32T_H + 14T_X + 24T_C$
P3	$2T_H + 3T_X + 4T_C$	$4T_H + 6T_C + 2T_X$	$6T_H + 4T_X + 7T_C$	$6T_H + 4T_X + 7T_C$
Total	$22T_H + 12T_X + 28T_C$	$15T_H + 40T_C + 7T_X$	$21T_H + 17T_X + 26T_C$	$44T_H + 18T_X + 35T_C$

TABLE 11.4

Comparative Analysis of the Proposed Method with Some Existing Techniques in Terms of Security and Functional Requirements

Security Constraint	Mishra et al.	Chaudhury et al.	Chaturvedi et al.	Proposed
Stolen smart card attack	✗	—	✔	✔
Impersonation attack	✔	✗	—	✔
Insider attack	✔	✔	✗	✔
Password guessing attack	✔	✗	✔	✔
Replay attack	✗	✔	✗	✔
Denial of service attack	✔	—	—	✔
Stolen verifier	—	✔	✗	✔
Forward secrecy	✗	✔	✔	✔
Session key security	✗	—	✔	✔
Mutual authentication	✗	✔	✔	✔

security measures, the algorithm resists major attacks and satisfies major functionality requirements. From the security and functional requirements comparison in Table 11.4, we get an idea as to how the proposed technique compares to other similar techniques with respect to the satisfaction of security and functionality requirements. Here, we see that the proposed technique is superior than the other state-of-the-art techniques, and whatever little loss there is in terms of the computational complexity, it makes up for it in resisting various kinds of attacks. Thus, when deploying this scheme, one can be assured of the fact that the popular attacks won't affect it and thus any kind of sensitive information transmitted will be secure even in an insecure network.

11.5   CONCLUSION

In this research article, a RUA scheme for telecare medical information systems is proposed. The technique employs a smart card, and RSA algorithm plays a key role to provide security. We have discussed the basis for our algorithm and also the various interactions which take place through the process flow in detail. The strengths and weaknesses of the algorithm is also discussed in detail and comparison with some state-of-the-art schemes in the same field implemented with other techniques are performed. From all the analysis, it can be concluded that the proposed RUA scheme is a flexible and secure one and can be used for telecare medical information systems. The scheme is secure as it satisfies all major security attacks that is possible on such a system, and is flexible as the scheme allows the user to choose a medical service of his/her choice and acts accordingly rather than forcing the user to act according to the algorithm. Thus, the algorithm can be used for security in a telecare medical information system.

ACKNOWLEDGEMENTS

We express our heartiest thanks to Bibhas Chandra Dhara, professor of information technology, Jadavpur University, Saltlake Campus, for his suggestions and encouragement. We thank our alma mater, the Department of Information Technology of the Indian Institute of Engineering Science and Technology, Shibpur, and Department of Information Technology of Jadavpur University, Saltlake Campus, Kolkata, for providing us with the infrastructure required to perform experiments and test the hypothesis regarding the proposed algorithm. Finally, we express our sincere thanks to all the authors who have contributed in this field. Their contributions have left behind a rich literature of which one can easily refer to while building a scheme of his own.

REFERENCES

1. Leslie Lamport (1981) Password authentication with insecure communication. *Communications of the ACM*, 24(11), p 770–772.
2. C-C. Chang, T-C, Wu (1991) Remote password authentication with smart cards. *IEE Proceedings E (Computers and Digital Techniques)*, 138(3), p 165–168.
3. Adi Shamir et al. (1984) Identity-based cryptosystems and signature schemes. In *Crypto*, vol. 84, Springer, Vol. 196, p 47–53.
4. Min-Shiang Hwang, Li-Hua Li (2000) A new remote user authentication scheme using smart cards. *IEEE Transactions on Consumer Electronics*, 46(1), p 28–30.
5. Taher ElGamal (1985) A public key cryptosystem and a signature scheme based on discrete Logarithms. *IEEE Transactions on Information Theory*, 31(4), p 469–472.
6. Manik Lal Das, Ashutosh Saxena, Ved P. Gulati (2004) A dynamic id-based remote user authentication scheme. *IEEE Transactions on Consumer Electronics*, 50(2), p 629–631.
7. Yan-yan Wang, Jia-yong Liu, Feng-xia Xiao, Jing Dan (2009) A more efficient and secure dynamic ID-based remote user authentication scheme. *Computer Communications*, 32(4), p 583–585.
8. Cheng-Chi Lee, Che-Wei Hsu (2013) A secure biometric-based remote user authentication with key agreement scheme using extended chaotic maps. *Nonlinear Dynamics*, 71(1–2), p 201–211.
9. Chen Chin-Ling, Yong-Yuan Deng, Yung-Wen Tang, Jung-Hsuan Chen, Yu-Fan Lin (2018) An improvement on remote user authentication schemes using smart cards. *Computers*, 7(1), p 9.
10. Manish Patel Shingala, Nishant Chintan Doshi, N. Doshi (2018) An improve three factor remote user authentication scheme using smart card. *Wireless Personal Communications*, 99(1), p 227–251.

11. Marimuthu Karuppiah, Ashok Kumar Das, Xiong Li, Saru Kumari, Fan Wu, Shehzad Ashraf Chaudhry, R. Niranchana (2018) Secure remote user mutual authentication scheme with key agreement for cloud environment. *Mobile Networks and Applications*, p 1–17.

12. Roy Sandip, Chatterjee Santanu, Mahapatra Gautam (2018) An efficient biometric based remote user authentication scheme for secure Internet of things environment. *Journal of Intelligent and Fuzzy Systems*, 34(3), p 1403–1410.

13. Chandrakar Preeti, Om Hari (2018) An efficient two-factor remote user authentication and session key agreement scheme using Rabin cryptosystem. *Arabian Journal for Science and Engineering*, 43(2), p 661–673.

14. Chunyi Quan, Jaewook Jung, Hakjun Lee, Dongwoo Kang, Dongho Won (2018) Cryptanalysis of a chaotic chebyshev polynomials based remote user authentication Scheme. Information Networking (ICOIN) International Conference on, 2018, p 438–441.

15. Chu-Hsing Lin, Yi-Yi Lai (2004) A flexible biometrics remote user authentication scheme. *Computer Standards and Interfaces*, 27(1), p 19–23.

16. Shehzad Ashraf Chaudhry, Husnain Naqvi, Mohammad Sabzinejad Farash, Taeshik Shon, Muhammad Sher (2015). An improved and robust biometrics-based three factor authentication scheme for multiserver environments. *The Journal of Supercomputing*, 74(8), p 3504–3520. Springer.

17. Xiong Niu Li, Jianwei Kumari, Saru Wu, Fan Choo, Kim Kwang Raymond (2018) A robust biometrics based three-factor authentication scheme for global mobility networks in smart city. *Future Generation Computer Systems*, 83, p 607–618.

18. Qi Chen Jiang, Zhiren Li, Bingyan Shen, Jian Yang, Jianfeng Li Ma (2018) Security analysis and improvement of bio-hashing based three-factor authentication scheme for telecare medical information systems. *Journal of Ambient Intelligence and Humanized Computing*, 9(4), p 1061–1073.

19. Yi-Pin Liao, Shuenn-Shyang Wang (2009) A secure dynamic id based remote user authentication scheme for multi-server environment. *Computer Standards & Interfaces*, 31(1), p 24–29.

20. Hsiang Han-Cheng, Shih Wei-Kuan (2009) Improvement of the secure dynamic ID based remote user authentication scheme for multiserver environment. *Computer Standards & Interfaces, Elsevier*, 31(6), p 1118–1123.

21. Cheng-Chi Lee, Tsung-Hung Lin, Rui-Xiang Chang (2011) A secure dynamic id based remote user authentication scheme for multi-server environment using smart cards. *Expert Systems with Applications*, 38(11), p 13863–13870.

22. Xiong Li, Jian Ma, Wendong Wang, Yongping Xiong, Junsong Zhang (2013) A novel smart card and dynamic id based remote user authentication scheme for multi-server environments. *Mathematical and Computer Modelling*, 58(1–2), p 85–95.

23. Chuang Ming-Chin, Chen Meng Chang (2014) An anonymous multi-server authenticated key agreement scheme based on trust computing using smart cards and biometrics. *Expert Systems with Applications*, 41(4), p 1411–1418. Elsevier.

24. Dheerendra Mishra, Jangirala Srinivas, Sourav Mukhopadhyay (2014) A secure and efficient chaotic map-based authenticated key agreement scheme for telecare medicine information systems. *Journal of Medical Systems*, 38(10), p 120.

25. Li-Hua Li, Luon-Chang Lin, Min-Shiang Hwang (2001) A remote password authentication scheme for multiserver architecture using neural networks. *IEEE Transactions on Neural Networks*, 12(6), p 1498–1504.

26. Iuon-Chang Lin, Min-Shiang Hwang, Li-Hua Li (2003) A new remote user authentication scheme for multi-server architecture. *Future Generation Computer Systems*, 19(1), p 13–22.

27. Debiao He, Ding Wang (2015) Robust biometrics-based authentication scheme for multiserver environment. *IEEE Systems Journal*, 9(3), p 816–823.

28. Vanga Odelu, Ashok Kumar Das, Adrijit Goswami (2015) A secure biometrics-based multi server authentication protocol using smart cards. *IEEE Transactions on Information Forensics and Security*, 10(9), p 1953–1966.
29. Yanrong Lu, Lixiang Li, Haipeng Peng, Yixian Yang (2015) A biometrics and smart cards based authentication scheme for multi-server environments. *Security and Communication Networks*, 8(17), p 3219–3228.
30. Toan-Thinh Truong, MINH-Triet Tran, Anh-Duc Duong, Isao Echizen (2017) Provable identity based user authentication scheme on ECC in multi-server environment. *Wireless Personal Communications*, 95(3), p 2785–2801. Springer.
31. Yan Zhao, Shiming Li, Liehui Jiang (2018) Secure and efficient user authentication scheme based on password and smart card for multiserver environment. *Security and Communication Network,* vol. 2018, p 1–13.
32. Lee Ren-Guey, Heng-Shuen Chen, Chung-Chih Lin, Kuang-Chiung Chang, Jyh-Horng Chen (2000) Home telecare system using cable television plants-an experimental field trial. *IEEE Transactions on Information Technology in Biomedicine : A Publication of the IEEE Engineering in Medicine and Biology Society*, 4(1), p 37–44.
33. Woodward Bryan, Robert S. H. Istepanian, C. I. Richards (2001) Design of a telemedicine system using a mobile telephone. *IEEE Transactions on Information Technology in Biomedicine : A Publication of the IEEE Engineering in Medicine and Biology Society*, 5(1), p 13–15.
34. Z. Y. Wu, Y. C. Lee, F. Lai, H. C. Lee A secure authentication scheme for telecare medicine information systems. *Journal of Medical Systems*, 36(3), p 1529–1535.
35. He Jianhua Debiao, Zhang Chen Rui, Z. Rui (2012) A more secure authentication scheme for telecare medicine information systems. *Journal of Medical Systems*, 36(3), p 1989–1995.
36. J. Wei, X. Hu, W. Liu. An Improved Authentication Scheme for Telecare Medicine Information Systems. *Journal of Medical Systems*, 36(6), p 3597–3604.
37. F. Wen, D. Guo (2014) An improved anonymous authentication scheme for telecare medical information systems. *Journal of Medical Systems* 38(5), p 26.
38. Q. Xie, W. Liu, S. Wang, L. Han, B. Hu, T. Wu (2014) Improvement of a uniqueness-and-anonymity-preserving user authentication scheme for connected health care. *Journal of Medical Systems*, 38(9), p 91.
39. Xu Lili, Fan Wu (2015) Cryptanalysis and improvement of a user authentication scheme preserving uniqueness and anonymity for connected health care. *Journal of Medical Systems*, 39(2), p 10.
40. Q. Xie, B. Hu, N. Dong, D. S. Plo Wong (2014) Anonymous three-party password-authenticated key exchange scheme for telecare medical information systems. 9(7), p e102747.
41. D. Mishra, S. Mukhopadhyay, A. Chaturvedi, S. Kumari, M. K. Khan (2014) Cryptanalysis and improvement of Yan et al.s biometric-based authentication scheme for telecare medicine information systems. *Journal of Medical Systems*, 38(6), p 24.
42. X. Yan, W. Li, P. Li, J. Wang, X. Hao, P. J. Gong (2013) A secure biometrics-based authentication scheme for telecare medicine information systems. Journal of Medical Systems, 37(5), p 16.
43. D. Guo, Q. Wen, W. Li, H. Zhang, Z. Jin (2015) An improved biometrics-based authentication scheme for telecare medical information systems. *Journal of Medical Systems*, 39(3), p 20.
44. Debasis Giri, Tanmoy Maitra, R. Amin, P. D. Ruhul Srivastava (2015) An efficient and robust rsa-based remote user authentication for telecare medical information systems. *Journal of Medical Systems*, 39(1), p 145.
45. R. Amin, G. P. Biswas (2015) An improved rsa based user authentication and session key agreement protocol usable in tmis. *Journal of Medical Systems*, 39(8), p 79.

46. R. Amin, T. Maitra, D. Giri, P. D. Srivastava (2017) Cryptanalysis and improvement of an RSA based remote user authentication scheme using smart card. *Wireless Personal Communications*, 96(3), p 4629–4659.

47. Zhang Liping, Shaohui Zhu, Shanyu Tang (2016) Privacy protection for telecare medicine information systems using a chaotic map-based three-factor authenticated key agreement scheme. *IEEE Journal of Biomedical and Health Informatics*, 21(2), p 465–475.

48. Jongho Moon, Younsung Choi, Kim Jiye, Won Dongho (2016) An improvement of robust and efficient biometrics based password authentication scheme for telecare medicine information systems using extended chaotic maps. *Journal of Medical Systems*, 40(3), p 70.

49. X. Li, F. Wu, M. K. Khan, L. Xu, J. Shen, M. Jo (2018) A secure chaotic map-based remote authentication scheme for telecare medicine information systems. *Future Generation Computer Systems*, 84, p 149–159.

50. T. Serraj, M. C. Ismaili, A. Azizi (2017) A improvement of SPEKE protocol using ECC and HMAC for applications in telecare medicine information systems. In *Europe and Mena Cooperation Advances in Information and Communication Technologies*, Cham: Springer, p 501–510.

51. M. Qi, J. Chen, Y. Chen (2018) A secure biometrics-based authentication key exchange protocol for multi-server TMIS using ECC. *Computer Methods and Programs in Biomedicine*, 164, p 101–109.

52. A. Irshad, M. Sher, O. Nawaz, S. A. Chaudhry, I. Khan, S. Kumari (2017) A secure and provable multi-server authenticated key agreement for TMIS based on Amin et al. scheme Amin et al. scheme. *Multimedia Tools and Applications*, 76(15), p 16463–16489.

53. S. S. Sahoo, S. Mohanty (2018) A lightweight biometric-based authentication scheme for telecare medicine information systems using ECC. 9th International Conference on Computing, Communication and Networking Technologies (ICCCNT), IEEE, p 1–6.

54. Y. Ji, J. Zhang, J. Ma, C. Yang, X. Yao (2018) BMPLS: Blockchain-based multi-level privacy-preserving location sharing scheme for telecare medical information systems. *Journal of Medical Systems*, 42(8), p 147.

55. X. Zhu, J. Shi, C. Lu (2019) Cloud health resource sharing based on consensus-oriented blockchain technology: Case study on a breast tumor diagnosis service. *Journal of Medical Internet Research*, 21(7), p e13767.

56. A. H. Mohsin, A. A. Zaidan, B. B. Zaidan, O. S. Albahri, A. S. Albahri, M. A. Alsalem, K. I. Mohammed (2019) Based medical systems for patients authentication: Towards a new verification secure framework using CIA standard. *Journal of Medical Systems*, 43(7), p 192.

57. Tan Zuowen (2018) Secure delegation-based authentication for telecare medicine information systems. *IEEE Access*, 6, p 26091–26110. IEEE.

58. Amin Ruhul, Islam S. K. Hafizul, Gope Prosanta, Choo Kim-Kwang Raymond, Tapas Nachiket (2018) Anonymity preserving and lightweight multi-medical server authentication protocol for telecare medical information system. *IEEE Journal of Biomedical and Health Informatics*, IEEE, vol. 23(4) p 1749–1759 (2019)

59. Chatterjee Santanu, Roy Sandip, Das Ashok Kumar, Chattopadhyay Samiran, Kumar Neeraj, Reddy Alavalapati Goutham, Park Kisung, Park Youngho (2017) On the design of fine grained access control with user authentication scheme for telecare medicine information systems. *IEEE Access*, vol. 5, p 7012–7030

60. Xiong Hu, Tao Junyi, Yuan Chen (2017) Enabling telecare medical information systems with strong authentication and anonymity. *IEEE Access*, 5, p 5648–5661. IEEE.

61. F. M. Salem, R. Amin (2019) A privacy-preserving RFID authentication protocol based on El-Gamal cryptosystem for secure TMIS. *Information Sciences*.

62. Z. Zhou, P. Wang, Z. Li (2018) A quadratic residue-based RFID authentication protocol with enhanced security for TMIS. *Journal of Ambient Intelligence and Humanized Computing*, p 1–13.

63. C. T. Li, D. H. Shih, C. C. Wang (2018) Cloud-assisted mutual authentication and privacy preservation protocol for telecare medical information systems. *Computer Methods and Programs in Biomedicine*, 157, p 191–203.

64. Garai, I. Pntek, A. Adamk (2019) Cognitive cloud-based telemedicine system. In *Cognitive Infocommunications, Theory and Applications*, Springer, p 305–328.

65. Shehzad Ashraf Chaudhry, Khalid Mahmood, Husnain Naqvi and Muhammad Khurram Khan (2015) An improved and secure biometric authentication scheme for telecare medicine information systems based on elliptic curve cryptography. *Journal of Medical Systems*, 39(11), p 175. Springer.

66. Ankita Chaturvedi, Dheerendra Mishra, Sourav Mukhopadhyay (2017) An enhanced dynamic ID-based authentication scheme for telecare medical information systems. *Journal of King Saud University-Computer and Information Sciences*, 1(1), 29, p 54–62. Elsevier.

67. R. Madhusudhan, R. C. Mittal (2012) Dynamic id-based remote user password authentication schemes using smart cards: A review. *Journal of Network and Computer Applications*, 35(4), p 1235–1248.

68. Dheerendra Mishra, Ashok Kumar Das, Sourav Mukhopadhyay (2014) A secure user anonymity-preserving biometric-based multi-server authenticated key agreement scheme using smart cards. *Expert Systems with Applications*, 41(18), p 8129–8143.

69. Dominique Paret Roderick Riesco (2005) *RFID and Contactless Smart Card Applications*. Wiley Online Library, USA.

70. William Stallings (2006) *Cryptography and Network Security: Principles and Practices*. Pearson Education. India

71. Jawahar Thakur, Nagesh Kumar (2011) Des, aes and blowfish: Symmetric key cryptography algorithms simulation based performance analysis. *International Journal of Emerging Technology and Advanced Engineering*, 1(2), p 6–12.

72. Chwei-Shyong Tsai, Cheng-Chi Lee, Min-Shiang Hwang (2006) Password authentication schemes: Current status and key issues. *IJ Network Security*, 3(2), p 101–115.

73. Ding Wang, Debiao He, Ping Wang, Chao-Hsien Chu (2015) Anonymous two-factor authentication in distributed systems: Certain goals are beyond attainment. *IEEE Transactions on Dependable and Secure Computing*, 12(4), p 428–442.

74. Xiong Li, Jianwei Niu, Saru Islam Kumari, SK Hafizul, Fan Wu, Muhammad Khurram Khan, Ashok Kumar Das (2016) A novel chaotic maps-based user authentication and key agreement protocol for multi-server environments with provable security. *Wireless Personal Communications*, 89(2), p 569–597.

75. Stefan Mangard, Elisabeth Oswald, Thomas Popp (2008) *Power Analysis Attacks: Revealing the Secrets of Smart Cards*. Springer Science & Business Media 31, USA

76. Shehzad Ashraf Chaudhry, Husnain Naqvi, Mohammad Sabzinejad Farash, Taeshik Shon Muhammad Sher (2018) An improved and robust biometrics-based three factor authentication scheme for multiserver environments. *The Journal of Supercomputing*, 74(8), p 3504–3520.

77. Xiong Niu Li, Jianwei Kumari, Saru Wu, Fan Sangaiah, Arun Kumar Choo, Kim-Kwang Raymond (2018) A three-factor anonymous authentication scheme for wireless sensor networks in Internet of things environments. *Journal of Network and Computer Applications*, 103, p 194–204.

80. Khalid Chaudhry Mahmood, Shehzad Ashraf Naqvi, Husnain Kumari, Saru Li, Xiong Sangaiah, Arun Kumar (2018) An elliptic curve cryptography based lightweight authentication scheme for smart grid communication. *Future Generation Computer Systems*, 81, p 557–565.

81. Saru Kumari, Marimuthu Karuppiah, Ashok Kumar Das, Xiong Li, Fan Wu, Vidushi Gupta (2018) Design of a secure anonymity-preserving authentication scheme for session initiation protocol using elliptic curve cryptography. *Journal of Ambient Intelligence and Humanized Computing*, 9(3), p 643–653.

82. Tian-Fu Lee (2018) Provably secure anonymous single-sign-on authentication mechanisms using extended Chebyshev chaotic maps for distributed computer networks. *IEEE Systems Journal*, 12(2), p 1499–1505.

83. Xiong Li, Wu Fan, Muhammad Khurram Khan, Lili Xu, Jian Shen, Minho Jo (2018) A secure chaotic map-based remote authentication scheme for telecare medicine information systems. *Future Generation Computer Systems*, 84, p 149–159.

84. Ashok Kumar Das (2015) A secure user anonymity-preserving three-factor remote user authentication scheme for the telecare medicine information systems. *Journal of Medical Systems*, 39(3), p 30.

85. W. Liu, Q. Xie, S. Wang, B. Hu (2016) An improved authenticated key agreement protocol for telecare medicine information system. *SpringerPlus*, 5(1), p 555.

12 Illegitimate EPR Modification

A Major Threat in IoT-Based Healthcare System and Its Remedy through Blind Forensic Measures

Suchismita Chinara, Ruchira Naskar, Jamimamul Bakas, and Soumya Nandan Mishra

CONTENTS

12.1   INTRODUCTION

The term 'Internet of Things (IoT)' (Lee and Lee 2015) was introduced by Kevin Ashton in 1999. IoT is the system of connecting devices with one another for sharing information with the help of the Internet. On the other hand, it can be defined as *things* having identities and virtual personalities in smart environments using intelligent interfaces to connect and communicate within social, medical, environmental, and end user context. These devices are connected to the Internet with different technologies like cellular networks and M2M technologies like LTE, Wi-Fi, and 5G. Various gleaning technologies like artificial intelligence, pervasive computing, and embedded devices make the devices smarter, enabling them to be applied in various applications starting from smart cities to smart healthcare. By 2020, it is expected that around 50 billion IoT devices will be connected globally to the Internet.

As far back as the beginning of the Internet in 1989, connecting 'things' in the Internet started broadly. In 1990, John Romkey invented a toaster that was connected to the Internet considering it as the first IoT device (Romkey 2016). TCP/IP networking was used to connect the toaster to the computer. However, human intervention was required to insert the slice of bread into the toaster. To automate the process, Romkey and Hackett invented a robot system that helped to insert the bread without any human help. In 1991, the Trojan Room coffee pot was seen as the starting inspiration for connecting a web camera to the Internet (Stafford-Fraser 1995). A camera was set up in that place to take live pictures of the coffee machine which was displayed to all users on their desktops to keep track of the quantity of coffee left in the pot. This saved user's time in going to the room for coffee and finding the machine empty. In 1994, Steve Mann invented a portable device called 'Wearcam' which could be fitted inside the wearer's clothing (Mann 1996). The device has all the features that a normal multimedia computer can have like microphones, video processing capabilities, cameras, and its own IP address. The Wearcam device was used for two applications, namely Personal Visual Assistant (PVA) and Visual Memory Prosthetic (VMP). The former is based on spatial visual filtering and the latter is based on temporal visual filtering. Both these devices fitted with cameras were worn over the eyes to capture the live video of the surroundings around the person and then feeding the captured video to the World Wide Web (www). In 1997, Paul Saffo presented an article on sensors. The article depicts some underlying technologies like micro machines, piezo materials, microelectromechanical systems (MEMS), VLSI video, and some other technologies responsible for the growth of sensors. In 2000, LG announced the world's first Internet refrigerator that can track the food items inside it and check whether it is replenished or not (Osisanwo, Kuyoro, and Awodele 2015). In 2003, radio frequency identification (RFID) (Want 2006) was used at an enormous dimension in US armed forces in their Savi programme. In 2005 media like *The Guardian*, Logical American, and *The Boston Globe* referred to numerous articles about IoT and its future course. Also in this year, Rafi Haladjian and Olivier Mevel created a Wi-Fi–enabled robotic rabbit called Nabaztag that can predict weather conditions; measure air quality; and read texts, emails, and news headlines from RSS feeds (Kuyoro, Osisanwo, and Akinsowon 2015). In 2008, the Internet Protocol for Smart Objects (IPSO) Alliance organization was founded for promoting the use of Internet Protocol (IP) in 'smart objects' and to empower the Internet of Things. Also in the same year, 'white spectrum' was approved by the US Federal Communications Commission (FCC). Finally in 2011, IPv6 was launched in the objective to connect more devices to the Internet. The beginning of IPV6 and the low power transmission protocols (LPWAN) have brought the revolution in IoT systems.

The application of IoT is widespread to every domain of human life (Afzal et al. 2019). A few of them are in industry, in agriculture, in traffic surveillance, in smart homes, in waste management, in smart cities, and most importantly in smart healthcare. The history of healthcare throws light on sensors being used long back in healthcare. Tiny sensors embedded into medical equipment were used to get insight into physiological health statuses that are difficult for detection, diagnosis, and treatment. The merging of IoT with sensor technology has enhanced the scope of the healthcare domain to a great extent. The IoT system involves advanced data processing methodology, transmission technology, and a machine-learning approach to make the healthcare smart and secure.

The rest of this chapter is organized as follows. In Section 12.2, we present a framework of IoT-based healthcare system, and its benefit in real life followed by some challenges in the IoT-based healthcare system environment. In Section 12.3, we present the major security challenges in a IoT-based healthcare system. In Section 12.4, we present a modern-day forensic solution, as well as traditional security approaches, to address the security challenges presented in Section 12.3. Here we mainly focus on the most severe form of security attack against electronic patient records, which is the data modification attack. In Section 12.5, we conclude the chapter along with a discussion on the open problems in the domain of IoT in healthcare.

12.2   IOT-BASED HEALTHCARE FRAMEWORK

Medical services today are costlier than ever, and as the world population is increasing, the number of chronic diseases is on the rise. Along with this, in some parts of the world, healthcare is getting out of reach for a major portion of the society, owing to the lack of efficient and reliable communication systems, and people in such locations are inclined to chronic diseases. While technology cannot stop the population from ageing, it can at least make healthcare easier in terms of accessibility. For example, as we all know the medical diagnosis adds to a large part of a hospital bill, technology could move the routine medical check-up from a hospital to a patient's home. The technology could function more competently and provide better treatment.

Figure 12.1 shows the smart healthcare applications in a nursing home, old age home, and smart home. The working principle of smart healthcare remains the same

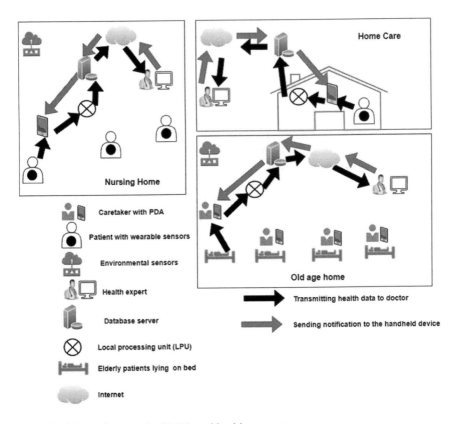

FIGURE 12.1 A framework of IoT-based healthcare system.

for all applications. The patients or elderly persons are attached with wearable sensors that constantly measure the physiological parameters. The activations of the sensors can be done by the smart phones owned by the patients or their family members. There exists local processing units (LPUs) that connect to the smart phone and the low-power data transmission takes place from the wearable sensors via the smartphones. Further, the patient data is forwarded to the database server or cloud using wireless transmission technologies like Wi-Fi, a cellular network, or any 5G technology. Most of the time, the server has the capability to perform automatic disease diagnosis by using advanced algorithms or machine-learning technologies.

As it can be seen in Figure 12.1, the server or the cloud can send any urgent notification to the health expert or the health expert can fetch the details from the server when required. Communication is possible from the patient to the health expert and vice versa. This architecture is very useful for elderly people living at home or in old age homes as well as they need continuous attention and constant monitoring. Patients of nursing homes are also beneficiaries of such smart healthcare.

In home care applications, the sensors attached on a patient's body senses health-related parameters and transmits it to the patient's family member or nurse. The sensors can directly communicate with the home local station through ZigBee modules. The application programme developed at the back-end network helps in analyzing the patient's data. In the in-hospital application, the sensors are deployed in the same way as in a nursing home and home care applications. The difference lies in the fact that here a group of patients are monitored by nurses using their PDAs (Kumar and Lee 2012; AbdElnapi et al. 2018).

There are unparalleled benefits for the quality and efficiency of treatments. Some benefits of smart healthcare systems are described as follows:

- Smart devices connected through the Internet do real-time monitoring of the patients, saving lives in case of emergencies like diabetes, heart failure, asthma, etc. The smart phone app collects information from the smart devices and transfers it to the doctor or physician. Various health parameter data like oxygen, blood sugar level, blood pressure, and ECG are collected by the IoT device and stored in clouds.
- Tracking and alert: In applications where a patient's life is at risk an on-time alert is critical. IoT smart devices capture the patient's vital data and transmit it to the doctors for tracking in real time and sending notifications through mobile apps and other devices. Thus the report gives important data about the patient's health conditions in a particular place and time. Hence, IoT helps in real-time monitoring, tracking, and alerting with better accuracy and treatment by the doctors.
- Remote medical assistance: There may arise a situation when a patient needs to contact a doctor who is located many kilometres away from the patient. In such cases, a doctor can check patient's conditions through a smart mobile app and identify the patient's location using GPS. Also through mobile communication the doctor can send some advice on taking medicines. In this way patient care can be done remotely.
- Research IoT: Devices have the capability to collect a large amount of data about a patient's health condition, which can be used for the research process. This data would have taken years to collect manually. This data can be used as a support to statistical study on health-related research. In this way IoT saves time in the collection process and has a great impact in the area of medical research.

Although there are many benefits of smart healthcare systems, there are several challenges. Some challenges are discussed in the following:

- In the domain of IoT, data security and privacy is one of the major challenges. IoT smart devices collect and send data in real time, however most IoT devices don't possess proper standard and data protocols, which allow attackers or cyber criminals to hack the healthcare system and gain personal data about both patients and doctors. The criminals misuse the patient's information to buy drugs, medical devices and create fake IDs.
- IoT devices are of a different variety causing hindrance while deploying multiple devices of IoT in the healthcare sector. All these devices have different communication protocols resulting in an improper process of aggregating data.

- As mentioned in the earlier point, aggregating data becomes difficult due to different communication standards of the protocols. However, the density of data is so large that the doctor faces difficulty in making quality decisions regarding a patient's health condition.
- In the current state-of-the-art, there is a huge scope to improve healthcare services for common mass hugely, with the help of IoT services. This is particularly true for developing nations across the world. With such services, the cost involved in medical diagnoses and treatment are expected to reduce manifolds. Hence IoT in healthcare is a rapidly growing area which has attracted huge research interest all over the world.

Now we will discuss some of the applications in healthcare:

- Hearables: These devices are designed for people suffering from hearing loss. The smart phone made it possible for Bluetooth to be compatible with hearables. Doppler Labs is one example.
- Ingestible sensors: These sensors are pin sized, and monitor a patient's body and will send a warning if there is any abnormality. Even these sensors can be an aid for a diabetic patient, as they would help in giving early warnings of the disease.
- Moodables: These are state-of-mind-enhancing devices which help in giving our temperament for the duration of the day. They send a power current to our cerebrum which lifts our state of mind.
- Audemix is an IoT device which eases the work of a doctor by reducing the manual conversation with his/her patient during charting. The device comprises of a voice command that captures patient data. This saves around 15 hours per week for a doctor.

12.3 SECURITY CHALLENGES IN IOT HEALTHCARE

IoT in healthcare provides enormous benefits to patients today, allowing them to be closely monitored, diagnosed, and even treated by doctors and physicians, even in the case of remote extreme locations. IoT in the medical sector has also enabled the usage of patient records and diagnostic statistics, in the medical research domain, parallel to treatment of patients by doctors. Doctors, nurses, and medical staff can collect information automatically from the medical devices attached to human body and take proper disease treatment decisions for a patient.

However, the most critical threat to this framework of IoT applications in healthcare is the security concern that arises pertaining to security and privacy protection of patient data stored/transmitted in electronic form. In such scenarios, since electronic patient records (EPRs) are communicated over insecure channels, they are vulnerable to illegitimate intrusions and privacy breach attacks.

More often than not, the security of medical devices connected over public channels such as the Internet, are neglected by medical companies. Patients are at high risk of being injured or killed by a zero-day exploit without proper detection. The hijacking of a large amount of medical devices in recent years (James and Simon

2017; Meggitt 2018) has forced the institution of special guidelines for manufacturers of such devices, which instruct strict adherence to cyber security rules, addressing present-day major cyber vulnerabilities (Mukhopadhyay and Suryadevara 2014; Khan and Salah 2018).

Out of a host of security challenges and vulnerabilities existing in the IoT healthcare system today, we discuss a few of the major ones in the following section, followed by ways to combat them in the form of modern-day forensic solutions.

12.3.1 Security Attacks in IoT-Based Healthcare System

12.3.1.1 Eavesdropping Attack

In healthcare, when a patient's data needs to be sent from his/her body to the doctor, a cyber attacker can eavesdrop the patient's data resulting in a patient's privacy breach. This data can be misused later by the attacker by posting it on social media like Facebook and Twitter, as shown in Figure 12.2.

Monitoring and eavesdropping a patient's data can be a common threat to a patient's privacy. An attacker can easily identify a patient's health-related data from the communication channels by snooping patient's vital signs. The messages can be more easily captured if the attacker has a powerful receiver antenna. The captured messages by the attacker help him/her to identify the patient's physical location and might try to harm him/her. The messages can also contain useful information like message ID, source, and destination address helping the attacker to gain relevant information.

12.3.1.2 Data Modification Attack

In wireless medical applications, the body sensors sense the vital parameters from the human body and transmit these data to the caregiver or hospital database. While sending this data, an attacker can access the data (patient's medical information) and alter this data. Later, he/she can transmit the altered data to the hospital database or nurse, risking the patient's life. When the data is in transit, two types of attacks are possible: (1) interception and (2) message (data) modification.

In an interception an attacker tries to attack a wireless medical sensor network helping him/her to access the sensor node information like node ID, sensor node type, and cryptographic keys. In message modification, the patient's vital data is attacked and tampered by the attacker. This can mislead the involved users in the

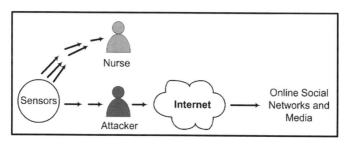

FIGURE 12.2 Eavesdropping attack.

whole process. For example, an attacker can attack and modify the data transmitted by a cardiograph sensor to a hospital staff causing an overdose of the medicine being provided to the patient. Further, the altered data can invoke false alarms or can provide fake results of the patient's health condition.

12.3.1.3 Masquerade and Replay Attack

In the healthcare application, a node can be rogued by an attacker to misroute packets intentionally while transmitting a patient's data to the remote server. An attacker can get access to the victim node which generates the possible masquerade. The masquerade node can now act as a real node causing disruption of the application by sending a false alarm to remote sites. The masquerade node can capture a patient's vital physiological data and later it can pose replay threats to real-time healthcare applications. So instead of operating on a patient's newly received data, the masquerade node replays the old messages again and again, causing mistreatment or overtreatment of patients.

12.3.1.4 Impersonation Attack

Through this attack an attacker can gain the personal information of a patient. In healthcare, the patient moves around the hospital. So it becomes important for the doctor or nurse to keep track of the patient's location in order to serve them in case of emergency. An attacker can interfere with the radio signals, thereby gaining information about the patient's location which leads to hacking a person's personal data.

Also, an attacker can gain information related to the health status of a patient exercising in a health club. Data captured by the sensor can be used to find the current activity of a patient and the attacker can alter this data to send false messages or wrong exercises to the patient causing harm to him/her. For example, the wireless sensor network (WSN) is monitoring an athlete exercising in a health club. The sensors attached to his/her body senses the heart rate, sensing time with location, and sends health-related feedback to the base station. An attacker can alter this data and can ruin the athlete's career by bringing the athlete under suspicion in doping tests.

12.3.1.5 Vulnerabilities in IoT Devices

Around 68,000 medical systems were exposed online by two security researchers in 2015, out of which 12,000 systems belonged to one medical organization. It was found that these systems connected through the Internet were operating on a very old version of Windows XP having lot of vulnerabilities in it.

IoT has risen the risk in data security and liability in the healthcare sector. A patient's heart condition can now be easily monitored by a doctor through programming the implantable cardioverter–defibrillator (ICD) device. The device helps in delivering heart rhythm data of a patient to the doctor and sending electric shocks to get information about heart rate. A malicious attacker can interrupt the device by malfunctioning it causing risk to the patient's life.

Hackers can cause harm to an individual by tampering with the medical IoT devices. They hack the devices to steal information and get to know the architecture of the larger healthcare system. 'Medjacking', where a device gets hijacked by an attacker, is the most vulnerable attack in most healthcare organizations. There are various examples where medjacking has caused serious harm to hospitals. In a hospital, the passwords were stolen by infecting a blood gas analyzer using malware and the data was sent to Eastern Europe. In another case, the image storage system of the radiology department was hacked to access the main network and the hacked data was sent to a particular place located in China. In another hospital, a drug pump was hacked to gain information by hijacking the hospital network. Hackers find it easy to steal large amounts of data from healthcare devices due to the devices' lack of security.

Another example of a medical device is insulin pumps that are attached on the patient's body to inject insulin. In 2008, a device was launched called Animas OneTouch Ping which is worn under clothing. The device came with a wireless remote control that can be used by a patient to pump a dose of insulin. An attacker can hack the insulin pumps to overdose patients with insulin. The hackers cannot be prevented from accessing the device, but the patient can be warned when such a situation arises.

Another device called an infusion pump is also exposed to various cyber bugs. The device delivers proper nutrients and medications to the patient. A malicious attacker can access and alter the drug dosage to harm the patient.

In this chapter, we focus only on the data modification attack in IoT healthcare.

12.3.2 Data Modification Attack in IoT Healthcare

In an IoT-based healthcare system, the experts (doctors) diagnosis a patient based on the EPR (Håland 2012), which contains various medical information like blood pressure, sugar level, magnetic resonance imaging (MRI), CT scan, X-ray, ultrasonography reports, etc. of a patient. The doctor gives treatment to the patient based on this medical information. So, this medical information is very sensitive for the patients as well as doctors. Because it might be possible that some bad wisher of the patient modified the EPR during the transmission through am insecure channel in IoT to mislead the doctor for wrong treatment is a life-threating issue for the patient. An MRI image of the brain and its tampered versions are shown in Figure 12.3. Figure 12.3a is an authentic MRI image, where a tumour is found in left side. Figure 12.3b

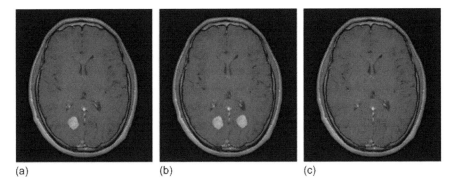

(a)   (b)   (c)

FIGURE 12.3   (a) MRI image. (b and c) Tampered MRI image after copy–move attack.

represents a simple copy–move attack MRI image, where the tumour region are copied and pasted into another side in the same image. From Figure 12.3a, it looks like there are two tumours in the MRI image. Whereas a tumour is not found in Figure 12.3c due to another copy–move attack, where some normal regions are copied and pasted over the tumour region to hide the tumour. So, the authenticity of the EPR needs to be checked before starting the treatment process. In Section 12.4 of this chapter, we present some solutions to detect the authenticity of EPR.

12.3.3   CHALLENGES OF TODAY'S HEALTHCARE

In recent years, even though the demand for healthcare services has risen, people still follow the traditional model of hospital-centric healthcare, where patients visit doctors for treatment. Patients frequently visit hospitals to get checked by doctors for managing chronic diseases, checking disease progression, and taking required precautions through treatments. In most hospitals, a disease- and physician-centred model is used that does not include patients as a part of the healthcare process. Some of the challenges in today's healthcare process are mentioned next (Amaraweera and Halgamuge 2019):

- It becomes difficult for a doctor or physician to make proper treatment within the limited time for each patient due to an increasing population with disabilities and illness. The daily routine of a patient, like diet, sleep, physical exercise, and social life, is ignored by physicians due to short screening time. All these activities are important for diagnosis and treatment.
- There is a lack of devices and equipment for a physician to check a patient's health condition and give proper treatment like exercise, medication, and diet. These things holds equal importance for a person to get proper medical care.
- The number of older people is growing at a rapid pace around the world. The medical care must provide enough resources and facilities for this elderly population.
- Most of the world's population lives in an urban environment. Due to increasing population in the big cities, demand for accessing healthcare facilities rises to serve the population. Along with that, contagious diseases spread in the cities with dense population.
- With increasing population and illness, the demand for healthcare facilities increases, thereby increasing the demand for jobs in healthcare surgeons, workers, staff, dentists, and medical caretakers who are responsible to enhance the medical ecosystem in rural and urban areas. Infrastructure of the hospitals and healthcare sector can be expanded to address the mentioned challenge.
- The rise in cost for treatment and medicine for the patients is one of the major challenges that healthcare is facing today. IoT comes to the rescue enabling a variety of medical devices to connect and collect data. This data is stored in clouds, which can be later used by the hospital, doctor, and analysis labs for treatment purposes. The different applications of healthcare

have different service quality, latency, and storage requirements, splitting the network layer into two sublayers, i.e. fog and cloud layer. Local buffering is handled by the fog layer, and data management, application services, and connectivity to the fog layer are handled by the cloud layer. Moreover, the security in devices need to be increased by providing secured hardware, routing protocols, and giving proper training to the workers to manage secure data.

12.4 SECURITY SOLUTIONS FOR DATA MODIFICATION IN IOT HEALTHCARE SYSTEM

In this section we shall present state-of-the-art solutions for data (EPR) modification detection, especially in digital medical images such as MRI, CT scan, and X-ray, in the form of active vis-à-vis passive security measures. The active measures include traditional security solutions such as watermarking (Cox et al. 2000), fingerprint-embedding (Paul et al. 2016), and hash creation (Harran, Farrelly, and Curran 2018). The passive measures comprise a newer form of security solution known as *digital forensics* (Reith, Carr, and Gunsch 2002).

First, we shall present and discuss the active security solutions which are in use in the current state of the art, mainly for EPR data modification detection. Later, we shall focus on the passive forensic techniques for EPR protection and modification detection pertaining to IoT security.

12.4.1 ACTIVE SOLUTION: DIGITAL WATERMARKING IN MEDICAL IMAGES

Digital watermarking (Cox et al. 2000) is the process of embedding a piece of external information, known as a *watermark* (hash or fingerprint), into the data to be secured (such as a digital images, text, audio, or video). Generally, the procedure of digital watermarking and subsequent data authentication using the same, consists of the following four broad modules/steps: (1) watermark generator, (2) watermark embedder, (3) watermark extraction, and (4) authentication, as shown in Figure 12.4.

The generator generates the watermark (hash or fingerprint) of a specified application based on the original data and key, which is embedded within the data to be secured by the watermark embedder, sometimes based on an embedding key.

Later, the embedded watermark is extracted by the watermark extraction module, which is used for authentication verification of data. The watermark embedded data

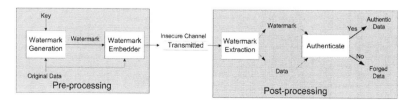

FIGURE 12.4   General block diagram of digital watermarking.

are transmitted over insecure/public channels. The receiver receives the transmitted data and performs watermark extraction. The original data is also used to compute the same watermark by the receiver. The extracted and computed watermarks are matched for data authentication at the receiver end. If those do not match, the received data is considered to be forged. In the following paragraph, we present a method (Memon and Gilani 2011) to check the authenticity of a medical image (CT scan) using the watermarking process.

In the watermarking technique, watermark(s) generation is a key step, and the watermark(s) generation process is varied from application to application. Memon and Gilani (2011) used three different key elements (message authentication code (MAC), hospital logo, and patient information) to generate watermark(s) for a chest CT scan medical image. The generated watermark(s) are then embedded into the CT scan images. A block diagram depicting the watermark generation and embedding processes (Memon and Gilani 2011) is shown in Figure 12.5.

12.4.1.1 MAC Generation Using Hash Function

Message authentication is a crucial technique in the data security domain. It is used to verify the source of received data as well as authenticity of the data. The message authentication code (MAC) is the key element of any authentication technique. There exists many techniques to generate the MAC from a piece of data; one of those is hashing functions (Deepakumara, Heys, and Venkatesan 2001). The hash function (Bakhtiari et al. 1995) is a mathematical function that maps a large (may be varied

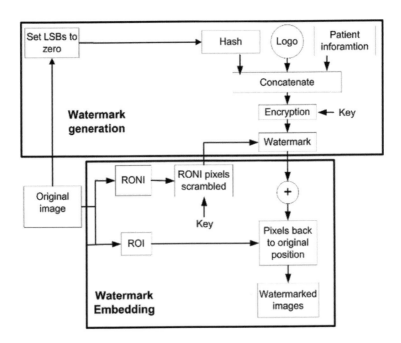

FIGURE 12.5   Block diagram of the watermark generation and embedding process (Memon and Gilani 2011).

in size) amount of data into small (fixed sized) data, which is known as the hash of it. The hash serves as a data authenticator.

A number of hash functions are available in the current state of the art, such as RIPEMD160, SHA384, SHA224, SHA1, MD2, and MD5 (Satoh and Inoue 2007). However, message digest-5 (MD5) and secure hash algorithm (SHA-1) are two of the most well-known hashing algorithms used to calculate the MAC till today. The MD5 hash algorithm (Rivest 1992) is used by Memon and Gilani (2011) to calculate the MAC. Along with the calculated MAC, Memon and Gilani used the hospital logo and patient information to generate the watermark as follows:

1. Convert grey level logo (image) of size $m \times n$ into a binary vector, L, where $L = \left[l(i), l \in \{0,1\}, 1 \le i \le m \times n \right]$.
2. Generate ASCII code for each character of patient information text file.
3. Convert ASCII code into its corresponding binary code and make a binary vector, P, such that $P = \left[p(i), p \in \{0,1\}, i \le i \le M \right]$, where $M = C \times b$. C is the number of characters used to record the patient data, and b denotes the number of bits used to represent each ASCII code.
4. Set the least significant bit (LSB) of each pixel in the input image to zero.
5. Compute the hash function using the MD5 algorithm, which generates S characters long string as the output.
6. Convert each string obtained from step v into a binary vector, D, such that $D = \left[d(i), d \in \{0,1\}, 1 \le i \le S \times b \right]$.
7. Now concatenate all the binary vectors L, P, and D into the new vector W, which forms the watermark for this scheme. Also, L, P, and D individually represent the watermarks for the logo, patient information, and original image, respectively.

12.4.1.2 Watermark Preprocessing

If the attacker possesses prior information about the watermark, he/she can easily modify the watermark. Hence, the watermark W needs to be preprocessed before being directly embedded. The method used by Wu et al. (2005) to preprocess the watermark is as follows:

1. Generate a pseudo-random binary vector B using a secret key k. The size of B is the same as W and B is represented as follows:

$$B = \left[b(i), b \in \{0,1\}, 1 \le i \le N \right] \qquad (12.1)$$

1. where N is the size of watermark W in bits.
2. Perform the XOR operation between W and B to obtain the final watermark:

$$W^* = W \oplus B \qquad (12.2)$$

Finally, watermark W^* is embedded into the host image as shown in Figure 12.5. In the following section, we describe the watermark embedding procedure.

12.4.1.3 Watermark Embedding Process

Generally, a medical image can be separated into two regions: (1) region of interest (ROI), which contains more useful information about the patient; and (2) region of non-interest (RONI), which does not contain any significant information. Basically, watermark(s) are embedded into RONI to give better security without compromising the sensitive diagnostic information (Navas, Thampy, and Sasikumar 2008). So, first we need to separate RONI from ROI in a medical image before embedding the watermarks. Generally, ROI is considered to be a rectangle (Navas, Thampy, and Sasikumar 2008; Smitha and Navas 2007) in a medical image. However, in some cases ROI is not exactly a rectangle, for example, in the CT scan image of the human chest area where the shape of the lung parenchyma on lung CT is elliptical by nature. Memon and Gilani (2008) extracted the ROI by drawing a logical ellipse in a CT scan image. They performed a segmentation algorithm to separate the ROI and RONI in a chest CT scan image. After separating the ROI and RONI of an image, the watermark(s) are embedded into the RONI portion. The steps of watermark embedding process (Memon and Gilani 2011) can be summarized as follows:

1. Generate a watermark as described in Section 12.4.1.1.
2. Encrypt the watermark W using Equation 12.2.
3. Perform an ROI extraction to separate the ROI and RONI.
4. Jumble pixels in RONI using a key.

Embed the generated watermark, W^* with jumbled pixels in LSBs of RONI.

6. Perform a re-jumble operation in the pixels of RONI to get the actual position of the pixels.
7. Merge ROI and RONI to obtain the watermarked images.

12.4.1.4 Watermark Extraction and Authentication Process

The receiver receives the watermark embedded image and extracts the watermark(s) from the received image. The watermark extraction process is the reverse of watermark embedding. The block diagram of the watermark(s) detection and authentication process is shown in Figure 12.6.

To extract the watermark and check the authenticity of the received image, the following steps are performed (Memon and Gilani 2011):

1. Divide the watermark embedded image into ROI and RONI portions using a segmentation technique.
2. Jumble the pixels in RONI using the same key, which was used in the watermark embedding process for embedding.

Extract LSBs from the scrambled RONI pixels to estimate the preprocessed watermark W^*.

Decrypt the estimated watermark W^* using P (binary vector of patient information, described in Section 12.4.1.1) to estimate the watermark W.

5. Fragment the estimated watermark W into L', P', and D', where L', P', and D' represent the estimated watermark for the logo, patient information, and host image, respectively.
6. If the original watermark of the host image D matches with the estimated watermark D', then the received image is authentic. Else, the received image is detected as tampered.

This scheme comprises a traditional security mechanism followed for protection of EPR integrity. The watermark of patient information is used to securely store the patient information within the host image. This is because if the EPR information and the medical image are sent in two separate files, the attacker may trivially tamper one of these two files, which leads to incorrect patient diagnosis.

On the other hand, using the watermark of the hospital logo, we can check the authenticity of an image when only the RONI area of the image is tampered. However, if the attacker tampers the image in the ROI area, where the watermark has not been added, we cannot detect the forged image through the watermark of the hospital logo. In this case, using the embedded MAC, we can detect the image forgery, as shown in Figure 12.6.

However, the watermarking technique needs certain preprocessing steps, where watermark(s) are generated and embedded within a host image. For this purpose, we require special watermark embedding hardware chips and embedded software, which increase the cost of the overall process. Additionally, many times such techniques degrade the quality of data, due to hash or watermark embedding. Last, but not the least, it is not practical to assume that such a priori precautions would have

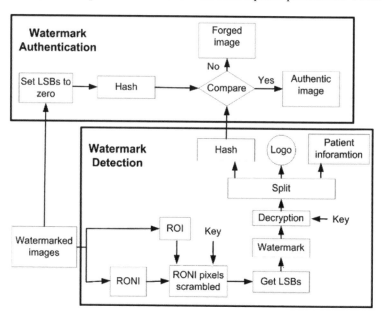

FIGURE 12.6 Block diagram of the watermark detection and authentication process (Memon and Gilani 2011).

been taken under all circumstances. To overcome these limitations, the new branch of passive security solutions, known as digital forensics, have emerged. Forensics solutions to preserve the integrity of EPR in IoT healthcare are discussed next.

12.4.2 PASSIVE SOLUTION: FORENSIC SOLUTION FOR MEDICAL IMAGE AUTHENTICITY DETECTION

Forensic security measures are comprised of passive threat detection mechanism in the IoT domain, which completely rely on post-processing-based data analysis and investigation. Hence such solutions are free from the assumption of precautionary measures adopted for attack prevention. Rather their operation is completely post-processing based. Once an attack has already been delivered, such solutions would provide an efficient way to detect the presence of an attack. A block diagram representing the working principle of passive forensic techniques is presented in Figure 12.7.

In this chapter, we present a forensic solution to detect illegitimate modifications in healthcare images such as MRI or CT scan images. Such images are highly prone to a certain form of attack, known as region duplication or copy–move forgery, where a specific portion of the image is copied and moved to some other region within the same image, with the malicious intention to repeat or obscure sensitive objects within the image. Healthcare images mostly consisting of regular/uniform textures or patterns throughout are highly prone to this form of malicious modification.

In this chapter, we present a recent image key-point-based scheme for detection of copy–move forgery in healthcare images, proposed by Dixit and Naskar (2019).

12.4.2.1 Key-Point-Based Copy–Move Forgery Detection in Images

Dixit and Naskar (2019) initially used Scale Invariant Feature Transform (SIFT) (Lowe 2004) to extract the image key-points, along with removing the redundant key-points by using maximally stable extremal regions (MSERs) (Matas et al. 2004) of the image. After this, a two-level matching operation is performed to localize the forged regions within an image. In the first level, a key-point matching operation is performed based on a generalized 2 nearest neighbour (g2NN) test (Amerini et al. 2013), followed by clustering the matched key-points utilizing agglomerative hierarchical clustering (Amerini et al. 2011) for detecting duplicate image regions. To optimize the false matches, a second-level matching operation is performed based on a graph similarity approach. In the following sections, we describe these steps in detail.

12.4.2.2 Preprocessing, Feature Extraction, and Selection

The amount of high contrast key-points of an image directly affects the performance of a key-point-based copy–move (region duplication) detection scheme (Amerini et al. 2011; Amerini et al. 2013; Pan and Lyu 2010). Fewer high-contrast extracted key-points

FIGURE 12.7 Block diagram of digital forensics process.

degrade the performance. So increasing the number high-contrast extracted key-points (Dixit and Naskar 2019), the test image is converted into an opponent colour space (Van De Sande, Gevers, and Snoek 2009) using the following equation:

$$O_1 = \frac{R-G}{\sqrt{2}}$$

$$O_2 = \frac{R+G-2B}{\sqrt{6}} \qquad (12.3)$$

$$O_3 = \frac{R+G+B}{\sqrt{3}}$$

where O_1 and O_2 are the colour-containing information channels, and O_3 is the intensity (luminance) channel. R, G, and B are the red, green, and blue colour channel pixel intensities of an RGB image. So, after converting the RGB image into the opponent colour space, the SIFT feature (Lowe 2004) is extracted from each opponent channels. It is possible that the extracting SIFT may contain a number of redundant SIFT features, which increase the computation cost during the feature matching. To remove the redundant SIFT feature, Dixit and Naskar (2019) divided the test image into MSERs, following Matas et al. (2004). Hence, the SIFT features are bundled in each MSER. SIFT features are considered those that are falling into at least one MSER, and the remainder are discarded. Similarly, MSERs containing at least one SIFT are considered, and the remainder are discarded. Experimental results (on Ardizzone, Bruno, and Mazzola 2015] dataset) of extracted and selected SIFT feature and MSER are shown in Figure 12.8a and b, respectively (SIFT feature and MSER are denoted by '+' and an ellipse, respectively). A 128-dimensional feature vector is used to describe each MSER.

12.4.2.3 First Level of Matching: Feature Matching and Clustering of Matched Key-Points

After performing the previous step, a set of bundle features is extracted, i.e. $F = \left(f_1, f_2, f_3, \ldots \ldots f_n \right)$, where $f_i = \left(k_1, d_1 \right)$ consists of key-point coordinates

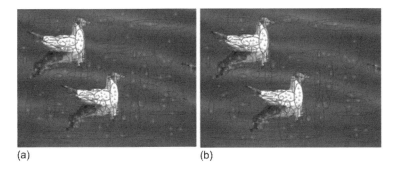

(a)   (b)

FIGURE 12.8 (a) SIFT features and MSER. (b) Bundle features (Dixit and Naskar 2019). (Reprinted by permission from Springer Nature: *Multimedia Tools and Applications* [Dixit and Naskar 2019], copyright 2019.)

$k_i = (x, y)$ in the spatial domain and the feature descriptor, d_i, of the MSERs. Next, a similarity vector, $S_i = \{s_1, s_2, s_3, \ldots s_{n-1}\}$, is computed in ascending order utilizing the Euclidian distance (ED) between each key-point k_i and other $(n-1)$ key-points of an image based on the feature descriptors. To find the two similar features, the generalized 2 nearest neighbour (g2NN) (Amerini et al. 2011) test is used. If the ratio of Euclidian distance between the nearest and second-nearest neighbour descriptors (key-points) is less than threshold T, the two features are considered similar. All the steps of feature matching are as follows:

i. For each bundle feature f_i, compute the Euclidian distance vector S_i between key-points k_i and k_j, $1 \le j \le n$ and $j \ne i$. Hence, the size of the Euclidian distance vector for each bundle feature f_i is $(n-1)$.

ii. Compute a new Euclidian distance vector L_i, which contain the sorted the Euclidian distance vector S_i in ascending order.

iii. If key-point k_i satisfies the following condition, then key-point k_i and the key-point corresponding to $L_i(p)$ are matched.

$$\frac{L_i(p)}{L_i(p+1)} < T \tag{12.4}$$

where p is the position index of vector L_i.

iv. Otherwise, key-point k_i and the key-point corresponding to $L_i(p)$ are mismatched and rejected.

However, the preceding feature-matching algorithm is highly false positive, especially when the test image itself contains very similar texture regions. In other words, a number of matched key-points will be found within a same region in an image, because similarity of texture within regions is highly probable to be same. If we are able to create one or more groups among matched key-points, and two matched key-points belong to two different groups, then we can say those two key-points are duplicates of each other. To create groups among matched key-points, Dixit and Naskar (2019) used the agglomerative hierarchical clustering (Amerini et al. 2011) technique of matched key-points. The clustering technique works as follows:

1. Initially, each matched key-point act as an individual cluster. Hence, the total number of clusters is M. (M represents the number matched key-points.)

2. Compute a pairwise Euclidian distance matrix $Edist(C_i, C_j)$ for each cluster utilizing the centroid linkage method (Ding and He 2002), as follows:

$$Edist(C_i, C_j) = \left\| \bar{x}_{C_i} - \bar{x}_{C_j} \right\|_2 \tag{12.5}$$

where $Edist(C_i, C_j)$ represents the Euclidian distance between cluster C_i and C_j.

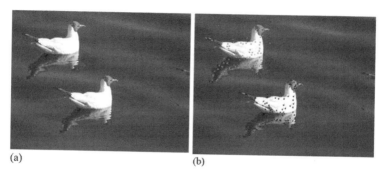

(a)   (b)

FIGURE 12.9 (a) Forged image. (b) Output after clustering (Dixit and Naskar 2019). (Reprinted by permission from Springer Nature: *Multimedia Tools and Applications* [Dixit and Naskar 2019], copyright 2019.)

$$\bar{x}_{C_i} = \frac{1}{p}\sum_{l=1}^{p} x_{C_i}(l) \text{ and } \bar{x}_{C_j} = \frac{1}{q}\sum_{l=1}^{q} x_{C_j}(l), \text{ where } p \text{ and } q \text{ denote the num-}$$

ber of key-points in cluster C_i and C_j respectively, and $x_{C_i}(l)$ and $x_{C_j}(l)$ represent the lth key-point in cluster C_i and C_j respectively.

3. Find the two clusters R and S, such that $Edist(R,S)$ is minimum for all pairs of clusters in the present cluster.
4. Merge the cluster R and S and form a new single cluster, and remove the corresponding data from $Edist$.
5. If all the matched key-points belong to a single cluster, then stop. Otherwise, repeat steps 2 to 5.

If more than one cluster is obtained, the test image is considered as an image forged by the copy–move attack. The result obtained after first-level matching followed by clustering of matched key-points is shown in Figure 12.9. Figure 12.9a represents a manually forged image, and the output after clustering is shown in Figure 12.9b. This first-level matching is highly prone to false matches due to the inherent property of the SIFT feature. To address this problem and optimize the false positives, in Dixit and Naskar (2019) perform a second level of matching, describe in the next section.

12.4.2.4 Second Level of Matching: Graph Similarity Analysis to Optimize False Positives

To reduce false positives, the graph similarity matching method is performed, whereby each cluster (or group) is considered as an attribute graph. Let a graph (represent a cluster) represented by $G = (V,E)$, where V and E are the set of vertices (here, key-points) and edges (here, distance between two key-points) respectively. An experimental result of graph formation after clustering (shown in Figure 12.10a) is shown in Figure 12.10b. Next, the graph similarity matching algorithm are performed, considering three node at a time of each graph, to compute similarity score. If the score of any two combinations with three nodes (one

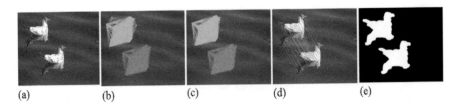

(a)   (b)   (c)   (d)   (e)

FIGURE 12.10 Experimental results (Dixit and Naskar 2019). (a) Result after clustering operation. (b) Formation of graph after clustering. (c) Graph after removal false positives. (d) Matched nodes. (e) Localization of forged region. (Reprinted by permission from Springer Nature: *Multimedia Tools and Applications* [Dixit and Naskar 2019], copyright 2019.)

from G_1 and another from G_2) are matched, those matched nodes are detected as duplicates of each other in an image. Otherwise, the test nodes are considered as false positives. An experimental result after removing the false positive is shown in Figure 12.10c. All the steps of graph similarity matching algorithm are given next:

1. For each cluster, create a graph $G = (V,E)$, where vertex V represents the key-points within a cluster, and edge E represents the distance between two corresponding key-points.
2. For each three-node combination of the ith graph G_i, compute the graph similarity score GS_i^t as follows:

$$GS_i^t = \frac{T_1 \times \sum_{k=1}^{3} \sum_{l=1}^{3} d_i^t(k) - d_i^t(l) + T_2 \times \sum_{k=1}^{3} \sum_{l=1}^{3} \sqrt{\left(x_i^t(k) - x_i^t(l)\right)^2 + \left(y_i^t(k) - y_i^t(l)\right)^2}}{T_1 + T_2}$$

(12.6)

where GS_i^t represents the graph similarity of the tth three-node combination of ith graph. $d_i^t(k)$ and $d_i^t(l)$ denote the descriptor of kth and lth nodes of the tth three-node combination of the ith graph, respectively.

$(x_i^t(k), y_i^t(k))$, and $(x_i^t(l), y_i^t(l))$ are the positioning coordinates of the kth and lth nodes of the tth thee-node combination of the ith graph, respectively. T_1 and T_2 are the user-defined threshold parameters ($T_1 = 1$ and $T_2 = 1$ are adopted in Dixit and Naskar [2019]).

3. Similarly (as step 2), compute the graph similarity score GS_j^t for each three-node combination of the jth graph G_j.
4. If $GS_i^p \approx GS_j^q$, then detect the pth and qth three-node combinations of G_i and G_j as duplicates.
5. Repeat steps 2 to 4 until all graphs are complete.

An experimental result after a second level of matching is shown in Figure 12.10d.

12.4.2.5 Duplicate Regions Detection and Localization

In order to identify the original region and localize the forged region among two detected forged region, Dixit and Naskar (2019) used a correlation map (Amerini et al. 2013) between the matched node image I_{match} and test image I_{test}. The correlation map gives the correlation between the I_{match} and I_{test}, and help us to generate a binary image map, which localized the forged regions of the test image, as shown in Figure 12.10e. The correlation map (Amerini et al. 2013) is computed using Equation 12.7:

$$C_{map}(p) = \frac{\sum_{j\in\omega(p)}\left(I_{test}(j)-\bar{I}_{test}\right)\times\left(I_{match}(j)-\bar{I}_{match}\right)}{\sqrt{\sum_{j\in\omega(p)}\left(I_{test}(j)-\bar{I}_{test}\right)^2\times\left(I_{match}(j)-\bar{I}_{match}\right)^2}} \quad \forall p \in I_{test} \quad (12.7)$$

where $\omega(p)$ represents a region with seven-neighbour pixels at centred pixel p, of the test image I_{test}.

$I_{match}(i)$ and $I_{test}(i)$ denotes the ith pixel intensities of I_{match} and I_{test}, respectively, and $\bar{I}_{match}$ and $\bar{I}_{test}$ denote the average pixel intensities of I_{match} and I_{test}, respectively.

The presented technique (Dixit and Naskar 2019) is able to detect the plain copy–move forgery and copy–move with geometric attack as well as post-processing after the copy–move forgery attacks in the digital image.

In this section, we presented an active and a passive technique for detection of data (EPR) modification attacks in IoT healthcare systems. Along with data modification detection, digital watermarking stores a patient's information embedded within the medical images during transmission over public communication channels. However, the disadvantage of such active techniques is that it requires specialized hardware or embedded software, and both the sender and receiver sides must be compatible with the technique technically. On the other hand, passive forensic techniques are fully post-processing based, and do not require any preprocessed information. So, the sender or receiver does not need technical compatibility. EPR data are received from various sources and diverse locations. So, it is challenging to come up with an IoT solution, where all senders and receivers are technically compatible. In this scenario, passive forensic techniques form an ideal solution for illegitimate EPR modification detection.

12.5 CONCLUSION

In this chapter, we have discussed IoT and its applications in our day-to-day lives. IoT-based healthcare is drastically changing the way healthcare services are provided to patients today. We presented a broad framework of an IoT-based healthcare system along with its benefits and major security challenges. One of the major security challenges in IoT healthcare is a data modification attack. We also presented the possible solutions for data (especially EPR) modification detection in the form of one active and one passive security mechanism.

In the active security measure, we have presented a watermarking scheme to detect image authenticity. The presented scheme is robust against Gaussian noise, median filtering, JPEG compression, copy–move, and histogram equalization attacks. Next, we have presented a security solution to illegitimate EPR healthcare image modification in the form of a recently published passive forensic technique. The presented forensic technique is validated on plain copy–move forgery, copy–move forgery with geometric transforms, as well as copy–move forgery with other post-processing operations (such as Gaussian noise, blurring, and brightness enhancement) attacks.

Besides a data modification attack, there are several other challenges in the domain of the IoT-based healthcare system. Scalability is one of the major challenges that the IoT system faces. The infrastructure of such systems should be designed in such a way that large-scale data may be received and processed in bulk amounts in real-time, seamlessly. Such data needs to be handled and maintained securely without compromising the performance of the system. Other major difficulties of IoT healthcare include the absence of EPR framework integration and lack of interoperability. Also, most of the healthcare data in IoT suffer from lack of common security practices or standards. These are the issues which need the immediate attention of the forensic community in relation to cybercrime and intrusion detection in IoT systems.

REFERENCES

AbdElnapi N M M, Omran N F, Ali A A, Omara F A (2018), A Survey of Internet of Things Technologies and Projects for Healthcare Services. In: 2018 International Conference on Innovative Trends in Computer Engineering (ITCE), Aswan, p 48–55.

Afzal B, Muhammad U, Shah G A, Ahmed E (2019), Enabling IoT Platforms for Social IoT Applications: Vision, Feature Mapping, and Challenges. *Future Generation Computer Systems*, 92, p 718–731. Elsevier.

Amaraweera S P, Halgamuge M N (2019), Internet of Things in the Healthcare Sector: Overview of Security and Privacy Issues. In: Mahmood Z. (eds) *Security, Privacy and Trust in the IoT Environment*, Springer, Cham, p 153–179.

Amerini I, Ballan L, Caldelli R, Del Bimbo A, Serra G (2011), A Sift-Based Forensic Method for Copy--Move Attack Detection and Transformation Recovery. *IEEE Transactions on Information Forensics and Security*, 6(3), p 1099–1110. IEEE.

Amerini I, Ballan L, Caldelli R, Del Bimbo A, Del Tongo L, Serra G (2013), Copy-Move Forgery Detection and Localization by Means of Robust Clustering with J-Linkage. *Signal Processing: Image Communication*, 28(6), p 659–669. Elsevier.

Ardizzone E, Bruno A, Mazzola G (2015), Copy--Move Forgery Detection by Matching Triangles of Keypoints. *IEEE Transactions on Information Forensics and Security*, 10(10), p 2084–2094. IEEE.

Bakhtiari S, Safavi-Naini R, Pieprzyk J (1995), Cryptographic Hash Functions: A Survey. *Technical Report*, Department of Computer Science, University of Wollongong, vol. 4, p 95–09.

Cox I J, Miller M L, Bloom J A (2000), Watermarking Applications and Their Properties. *Proceedings International Conference on Information Technology: Coding and Computing*, IEEE, Las Vegas, NV, p 6–10.

Deepakumara J, Heys H M, Venkatesan R (2001), FPGA Implementation of MD5 Hash Algorithm. In: Canadian Conference on Electrical and Computer Engineering. Canadian Conference on Electrical and Computer Engineering. Conference Proceedings (Cat. No.01TH8555), Toronto Ontario. Canada, vol. 2, p 919–924.

Ding C, He X (2002), Cluster Merging and Splitting in Hierarchical Clustering Algorithms. In: IEEE International Conference on Data Mining, Proceedings, IEEE, Maebashi City, Japan, p 139–146.

Dixit R, Naskar R (2019), Region Duplication Detection in Digital Images Based on Centroid Linkage Clustering of Key-Points and Graph Similarity Matching. *Multimedia Tools and Applications, Springer*, 78(10), p 13819–13840.

Håland E (2012), Introducing the Electronic Patient Record (EPR) in a Hospital Setting: Boundary Work and Shifting Constructions of Professional Identities. *Sociology of Health & Illness, Wiley*, 34(5), p 761–775.

Harran M, Farrelly W, Curran (2018), A Method for Verifying Integrity & Authenticating Digital Media. *Applied Computing and Informatics, Elsevier*, 14(2), p 145–158.

James A, Simon M B (2017), MEDJACK. 3 Medical Device Hijack Cyber Attacks Evolve. In: RSA Conference, San Francisco.

Khan M A, Salah K (2018), IoT Security: Review, Blockchain Solutions, and Open Challenges. *Future Generation Computer Systems*, 82, p 395–411. Elsevier.

Kumar P, Lee H J (2012), Security Issues in Healthcare Applications Using Wireless Medical Sensor Networks: A Survey. *Sensors*, 12(1), p 55–91.

Kuyoro S, Osisanwo F, Akinsowon O (2015), Internet of Things (IoT): An Overview. In: 3rd International Conference on Advances in Engineering Sciences & Applied Mathematics, p 53–58, London, UK.

Lee I, Lee K (2015), The Internet of Things (IoT): Applications, Investments, and Challenges for Enterprises. *Business Horizons*, 58(4), p 431–440. Elsevier.

Lowe D G (2004), Distinctive Image Features from Scale-Invariant Keypoints. *International Journal of Computer Vision, Springer*, 60(2), p 91–110.

Mann S (1996), Wearable, Tetherless Computer–Mediated Reality: WearCam as a Wearable Face–Recognizer, and Other Applications for the Disabled. TR 361, M.I.T. Media Lab Perceptual Computing Section, Cambridge, Ma, February 2 1996. Also appears in AAAI Fall Symposium on Developing Assistive Technology for People with Disabilities, 9–11 November 1996, MIT.

Matas J, Chum O, Urban M, Pajdla T (2004), Robust Wide-Baseline Stereo from Maximally Stable Extremal Regions. *Image and Vision Computing*, 22(10), p 761–767. Elsevier.

Meggitt S (2018), MEDJACK Attacks: The Scariest Part of the Hospital [scholarly project], Tufts University, Massachusetts, Retrieved July, 2019 from http://www.cs.tufts.edu/c omp/116/archive/fall2018/smeggitt.pdf.

Memon N A, Gilani S A M (2008), NROI Watermarking of Medical Images for Content Authentication. In: 2008 IEEE International Multitopic Conference, Karachi, p 106–110.

Memon N A, Gilani S A M (2011), Watermarking of Chest CT Scan Medical Images for Content Authentication. *International Journal of Computer Mathematics*, 88(2), p 265–280. Taylor & Francis.

Mukhopadhyay S C, Suryadevara N K (2014), Internet of Things: Challenges and Opportunities. In: Mukhopadhya S. (eds) *Internet of Things*, Smart Sensors, Measurement and Instrumentation, vol 9. Springer, Cham, Switzerland.

Navas K A, Thampy S A, Sasikumar M (2008), EPR Hiding in Medical Images for Telemedicine. *International Journal of Biomedical Sciences, Citeseer*, 3(1), p 44–47.

Osisanwo F, Kuyoro S, Awodele O (2015), Internet Refrigerator--A Typical Internet of Things (IoT). In: 3rd International Conference on Advances in Engineering Sciences & Applied Mathematics (ICAESAM'2015), London (UK).

Pan X, Lyu S (2010), Region Duplication Detection Using Image Feature Matching. *IEEE Transactions on Information Forensics and Security*, 5(4), p 857–867. IEEE.

Reith M, Carr C, Gunsch G (2002), An Examination of Digital Forensic Models. *International Journal of Digital Evidence*, 1(3), p 1–12.

Rivest R (1992), *The Md5 Message-Digest Algorithm*. The Md5 Message-Digest Algorithm. RFC 1321, MIT Laboratory for Computer Science and RSA Data Security, Inc, Tech. Rep. DOI 10.17487/RFC1321.

Romkey J (2016), Toast of the IoT: The 1990 Interop Internet Toaster. *IEEE Consumer Electronics Magazine*, 6(1), p 116–119. IEEE.

Satoh A, Inoue T (2007), ASIC-Hardware-Focused Comparison for Hash Functions MD5, RIPEMD-160, and SHS. *Integration, the VLSI Journal*, 40(1), p 3–10. Elsevier.

Smitha B, Navas K A (2007), Spatial Domain-High Capacity Data Hiding in ROI Images. In: 2007 International Conference on Signal Processing, Communications and Networking, IEEE, Chennai, India, p 528–533.

Stafford-Fraser Q (1995), The Trojan Room Coffee Pot: A (non-technical) biography, Retrieved July 2019 from https://www.cl.cam.ac.uk/coffee/qsf/coffee.html.

Van De Sande K, Gevers T, Snoek C (2009), Evaluating Color Descriptors for Object and Scene Recognition. *IEEE Transactions on Pattern Analysis and Machine Intelligence*, 32(9), p 1582–1596. IEEE.

Want R (2006), An Introduction to RFID Technology. *IEEE Pervasive Computing*, 5(1), p 25–33. IEEE.

Wu X, Hu J, Gu Z, Huang J (2005), A Secure Semi-Fragile Watermarking for Image Authentication Based on Integer Wavelet Transform with Parameters. In: Australian Information Security Workshop, New Castle, Australia, p 75–80.

YU P L, Sadler B M, Verma G, Baras J S (2016), Fingerprinting by Design: Embedding and Authentication. In: Wang C, Gerdes R, Guan Y, Kasera S (eds) *Digital Fingerprinting*, Springer, New York, p 69–88.

13 IoT
Foundations and Applications

Suchismita Chinara, Ranjit Kumar,
and Soumya Nandan Mishra

CONTENTS

13.1 INTRODUCTION

Before we start, let us imagine a few events in our day-to-day life. When you leave the house, you find that the umbrella in the corner has a blinking light indicating that there will be rain today, so it must accompany you. While getting ready to go to the office, the formal section of your wardrobe buzzes a ring indicating today is a client meeting day. You need to choose your outfit from the formal section. Now the question arises, does the umbrella know about the rain and does the wardrobe know about the client meeting? The answer is yes; the umbrella learns about the weather from your smart phone and provides the indication accordingly. Similarly, the wardrobe checks your email and learns about the meeting. Isn't it true that your life has become simpler and comfortable now? So what is the technology behind it? A similar concept was very popular a few years ago; it was named ubiquitous computing. It was the concept of computer science that made the availability of computing to any place and any time beyond desktop computing. In this context the devices may or may not be connected to the Internet to provide the contextual service. On the contrary, the objects in the *Internet of Things (IoT)* are connected to each other via the Internet to provide a similar service as the ubiquitous computing. Thus, IoT is a network of the physical and smart devices that are widely connected to each other and are approachable to one another over a range of wireless networks over the Internet.

The name 'Internet of Things' was not officially coined until the late '90s. One of the primary instances of IoT is from the mid-1980s and was a Coca Cola machine, situated at the Carnegie Melon University. Neighbourhood developers and techies would associate themselves to the Internet to the refrigerated apparatus to verify whether there was a beverage accessible and check whether it was cold before making the visit.

In 1999, Kevin Ashton, executive director of Auto-ID Labs at the Massachusetts Institute of Technology (MIT) and also known as the Father of IoT, was the first to portray the Internet of Things while making a presentation for Procter & Gamble. But the term did not get much attention till 2010. In 2011, 'Internet of Things' was included as a new emerging phenomenon by the market research company Gartner. Since the turn of the century, IoT has rolled out more attainments; starting from consumer goods to other industrial components, IoT usage also penetrates everyday objects. The prediction is that in the near future, IoT-based consumer goods will overtake that of the industrial goods as indicated in Figure 13.1.

The Internet of Things devices include computational devices, both operated mechanically and electrically, digitized machines, specified objects, animals, and

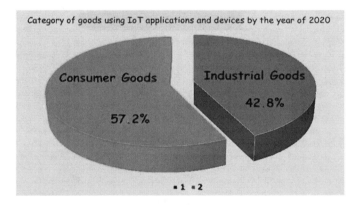

FIGURE 13.1   Prediction for IoT goods in consumer and industrial domain.

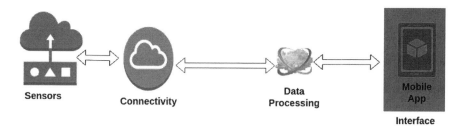

FIGURE 13.2 Major components of Internet of Things.

humans that are equipped with different and apparent identifiers. The objective of the IoT system is to control and automate everything around us, make things faster and real-time, and to gather more data for further analysis and prediction. The Internet of Things has the capacity to transfer information, resources, and valuable data over an integrated network without any man-to-man or man/human with computer interaction directly. The major components of IoT are:

- Sensors/ devices
- Connectivity
- Data processing
- User interface

This is indicated in Figure 13.2. *Sensors*/devices are embedded into the IoT devices to collect data from the surroundings. The sensors can be as simple as a temperature sensor or can be a complex motion sensor camera. An IoT device can be embedded with multiple homogeneous or heterogeneous sensors. Sometimes, the IoT device is also embedded with actuators. The umbrella in the earlier example has a sensor that checks the weather and the actuator that enables the blinker to blink as an outcome of the sensing analysis. Sensors ensure the physical identifiable objects are more responsive and can mould them to use their capability to ensure reclamation of data to the enforcement of the instruction given.

Connectivity is a major component of the IoT system that coordinates among multiple devices to work together. The sensed data of the IoT devices need to be forwarded to the cloud infrastructure or server for further analysis and processing. It requires reliable connectivity among the sensors and the cloud environment. Connectivity consists of the agreements and technologies which make use of the two physical objects to exchange information. With the boom in the telecom sector and the progress in global Internet usage and availability to all the devices, the connectivity issue has become cost-efficient. Several connectivity technologies are cellular network, satellite network, Wi-Fi, Bluetooth, and low-power wide-area network (LPWAN). Any of these technologies has its own specifications and constraints. Thus best connectivity in IoT is chosen while considering parameters like power consumption, range of connectivity, bandwidth, and compatibility.

The collected data by the IoT devices has no value if not processed for getting the required output. The data processing in IoT system can be a simple task of checking if the temperature reading in the AC/heater is within acceptable range. Sometimes it can be a much more complex task of identifying an intruder in your house through the video processing. *Data processing* involves data analysis algorithms to run efficiently and provide real-time output.

The *user interface* (UI) provides the bridge between the IoT system and the end user. The UI can use an alarm, an email, or a text notification to communicate with the end user. The user interface sometimes enables the user to track IoT systems. For example, the user can check video recordings of his house through the web server to identify the intruder captured on the camera installed in his house. The UI can also support complex tasks like switching off any appliance remotely or controlling the feeding medicine to the unattended elderly person at home.

13.2   CHALLENGES

In IoT systems, the devices are connected to the Internet to carry out information from and to the servers all the time. Thus, there exists the potential risk of hacking of such devices, concerns about surveillance, and fears concerning privacy and many such threats that have gathered public attention. Although IoT has the potential to enhance the lives of the human and is a boom in the 21st century, at the same time there are numerous challenges associated with the IoT framework that must be scrutinized. A few such challenges are described next.

13.2.1   SCALABILITY

As the number of connected things is increasing exponentially, the IoT network is required to be scalable. The fundamental principle of IoT is to connect anything over the Internet that is capable of collecting and sharing data among others (Gupta, Christie, and Manjula, 2017). Everything in the IoT network is uniquely recognized and personified virtually. However, there is no end limit for the number of things to get connected. So the framework and the model that need to be designed for the IoT network must be capable of accommodating any number of things into it. The future prediction is that humans will be the minority as the generators and receivers of the payload. Instead the channel will be flooded with the traffic generated by all kinds of things, resulting a much more complex Internet of Things. In a study conducted by INTEL, [] indicates that approximately 40% of the market share has been consumed by the manufacturing industry that includes robotics machinery and supply chain and equipment. Another 30% share is taken up by the healthcare industry that includes mobile health monitoring, electronic recordkeeping, disease prediction, and pharmaceutical safeguards. Other market shares are consumed by retail and security systems. A scalable IoT network should have the capability of moving from a smaller to a larger system, adapt to changes in the environment, and applications available in the changing environment.

13.2.2   TECHNOLOGICAL STANDARDIZATION

IoT devices surround everyone nowadays. Smart lights, automatic HVAC (heating, ventilation, and air conditioning) systems, voice-activated products, smart parking, traffic monitoring, etc. have become very common these days. However, the standardization of the products and protocols in the IoT system has not come at the same pace as the adoption of IoT in commercial and retail sectors. The absence of standardized norms and documents can make out the IoT devices as senseless activities. The low-bounded and cheap materials that are used for design of configurations have dreadful consequences. Standardization in the IoT system is a complex and challenging task (Gupta, Christie, and Manjula, 2017). It needs the highest precision in the specifications of making microcontrollers, sensors, actuators, and connectivity devices. The standardization requires a high level of collaboration among several international organizations as well. The Industrial Internet Consortium (IIC) is one such organization formed in 2014 to encourage the interconnectivity among the devices ("Industrial Internet Consortium," n.d.). IIC targets to bring together the multinational corporations (MNCs), governments, and academia to collaborate in designing the IoT test bed for the real-world applications. Another organisation Internet Task Force (IETF) that came to existence in 1986 is now slowly shifting towards IoT networks ("Internet Engineering Task Force," n.d.). They define protocols to support IPV6 over Low-Power Wireless Personal Area Networks (6LoWPAN) (Olsson 2014) and Routing Over Low-Power and Lossy Networks (ROLL) (Brandt and Porcu 2010). Yet another non-profit organization named the IoT Security Foundation (IoTSF) works with a motto, "Build secure, Buy secure, and Be secure" ("IoT Security Foundation," n.d.). It aims to create IoT security guidelines through different courses and trainings. IoT standardization covers the major domains like platform, connectivity, business model and the applications as shown in Figure 13.3.

13.2.3 INTEROPERABILITY

The Institute of Electrical and Electronics Engineers (IEEE) defines interoperability as the means of exchange of data between two or more systems or things. As per the

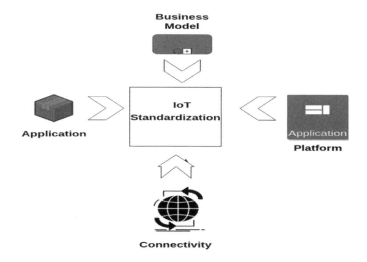

FIGURE 13.3 IoT domains that need standardization.

CIOReviewIndia report, interoperability enables 40% of the total economic value from IoT (Noura, Atiquzzaman, and Gaedke, 2017). The data that is exchanged between the different things have different languages, data formats, data models, and above all complex interrelationships. This results in the interoperability among the things being very complex. A new thing that gets connected to an existing IoT system may have unexpected data structures and protocols. For example, smart coffee kettle may communicate and share information with a human–robot asking it to pour the hot coffee into a cup and handover to the physically challenged person sitting in the wheelchair in the corner of the living room. The act may feel like a story to anybody, but IoT provides such a platform to today's world. This is all possible due to the interconnectivity and interoperability among heterogeneous devices. Interoperability can be at the device level, network level, syntactic level, or semantic level. Device-level interoperability deals with accessing devices through standardizing interfaces and the provision of adding any device/thing to the existing IoT platform. In the previous example, the coffee kettle may use ZigBee to get instructions from its server, the robot may use Wi-Fi to communicate with the outer world, and both of them may use Bluetooth to communicate with each other. So device-level interoperability enables these heterogeneous devices to understand and translate all these disparate communication technologies. Network-level interoperability deals with the protocol interoperability. There exists a set of routing protocols like RPL, CORPL, etc. to be used by the devices for network interface. The Fog of Things (Yu et al., 2018) also provides network interoperability in the cloud system. Software Defined Networking (SDN) is the new software-based solution to address network interoperability. Syntactic-level interoperability deals with the format and structure of the data used by the devices during the exchange of data and information. Web technologies like JSON, REST, and SOAP architecture provide greater interoperability. Semantic-level interoperability deals with the technologies needed for enabling the meaning of information to be shared by the devices (Noura, Atiquzzaman, and Gaedke, 2017).

13.2.4 SOFTWARE COMPLEXITY

Software complexity is yet another challenge in the IoT framework. To unify the interfacing devices in the IoT, and to standardize the protocols and specification of the devices, it is required to set up the infrastructure accordingly. This enhances the software complexity to a great extent.

13.2.5 DATA VOLUME AND INTERPRETATION

A huge amount of data is produced by IoT devices such as sensors, actuators, networks, etc.

Big data and its analysis are going to be the core study of the IoT system, while volume and its interpretation will be the major challenge (Irmak and Bozdal, 2017).

13.2.6 FAULT TOLERANCE

By now we understand that IoT is the interconnectivity among several heterogeneous devices. The connected devices may be sensors, actuators, gateway nodes, or any connectivity components. The devices may use different technologies individually. For example, some devices might use Wi-Fi technology, whereas some others might use IEEE 802.15.4 technology for their functioning. Thus bringing together varieties of technologies onto a single platform may result in device level or connectivity-level faults. Device-level fault deals with the failure in the sensors or actuators in the network. The faulty device must be replaced with new devices if possible or there must be backup devices to take up the role of the faulty ones. Connectivity-level fault is a major challenge. Usually the devices are connected wirelessly in the IoT network. The break in the connectivity may cause severe damage in certain cases like healthcare, disaster management, and emergency services. Data forwarding cannot be compromised in these applications. Researchers are working on fault-tolerant routing protocols (Chaithra and Gowrishankar, 2016) for the IoT network in order to continue providing the required services to end users. In certain sensitive IoT applications like healthcare, fault tolerance is a major challenge as the system needs to function as normal as possible even in the presence of any fault. Researchers have given the utmost attention in the design of fault-tolerant architecture (Gia et al. 2015), especially in the healthcare domain.

13.2.7 NETWORKING

In general, the topic of networking has great importance in the field of the Internet as it comprises many of the important factors which are used to manage the networks. In the domain of IoT, things cannot be predicted, as the movement of the object is not certain; movement changes from time to time and place to place as per the suitability of the users. The IoT objects also transmit signals from one network to another. This leads to the complication of dynamic gateways and difficulty keeping track of identifying and locating the devices which change position occasionally. The more we move towards a connected world, the challenges of making the systems, sensors, wearables, and devices connect in a single channel has become complex. To collect and cluster information from a dissimilar arrangement of hubs requires methods for connecting gadgets with a scope of handling capacities and interfaces together in a stable and balanced way.

13.2.8 PRIVACY AND SECURITY ISSUES

Privacy and security are essential pillars to the Internet and a major challenge to IoT. In the course of time, the inclination of IoT augments from millions of devices to tens of billions. The higher the number devices connected to the network, the higher is the chance for exploitation in safety susceptibility.

As authenticity, dependability, and confidentiality are significant perspectives, there are some prerequisites significant to unfair access to specific offices. The information systems are yet fragile and expensive in correlation of other created nation. From an Indian point of view, the distributed storage activity is still in the emerging stage (Abomhara and Køien, 2014).

13.3 IOT AND ITS APPLICATIONS

IoT, in general, represents a unique concept of sustainability of the devices of the interconnected network to analyze, collect, and use data across the globe and sharing across various platforms. The Internet of Things extends its connectivity from smart wearables to smart homes, from healthcare to traffic monitoring, and smart parking to smart cities. IoT is spreading the roots of benefits to the basic level of every trade and is gaining confidence in the aspects of our day-to-day lives by augmenting their presence to add comforts and simplifying the personal or daily routine tasks. Some of the applications of IoT are described in the following sections.

13.3.1 SMART HOMES

As the IoT system takes root, it comes along with its most popular and important application, i.e. smart home or the home automation system. A home is termed as a smart home when it provides a better living standard with the benefits of energy saving, safety, flexibility, and comfort. The number of people looking for a smart home is increasing rapidly day by day. The aim of having a smart home is to avail control of all appliances and electronic devices without much human intervention. There are enormous parameters that can be taken into consideration while designing a smart home. Temperature sensors, automatic HVAC, intruder detection system, and voice-detecting devices (echo bot) are few among them. Taking care of in-house elderly persons or physically challenged people is another major objective of having smart homes. Figure 13.4 depicts a smart home with features of smart foot mat for assisting in-house elderly people, security and alarm device, temperature

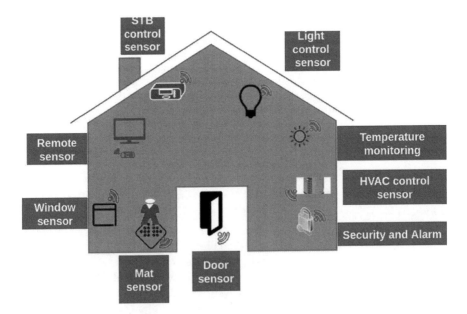

FIGURE 13.4 Smart home.

monitoring, light control, door control, environmental monitoring, HVAC control, etc. All these products are connected efficiently to make the life of the homeowner safe and comfortable.

Many homes have been implementing sensor-based devices to lessen burdens in their lifestyles. Even modern washing machines are programmed follow a series of parallel washes and spins according to the strength of washing needed for the garments. Similarly, refrigerators can be connected to smart phones to inform the owner about the availability of foodstuffs in it. Badabaji and Nagaraju (2018) designed a smart home system for controlling home appliances via a smart-phone-based application. It also includes real-time video streaming from the web camera for intruder detection at home. A server-based application at home can easily handle the data collected by the sensors at home and stored in a local database.

13.3.2 Smart Wearables

Like smart homes, smart wearables are another hot component among potential IoT applications. There exists several consumer wearables like Apple's smartwatch and Sony's Smart B-Trainer. These wearables have become so fancy and popular that every year thousands of different versions of these wearables come to market. Wearables also come in the form of assistive technology (AT) for accessibility. For example, Toyota has developed a wearable device to enhance the mobility of blind persons in indoor places like shopping malls and airports. This wearable can be worn around the shoulder and it takes input through a camera to recognize what is around the person. It instructs the blind person through sound or vibration. Similarly, Medtronic's glucose monitoring is another wearable that is worn under the skin to continuously keeps track of the body's glucose level.

These wearables can be incorporated into clothing or worn on the body as accessories. Health monitoring wearables offer biometric measurements like heart rate, glucose, temperature, and blood oxygen level. These wearables are able to sense, store, and track the biometric parameters. Sometimes these wearables help to predict the future occurrence of diseases like diabetes and the flu. Figure 13.5 denotes a few existing wearables that have become very popular.

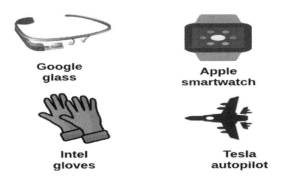

FIGURE 13.5 IoT gadgets.

Some advanced IoT-based wearables can even detect the presence of your jacket in the car and can adjust the car temperature accordingly. A few biometric measurement wearables have the provision of sharing or not sharing the sensed data with a third party. For example, you may want to share your biometric measurements with your physician who needs to check it regularly, whereas you might not want the same to be shared with your office colleagues. So the provision is kept open to either share or not to share the wearable sensed data to things (or person) around you. Sometimes, the wearables help in designing a smart home as well. It can adjust the room light intensity while the wearer is watching TV. It can even block the window light that may create glare on the TV or the backlight on the LCD can be adjusted. Above all the wearable can communicate with all other devices in the home to provide the viewer the best experience. A few brands that focus on wearable technology are Apple (Apple Watch), Google (Google Glass), Intel (gloves), Ralph Lauren (PoloTech Shirt), Tesla (Autopilot), Blackberry (smartwatch), Chui (universal key), Huawei (smartwatch with GPS), and Samsung (fitness tracker). The Kaa IoT platform enables wearable technology with ready-to-use IoT functions and applications. Kaa also provides scalable cloud functionality to ensure lossless communication between wearable devices and empower them with data analytics and visualization tools. Irrespective of the direction in which the wearables will evolve in future, they will always be integrated with IoT to provide a wide range of features.

13.3.3 SMART CITIES

IoT applications for smart cities span from water distribution to waste management, smart healthcare to environmental monitoring, pollution control to city security, and many more. The ultimate goal of a smart city is to eliminate the discomfort and problems of people of the city. Additionally, cities including smart supervision and surveillance, better automatic transportation, and managing the energy in the most efficient way minimizing environmental degradation are some of the practical examples considered in the construction of the smart cities. The whole world is investing billions and encouraging IoT technologies in developing smart cities. The popular term 'smart city' was launched in the White House in 2015 with an objective to facilitate technological collaboration between cities, federal agencies, universities, and private sectors. Kansas City, Missouri, has signed an agreement with Sprint and Cisco to build up the largest smart city while improving municipal services and providing various information about the citizens by gathering and analysing data on their behaviour in the city. China is also investing a huge amount in the IoT sector to bring up smart cities by 2020. Firms like Alibaba, Huawei, Lenovo, and Xiaomi are taking active participation in this mission. Fujisawa, a city in Japan, was constructed by Panasonic and is using IoT technology to make it a smart city. This city uses smart street light monitoring and rainwater recycling as a few smart features. The Malmo Green Digital City project of Sweden plans to make the whole city carbon free by smart environment monitoring system. The projects also aims to run the entire city on renewable energy by 2030. To summarize, the goal of any smart city is to provide smart urban infrastructure with the ability to collect and process data using the help of information and communications technology (ICT) such as a smart grid, smart

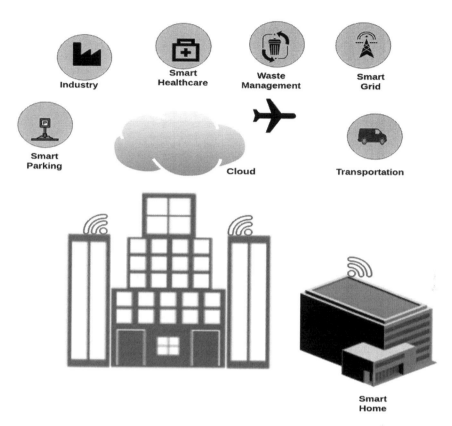

FIGURE 13.6 Smart cities.

metre, smart home, smart healthcare, and smart transportation. Figure 13.6 shows a few of the smart city aspects.

13.3.3.1 Environment Monitoring

The dimension of contamination in water, air, and food is expanding quickly because of parameters like enterprises, urbanization, dense population, and vehicle use which can influence human wellbeing. Deployment of wireless sensor networks (WSNs) can sense, store, analyze, and disseminate information regarding various natural resources like water level in lakes, streams, and sewages; gas concentration in the air of the city; and soil humidity and fertility. Similarly, the position changes that may occur due to a landslide, for example, can also be detected through sensing technology and IoT applications. Change in structures like dams and bridges can be detected and monitored for the safety of citizens (Lazarescu, 2013). Detection of infrared radiation is also part of environmental monitoring in any smart city. The places which are near to the nuclear power plants should be checked continuously with the help of sensors to measure the quality and safety checks and ensure the quality of life to the people nearby. IoT can be a very supporting technology for tracking and monitoring environmental features.

13.3.3.2 Waste Management

It is expected that by 2050, most of the human population will move towards urbanization, thus establishing more cities. As the population in any city grows, it also increases varieties of waste products. Thus waste management is highly essential in order to enhance the socioeconomic standard and environmental quality. A waste management model was prepared by Anagnostopoulos et al. (2017) which includes the steps of planning for waste collection, waste transport to specific locations, and recycling of the waste. Planning for waste collection involves design of an algorithm for route selection for the garbage truck and dynamic adaptation of routes by the garbage trucks. These days the truck picks up the garbage cans even if they are empty resulting in wasted man-hours. IoT devices inside the garbage cans will be able to connect to the computing server by any of the LPWAN technologies and can inform the server about the presence of garbage inside it. The computing server can collect all the information and optimize the route for the garbage collection by the trucks. Similarly, selection of specific locations for the type of waste needs smart decision-making. Recycling is a major component of waste management to have a sustainable environment. Ghose, Dikshit, and Sharma 2006) have an IoT-based solid waste recycle management information platform for efficient waste management. Figure 13.7 shows the waste management phases for a smart city. Anagnostopoulos et al. (2017) studied the waste management problem in the city of St. Petersburg, Russia, and provided solutions in order to save the huge amount of fuel consumed by waste collection trucks and to avoid the traffic that was caused by those trucks during peak hours.

13.3.3.3 Traffic Monitoring

The Internet of Things can be used to control the growing worldwide problem of traffic congestion, which causes unnecessary fuel consumption, wasted driver time, and increased driving stress. Traffic systems need special attention and smart solutions in case of medical emergency and accident vulnerability. Traffic problems, overcrowding, and unforeseeable travel time are common issues. IoT solves traffic congestion by using the dynamic location and control in the static monitoring state. It manages and responds to other vehicles in nearby areas to know their real-time location and predicts the movement of traffic and the number of vehicles passing through areas. By knowing the location and movement of traffic, one can select an alternative shorter route so that they don't get stuck in particular area. Some of the approaches where IoT can be very useful include diverting traffic in particular areas in case of any medical emergency such as the movement of the ambulance, natural calamity, the arrival of VIPs, and any sort of terrorist attack or similar situation. IoT implements Internet-based traffic control using the real-time location of commuters over an economical and innovative technology for the control of centralized smart-signalling-based traffic control (Prakash et al., 2018).

With the growth of the smartphone industry, it has become much more uncomplicated to keep track of the signals from IoT devices and to follow steps to decongest road traffic. These effective measures can be taken by the authorities of the city traffic police to divert the public from the measured location. It can be a great help for fire brigades to help the people in the affected area and to divert others from the

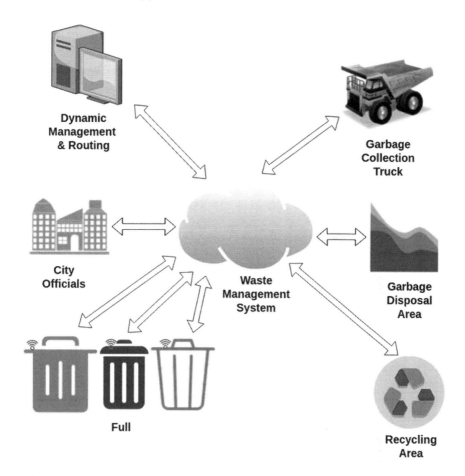

FIGURE 13.7 Waste management.

danger ensuring the safety of the public. Where the traffic has been an immense problem for a period of years, IoT has helped in collecting useful information from an arrangement of road sensors along with dozens of cameras that feed into a centralized system to control the city's traffic lights enabling the movement of the vehicles (Janahan et al. 2018) (Figure 13.8).

13.3.3.4 Smart Healthcare

Applications of IoT in healthcare is an aid to the people that brings change in the traditional techniques allowing people to draw attention to health monitoring and treatment. IoT applications in healthcare span from remote monitoring equipment to tiny sensors that are integrated to advanced equipment. It has the potential to improve the physician's treatment and the duration of interaction time between patients and caregivers. From personal fitness sensors to surgical robots, the IoT application can bring revolution in the healthcare domain. Some of the important features that can be included in IoT-based healthcare are (i) remote monitoring of health, (ii) real-time

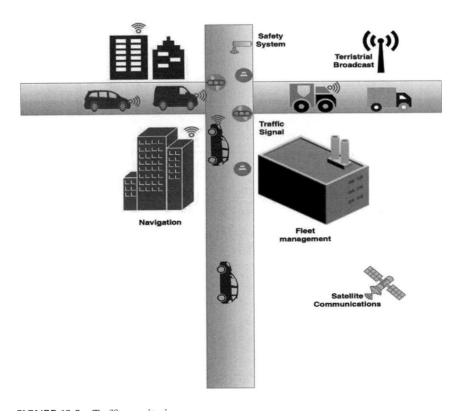

FIGURE 13.8   Traffic monitoring.

location tracking of the patient, (iii) prediction of occurrence of diseases beforehand, and (iv) critical and emergency care.

People in remote areas suffer a lot due to unavailability of health experts or means to communicate with them in emergency. IoT application helps the patient's physiological parameters to reach the doctor to enable diagnosis of the disease. The use of IoT along with ICT can allow the doctor to treat the patient in time. Remote monitoring is possible through medical equipment and paraphernalia equipped with sensors and IoT devices, e.g. monitoring equipment, wheelchairs, heart pumps, and nebulizers. GPS-based location tracking of the patient is also possible through IoT applications in case of emergency. Smart healthcare includes the prediction of occurrence of diseases by analyzing the family health history, and, in case of critical and emergency care, the hospitals receive the profile of the patient before their arrival so that the patient can receive essential care without delay. The use of IoT can make the connectivity among the patient and the caregiver while the former is moving in an ambulance in order to give continuous attention to the patient. Figure 13.9 depicts the IoT-based healthcare in a smart city.

13.3.3.5   Smart Grid

A smart grid uses intelligent and autonomous controllers, advanced algorithms, and data management techniques as efficient means of communication between power

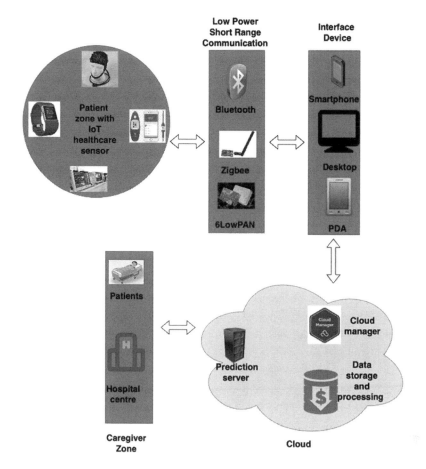

FIGURE 13.9   Smart healthcare.

utilities and consumers in order to have efficient and fair energy distribution in a smart city (Fang et al., 2011). It extracts information of the behaviours of consumers and electricity suppliers automatically for analysis and decision-making for energy distribution. IoT applications in a smart grid enhance efficiency, reliability, and economics in the energy sector. Figure 13.10 indicates the use of IoT in a smart grid.

13.3.3.6   Smart Retail

Applications of IoT solutions have captured the retail domain in order to increase purchases, reduce theft, make efficient inventory management, and provide pleasant shopping experience for consumers. Smart cities have many stores equipped with IoT-enabled technologies such as (i) predictive equipment maintenance, (ii) smart merchandise transportation, (iii) demand-driven warehouse, (iv) connected consumer, and (v) smart store (Figure 13.11).

Predictive equipment maintenance prevents equipment failure and defects, thus increasing the lifetime of equipment. Smart merchandise transportation includes

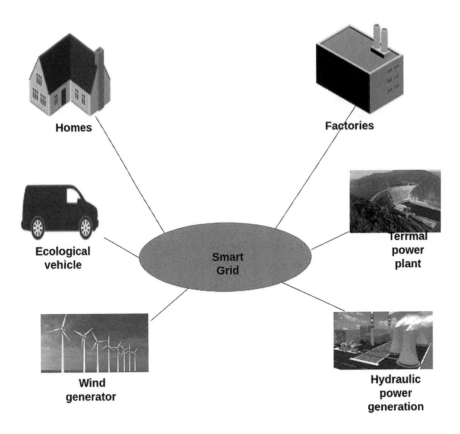

FIGURE 13.10   Smart grid.

GPS-based tracking of movement of merchandise, optimizing the merchandise route during transportation to reduce the cost of fuel and traffic congestion. A demand-driven warehouse requires warehouse automation in terms of a supply-and-demand plan. IoT technology enables monitoring of sales opportunities in real-time and tracking of missed in-store sales. Use of radio frequency identification (RFID) is a proven solution for inventory management in retail industry. In a smart store equipped with IoT-enabled devices, beacons play a key role for sending push notifications to the customers like information about discounts, attractive offers, and special events at the store. Beacons use low energy Bluetooth connection for these push notifications. The end result is better return on investment (ROI) to the store. Smart shelves in stores are equipped with RFID tags, RFID readers, and antennas. Through these, the smart shelves help to track the inventory, e.g. if any item is running low or any item is on the wrong shelf. In order to save the customer's waiting time during checkout and to save the cost of recruiting multiple employees by the store at the checkout counter, automated checkout can be a smart solution. It verifies the item automatically through a RFID reader, and automatically deducts money from customer's mobile payment app, providing a better shopping experience for all. Also, robots are nowadays replacing humans, saving the salary of store staff and increasing the volume of work in comparison to humans.

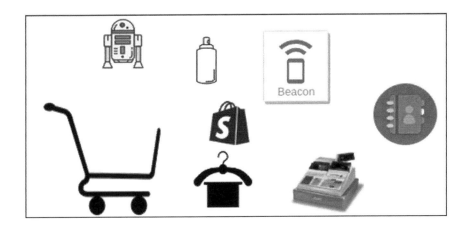

FIGURE 13.11 Smart retail.

13.3.4 SMART PARKING

Nowadays one of the foremost problems that arises along with traffic is parking because of the lack of ample parking space. The number of vehicles in a family is significantly more than the head count. If the trend continues, then the rise of parking demand in malls, offices, shops, airports, bus terminals, railway stations, etc. will be of great concern. In the advancement of technology, vehicles and transportation are convenient for low-income groups. Urbanized areas and offices are suffering congestion and parking problems; many people spend their time, energy, and fuel trying to find a parking space.

The problems of parking space can be fought in the view of IoT devices that can be tracked, monitored, and administered using computer networks and sensors connected over the Internet.

Smart parking is a strategy of parking vehicles which combines the technology and innovation in the human effort to minimize the use of fuel and time by finding a faster and easier way to park vehicles. The Internet devices use the sensing of devices to find out the occupancy of parking spaces. Cameras send signals to the devices to count the number of vehicles and free parking spaces.

Mobile apps would help users to register for parking services, and if the user specifies the entry and the exit time, then IoT devices along with sensors embedded in the pavements and cameras can allow a free parking space to the user and send the location to the mobile device where the user can tentatively book tickets for the parking lots and thus save a lot of time. For each parking area, infrared technology is used to detect the number of parking slots and the free parking slots are deployed in the screen and Internet, or a Wi-Fi module communicates between the user app and the affixed sensing element (Figure 13.12).

13.3.5 SMART AGRICULTURE

Use of sensors in the agriculture domain is quite an old concept. But in this traditional approach, it was difficult to get live sensor data. Sensors store the data in their local memory that was fetched at a later period for processing (Verdouw et al., 2016).

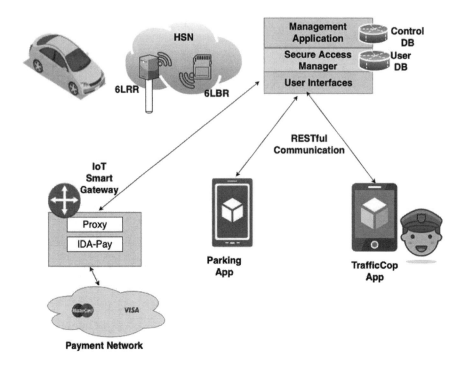

FIGURE 13.12 Smart parking.

But with the introduction of IoT into agriculture, advance sensors are used and they are connected via the cloud to provide real-time data for analysis and decision-making. It helps the farmer in reducing expenses and increasing the production with the help of accurate decisions based on the collected data. IoT applications in agriculture help in (i) understanding climatic conditions, (ii) doing precision farming, (iii) making a smart green house, and (iv) proper data analysis.

The population of the globe is yet to touch 10 billion by 2050, and nourishing food for such a gigantic population would be an ordeal. Embracing the applications of IoT and enhancing the procedure of smart agriculture could transparently reduce the burden of feeding the population.

Challenging the extremes of weather conditions and climate disruptions can dissuade people from practising extensive and intensive farming to meet the demand of the food industry.

Smart farming empowered with technologies of IoT will enable farmers and the cultivators to reduce waste and improve the amount of production. Farmers will also be able to check the use of fertilizers and limit the journeys of the machinery being used in irrigation (Figure 13.13).

13.3.6 FISH FARMING

Fish farming refers to the farming of a variety of marine species like shellfish, bait fish, molluscs, algae, sea vegetables, and fish eggs to breed. The water in ponds,

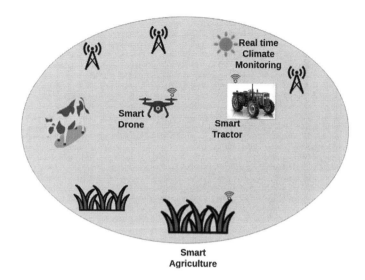

FIGURE 13.13 Smart agriculture.

rivers, oceans, lakes, etc. vary with respect to quality, pH value, temperature, acidity and alkaline, and other parameters. Though fish can adapt to a wide range of temperature changes, a sudden rise or fall in temperature can cause the death of fish due to respiratory arrest or paralysis. The amount of oxygen dissolved in water depends on its temperature. Thus temperature monitoring is a vital issue in fish farming. Though the applications of IoT have not penetrated much into the aquaculture domain, it has the potential to bring revolution in this industry. Use of IoT sensors to monitor aqua parameters enables remote fish farming (Figure 13.14). But one of the major challenges is to acquire high-performance and low-cost sensors, as these water-sensitive sensors are very expensive and difficult to connect to the IoT world (Dupont, Cousin, and Dupont, 2018). Similarly, deploying and connecting these sensors in water for communicating are complex tasks. Researchers are still working on IoT applications for fish farming in order to achieve affordable and easy-to-deploy technology (Dupont, Cousin, and Dupont, 2018; Janet, Balakrishnan, and Rani, 2019).

13.3.7 Disaster Management

Natural calamities, disasters, hurricanes, and floods have a traumatic effect on communities around the world and severely affect the natural beauty and natural resources, destroys homesteads and crops, and leads to the loss of human life. Natural disasters cannot be stopped, but IoT technologies can be extremely utilitarian for disaster awareness such as prediction of natural disasters and early warning systems to control the catastrophic effects (Sharma and Kaur, 2019). Whenever a disaster strikes the people, rescue teams, the government, and other local authorities need to coordinate a strategy based on the data and find qualified professional

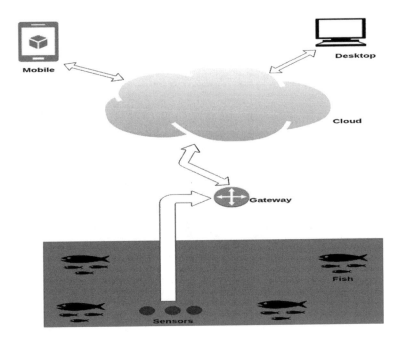

FIGURE 13.14   Fish farming.

workforce to direct the response. Government firms and other responsible agencies, which work in alliance with the rescue operation, should grasp the IoT and other network-driven technologies to obtain the timed accurate data that can be better utilized to take on the necessary operational work. Use of IoT and current technologies can help them more appropriately address the movement of the calamities to ensure maximum safety. In order to get a quick response with accuracy, government and emergency teams should organize a strong communication system with the help of the mobile devices (Figure 13.15).

Measuring shock waves by IoT-enabled technology can prevent the loss caused by earthquake. The accelerated movement of the ground should provoke the early warning of an earthquake. Depending upon the shock intensity, the devices can emit the warning levels which can be communicated with people to take the necessary steps. Monitoring the waves from the seismic monitoring stations as well the data from the sensors fixed in typical infrastructure similar to high-rise buildings, towers, and skyscrapers can be very useful in distinguishing the wave pattern at the time of earthquakes and in normal conditions. Similarly, forest fires can be detected by attaching the temperature sensors to trees to check the parameters of heat and compare with the natural conditions. Any disruptions in the levels of temperature, humidity, or water level, which can cause a hazardous effect on the survivability of the environment, can be signalled to the nearest record stations or forest officers where they can then make necessary arrangements to control the forest fire.

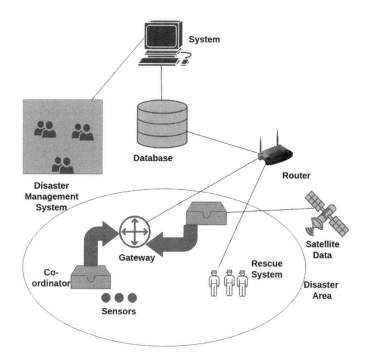

FIGURE 13.15   Disaster management.

13.4   CONCLUSION

This chapter concisely describes the evolution of the Internet of Things. The transition of ubiquitous computing through the Internet of Things has many challenges associated with it. This chapter describes the challenges of IoT including scalability, security, and standardization. In spite of all these challenges, the application of IoT has entered into almost every domain of life. Smart home, smart healthcare, smart city, and disaster management are some of the most important domains that makes life secure and comfortable.

This chapter will be helpful for the researchers to find out their domain of interest and study the challenges associated with it. The application of IoT is still growing into many other domains like fish farming, agriculture, and environmental monitoring. This chapter will enable researchers in these areas to do intensive study and find solutions to improve the social and economic conditions.

REFERENCES

Abomhara, Mohamed, and Geir M. Køien. 2014. "Security and Privacy in the Internet of Things: Current Status and Open Issues." In: *2014 International Conference on Privacy and Security in Mobile Systems (PRISMS)*, 1–8.

Anagnostopoulos, Theodoros, Arkady Zaslavsky, Kostas Kolomvatsos, Alexey Medvedev, Pouria Amirian, Jeremy Morley, and Stathes Hadjieftymiades. 2017. "Challenges and Opportunities of Waste Management in IoT-Enabled Smart Cities: A Survey." *IEEE Transactions on Sustainable Computing* 2(3): 275–89.

Badabaji, Swapna, and V. Siva Nagaraju. 2018. "An IoT Based Smart Home Service System." *International Journal of Pure and Applied Mathematics* 119(16): 4659–67.

Brandt, A., Buron,J. and Porcu,G, 2010. "Home automation routing requirements in low-power and lossy networks."

Chaithra, S., and S. Gowrishankar. 2016. "Study of Secure Fault Tolerant Routing Protocol for IoT." *International Journal of Scientific and Engineering and Research* 5(7): 1833–38.

Dupont, Charlotte, Philippe Cousin, and Samuel Dupont. 2018. "IoT for Aquaculture 4.0 Smart and Easy-To-Deploy Real-Time Water Monitoring with IoT." *Global Internet of Things Summit (GIoTS)* 2018: 1–5.

Fang, Xi, Satyajayant Misra, Guoliang Xue, and Dejun Yang. 2011. "Smart Grid—The New and Improved Power Grid: A Survey." *IEEE Communications Surveys and Tutorials* 14(4): 944–80.

Ghose, M. K., Anil Kumar Dikshit, and S. K. Sharma. 2006. "A GIS Based Transportation Model for Solid Waste Disposal--A Case Study on Asansol Municipality." *Waste Management* 26(11): 1287–93.

Gia, Tuan Nguyen, Amir-Mohammad Rahmani, Tomi Westerlund, Pasi Liljeberg, and Hannu Tenhunen. 2015. "Fault Tolerant and Scalable IoT-Based Architecture for Health Monitoring." In: *2015 IEEE Sensors Applications Symposium (SAS)*, 1–6.

Gupta, Anisha, R. Christie, and P. R. Manjula. 2017. "Scalability in Internet of Things: Features, Techniques and Research Challenges." *International Journal of Computational Intelligence Research* 13(7): 1617–27.

"Industrial Internet Consortium." n.d. https://www.iiconsortium.org.

"Internet Engineering Task Force." n.d. https://www.ietf.org.

"IoT Security Foundation." n.d. https://www.iotsecurityfoundation.org.

Irmak, Emrah, and Mehmet Bozdal. 2017. "Internet of Things (IoT): The Most Up-To-Date Challenges, Architectures, Emerging Trends and Potential Opportunities." *International Journal of Computer and Applications* 975: 8887.

Janahan, S. Kumar, M. R. M. Veeramanickam, S. Arun, Kumar Narayanan, R. Anandan, and Shaik Javed. 2018. "IoT Based Smart Traffic Signal Monitoring System Using Vehicles Counts." *International Journal of Engineering and Technology* 7(221): 309.

Janet, J., S. Balakrishnan, and S. Sheeba Rani. 2019. "IOT Based Fishery Management System." *International Journal of Oceans and Oceanography* 13(1): 147–52.

Lazarescu, Mihai T. 2013. "Design of a WSN Platform for Long-Term Environmental Monitoring for IoT Applications." *IEEE Journal on Emerging and Selected Topics in Circuits and Systems* 3(1): 45–54.

Noura, Mahda, Mohammed Atiquzzaman, and Martin Gaedke. 2017. "Interoperability in Internet of Things Infrastructure: Classification, Challenges, and Future Work." In: *International Conference on Internet of Things as a Service*, 11–8.

Olsson, Jonas. 2014. "6LoWPAN Demystified." *Texas Instruments*, 13.

Prakash, B, M. Naga, Sai Roopa, B. Sowjanya, and A. Pradyumna Kumar. 2018. "An Iot Based Traffic Signal Monitoring and Controlling System Using Density Measure of Vehicles." *International Journal of Research*, 5(12), 1173–1177.

Sharma, Meghna, and Jagdeep Kaur. 2019. "Disaster Management Using Internet of Things." In: *Handbook of Research on Big Data and the IoT, pp.211-222. IGI Global*

Verdouw, Cor, Sjaak Wolfert, Bedir Tekinerdogan, and others. 2016. "Internet of Things in Agriculture." *CAB Reviews: Perspectives in Agriculture, Veterinary Science, Nutrition and Natural Resources* 11(35): 1–12.

Yu, Ruozhou, Guoliang Xue, Vishnu Teja Kilari, and Xiang Zhang. 2018. "The Fog of Things Paradigm: Road Toward On-Demand Internet of Things." *IEEE Communications Magazine* 56(9): 48–54.

14 Physical Layer Security Approach to IoT

Rupender Singh and Meenakshi Rawat

CONTENTS

14.1 INTRODUCTION

Today the Internet of Things (IoT) has become the essential concept of every device making human life comfortable. IoT technology is capable of providing the ubiquitous connectivity and information congregate in various emerging applications such as smart cities, medical devices, vehicular technology, industrial environments, and many more [1–3]. Most of the physical devices can communicate with each other by connecting through different sensors [4]. In other words, IoT has provided a platform to these devices to communicate wirelessly [5, 6]. Because of broadcasting and the random nature of wireless technology, the security concerns for IoT applications are required to be addressed [7–13]. Thus, millions of dollars are spent every year to make IoT more secure. Figure 14.1 shows the spending on IoT security in last five years. Till now, traditional cryptographic algorithms such as an asymmetric encryption algorithm (RSA) or symmetric encryption algorithm (AES) are employed to ensure privacy and security [14]. Regrettably, these techniques were failed to associate with the physical phenomenon of wireless communication as these techniques are based on the assumption that the physical layer allows the transmission without error. Recently, many researchers have examined the random nature of wireless channels using information-theoretic security to guarantee secure communication between the transmitter and receiver. It is widely accepted across the world that the information-theoretic–based approach will provide secure communication over a physical-layer link in the strictest form. Table 14.1 shows the nomenclature for the abbreviations used throughout the chapter.

14.1.1 CONVENTIONAL SYSTEM MODEL FOR SECRECY

The perfect secrecy notion was first presented by Shannon [15] using the information-theoretic approach, as shown in Figure 14.2. Generally, the three nodes are

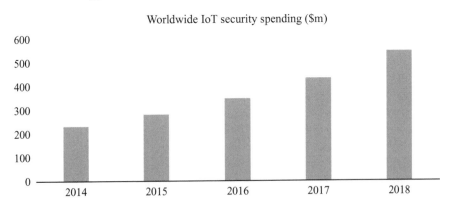

FIGURE 14.1 Worldwide spending on IoT security.

TABLE 14.1
Abbreviations

ABEP	Average bit error probability
AES	Advanced Encryption Standard
ASC	Average secrecy capacity
AWGN	Additive white Gaussian noise
BER	Bit error rate
CDF	Cumulative distribution function
CSI	Channel state information
DSR	Double shadowed Rician
EGC	Equal gain combining
IoT	Internet of Things
ITS	Intelligent transportation system
M2M	Mobile-to-mobile
MIMO	Multiple-input multiple-output
mm-Wave	Millimetre wave
MRC	Maximal ratio combining
MSE	Mean square error
PDF	Probability density function
PLS	Physical layer security
SC	Selection combining
SDoF	Secrecy degrees of freedom
SEE	Secrecy energy efficiency
SINR	Signal-to-interference-plus-noise ratio
SIMO	Single-input multiple-output
SNR	Signal-to-noise ratio
SOC	Secrecy outage capacity
SOP	Secure outage portability
SOR	Secure outage region
SPSC	Strictly positive secrecy capacity
SR	Secure region
RSA	Rivest-Shamir-Adleman
V2V	Vehicle-to-vehicle

mentioned as Alice, Bob, and Eve, where two nodes Alice and Bob out of three nodes are licit users, whereas third node Eve acts as an adversary. In this proposed secure system, both licit users share a non-reusable secret key k to encrypt the message U. A cryptogram C is used to encrypt the message U, which can be heard by the eavesdropper (Eve). Herein, if a posteriori probability of the encrypted message is equal to the a priori probability of U, then the perfect secrecy can be achieved. Hence, this can be written mathematically as

$$H(U/C) = H(U) \tag{14.1}$$

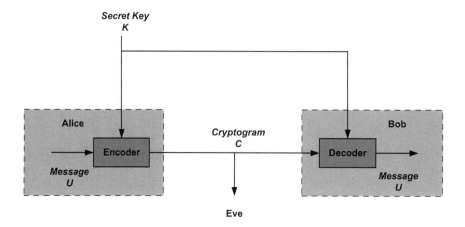

FIGURE 14.2   Secure system model proposed by Shannon.

where $H(U/C)$ denotes the eavesdropper's *equivocation* or conditional entropy of U given cryptogram C and $H(U)$ represents the entropy of the message U.

Equation 14.1 can be written using mutual information as

$$I(U,C) = 0 \qquad\qquad (14.2)$$

where $I(\circ;\circ)$ represents mutual information. As both U and C are mutually independent, so we have Equation 14.2.

In 1975, Wyner [16] laid the idea of a wiretap channel for perfect secrecy where two legitimate users called Alice and Bob, respectively, share their information secretly with each other while an adversary called Eve is trying to hear the secret information, as shown in Figure 14.3. In this model, the Alice–Eve channel can be estimated by assuming it as a degraded version of the Alice–Bob channel. The message

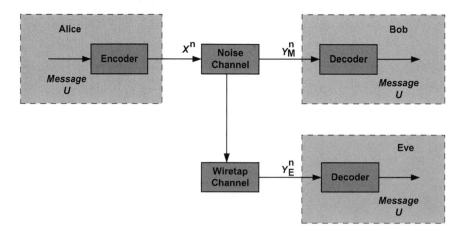

FIGURE 14.3   Wiretap system model proposed by Wyner.

$U^k = [U_1, U_2, U_3, ..., U_k]$ is encoded into code word $X^n = [X_1, X_2, X_3, ..., X_n]$ at the transmitter and is received by the legitimate receiver as $Y_M^n = [Y_{M_1}, Y_{M_2}, Y_{M_3}, ..., Y_{M_n}]$. This encoded message is also sent to the eavesdropper via the wiretap channel as degraded message $Y_E^n = [Y_{E_1}, Y_{E_2}, Y_{E_3}, ..., Y_{E_n}]$. They have provided the security performance in terms of secrecy capacity, which can be defined in terms of a extreme secure transmission rate from Alice to Bob, at which Eve is unable to obtain any information.

14.1.2 PRACTICAL WIRETAP CHANNEL SCENARIOS

Csiszár and Körner carried forward Wyner's work to non-degraded channels [17], as shown in Figure 14.4. In this system model, the transmitter sends the same confidential information to both the receiver and eavesdropper, while keeping the eavesdropper as ignorant of this confidential information as possible. Further works have provided analysis over additive white Gaussian noise (AWGN) and shown that secure communication can be obtained under the assumption of higher main channel capacity compared to the eavesdropper's channel capacity. In this case, if the Markov chain $R \to X \to Y_M Y_E$ holds, then the secrecy capacity C_s is defined as

$$C_S = \max_{R \to X \to Y_M Y_E} I(R, Y_M) - I(R, Y_E) \tag{14.3}$$

where R is defined as the auxiliary variable, which is purposely designed. Furthermore, it is assumed that the adversary's channel is a debased version of authentic channel, then the secrecy capacity reduces to

$$C_S = \max_{X \to Y_M \to Y_E} I(X, Y_M) - I(X, Y_E) \tag{14.4}$$

We can note that here channel prefixing is not needed due to Markov chain, thus the auxiliary variable R vanishes.

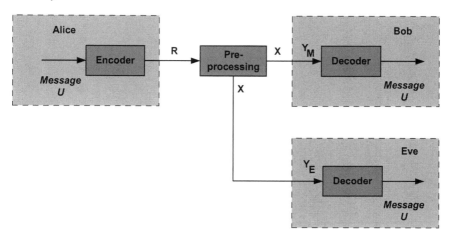

FIGURE 14.4 Practical wiretap model proposed by Csiszár and Körner.

Soon after, the secrecy capacity for the Gaussian discrete memoryless degraded wiretap channel was examined by Cheong and Hellman [18]. They presented that by maximizing the channel capacities C_M and C_E of the legitimate link and the wiretap link, respectively, it is easy to redefine the secrecy capacity as

$$C_S = C_M - C_E \qquad (14.5)$$

If the transmitter transmits power P with σ_M^2 and σ_E^2 noise variances in the Alice–Bob channel and adversary channel, respectively, then the secrecy capacity can be written as

$$C_S = \frac{1}{2}\log_2\left(1 + \frac{P}{\sigma_M^2}\right) - \frac{1}{2}\log_2\left(1 + \frac{P}{\sigma_E^2}\right) \qquad (14.6)$$

In general, it is expected that the adversary's channel capacity is smaller than the Alice–Bob channel capacity, so the secrecy capacity can be expressed as

$$C_S = \left[C_M - C_E\right]^+ \qquad (14.7)$$

where $\left[z\right]^+ = \max\left(0, z\right)$.

Moreover, many researchers have investigated the secrecy capacity in complex AWGN channels and presented the secrecy capacity in terms of channel gains h_M and h_E of the legitimate channel and eavesdropper's channel, respectively, as

$$C_S = \left[\log_2\left(1 + \frac{P\left|h_M\right|^2}{\sigma_M^2}\right) - \log_2\left(1 + \frac{P\left|h_E\right|^2}{\sigma_E^2}\right)\right]^+ \qquad (14.8)$$

14.1.3 MULTIPLE-INPUT MULTIPLE-OUTPUT (MIMO) SYSTEM

Recently, with the day-by-day developments in wireless communication techniques, communication systems are shifting from single-input single-output (SISO) to multiple-input multiple-output (MIMO) technology, where all nodes are equipped with multiple antennas. In this technology, multiple antennas at all nodes help to ameliorate the reliability and efficiency of the system. It is proven that the MIMO technique can play an important role for secure communication. Hero [19] was the first to propose the secure communication using the MIMO technique. A MIMO illegitimate channel was considered in this study, where both the legitimate users (Alice and Bob) and illegitimate receiver (Eve) are provided multiple antennas as shown in Figure 14.5. For this proposed system, the secrecy capacity C_s is formulated mathematically as

$$C_S = \max_{\rho_X \succ 0, tr(\rho_X) \prec P} \frac{\log\det\left(I + H_M\rho_X H_M^H\right)}{\log\det\left(I + H_E\rho_X H_E^H\right)} \qquad (14.9)$$

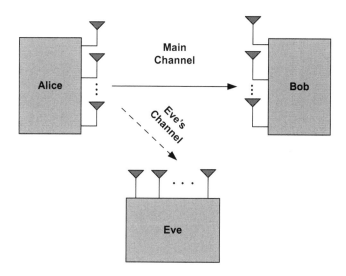

FIGURE 14.5   MIMO communication system with MIMO wiretap channel.

where ρ_X denotes the covariance matrix of transmitted signal X, P represents the power constraint, and $H_M = \mathbb{C}^{N_M \times N_T}$, $H_E = \mathbb{C}^{N_E \times N_T}$ are the MIMO channel gain of the legitimate channel and the adversary's channel, respectively.

14.2   RELATED WORK

So far, many works are devoted to studying the physical layer security (PLS) for IoT applications over the decade. Figure 14.6 shows the number of articles published in the last decade on IoT and IoT security. Exponential growth can be observed every year [19]. These studies have considered different wireless fading channels and provided the PLS secrecy analysis. The effect of short-term fading conditions such as Rayleigh, Nakagami-n, Nakagami-q, Weibull, κ-μ, α-μ, and α-η-κ-μ were explored [21–25]. However, none of these fading distributions can model the realistic emerging communication conditions, e.g. millimetre wave (mm-Wave) communications, device-to-device (D2D) communication, vehicle-to-vehicle (V2V) communication,

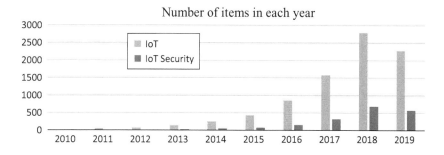

FIGURE 14.6   Number of items related to IoT and IoT security published as of July 2019 [20].

body area networks, and IoT. Therefore, practical realistic scenarios can be modelled by so-called composite fading conditions, which are capable of modelling the simultaneous occurrence of multipath and shadowing. In this context, Lei and co-workers [24, 26] considered generalized-K (GK) fading distribution to characterize both the channel links and conducted secrecy analysis. By utilizing mixture gamma distribution, the authors have investigated the secrecy performance of PLS and also derived the expressions for average secrecy capacity (ASC), probability of non-zero secrecy capacity (PNSC), and secure outage portability (SOP) [24]. A similar analysis was done for a single-input multiple-output (SIMO) system [27]. The PLS secrecy metric SOP for correlated channels were analyzed over Nakagami-m/gamma composite fading channels [28]. A generalized fading distribution, so-called κ-μ shadowed composite fading distribution, has also been examined [29, 30]. Sen et al. [29] analyzed the secrecy performance and derived the novel analytical expressions for SOP and SPSC in terms of the Meijer G-function, while Srinivasan and Kalyani [30] derived the SPSC by using the moment matching method.

14.3 CRYPTOGRAPHIC TECHNIQUES VERSUS PHYSICAL LAYER SECURITY

The fundamental difference between cryptographic techniques and physical layer security is illustrated in Figure 14.7. Cryptographic techniques are traditional approaches to protect the information against eavesdropping. In general, cryptographic encryption is done at the upper layer to provide data security. The principle idea of cryptographic encryption is to convert the transmitted information into ciphertext using a *secret key*. The data is extracted from ciphertext using the same secret key at the legitimate receiver. For instance, the asymmetric encryption algorithm Rivest-Shamir-Adleman (RSA) or the symmetric encryption algorithm Advanced Encryption Standard (AES) are the most popular cryptographic

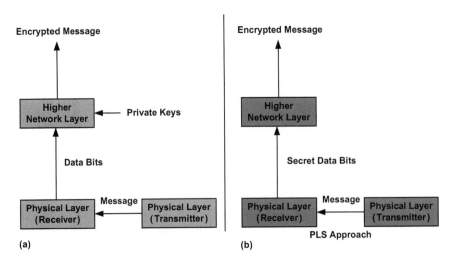

FIGURE 14.7 Cryptography versus PLS approaches [31].

techniques. Unfortunately, cryptography is vulnerable to eavesdropping because the eavesdroppers can use the benefits of the broadcast nature of wireless communication to intercept the secret key. In contrast, physical layer security provides better secure transmission by utilizing the random nature of wireless channels and using proper channel coding at the physical layer rather than the upper layer. It is interesting to note that the physical layer security ensures the achievable transmission secrecy regardless of the eavesdroppers computational power. Besides many advantages, physical layer security also provides a few disadvantages. It is very difficult to guarantee the secrecy with probability one, as physical layer security can be measured in terms of average information [32]. Moreover, the practical channels behave differently from theoretical channels.

14.4 CLASSIFICATION OF EAVESDROPPING

Illegitimate receivers are always trying to hear the secure information with their own abilities. On their abilities, eavesdropping in physical layer secrecy is classified into two types: (1) active eavesdropping and (2) passive eavesdropping.

14.4.1 ACTIVE EAVESDROPPING

In active eavesdropping, the legitimate transmitter perceives the channel state information (CSI) of the adversary and is able to detect the illegitimate receiver. The eavesdropper attacks the physical layer security by transmitting the deceive signals to confuse the legitimate receiver. The legitimate receiver intercepts this forged information, which results in deterioration in secrecy performance. These attacks are also referred as *masquerade attacks* [33]. In this scenario, the eavesdropper also has the ability to behave as a *jammer*. The illegitimate receiver sends noisy signals to the legitimate receiver, which makes transmitted information less trustworthy.

14.4.2 PASSIVE EAVESDROPPING

In passive eavesdropping, it is not possible to detect the illegitimate receiver as the legitimate transmitter fails to perceive the CSI of the eavesdropper. In this scenario, the eavesdropper can hear information without disrupting the communication between the legitimate users. Thus, we have to design signals in such a way to prevent eavesdropping.

14.5 PHYSICAL LAYER SECURITY PERFORMANCE METRICS

The secrecy performance of conventional encryption algorithms can be evaluated by measuring the total attacks during transmission. On the other hand, various physical layer security metrics can be utilized to characterize the secrecy for different channel conditions. However, the accuracy of the evaluation of these metrics can be limited by the knowledge of channel properties. Various scenarios of channel properties will be discussed in the following sections.

14.5.1 Channel State Information (CSI)

In the physical layer security, CSI plays an important role. CSI is also referred to as channel gain. CSI describes the scenario where the channel properties of the wireless communication link are available to the transmitter or receiver. This information about the channel provides the knowledge of signal transmission behaviour from the transmitter to the receiver. CSI also elucidates the analogous channel effects, e.g. scattering, fading, and power deterioration with distance. This is the method by which the channel can be estimated. CSI makes it possible to provide reliable communication for multi-antenna systems at high data rates by adapting instantaneous channel conditions. The CSI available to the transmitter is different from the CSI available to the receiver in physical systems. In general, the earlier condition is referred to as CSIT, while latter is referred to as CSIR.

Further, CSI is featured in two types: (1) instantaneous CSI and (2) statistical CSI.

14.5.1.1 Instantaneous CSI

The scenario when instant channel conditions are known is referred to as instantaneous CSI. Instantaneous CSI can be observed as knowing the impulse response of the filter. Under instantaneous CSI, the transmitted signal can adapt the impulse, while the received signal can be optimized to achieve minimum bit error rates (BERs) at high data rate transmission. The scenario is referred to as *full CSI* when instantaneous CSI of all wireless network nodes are available to other wireless network nodes.

14.5.1.2 Statistical CSI

Statistical CSI describes the statistical properties of the channel, e.g. the fading conditions, the spatial correlation, the line-of-sight (LOS) component, and the average channel gain are known. In the statistical CSI scenario, this information can be used to optimize the transmission as used in the case of instantaneous CSI. The scenario is referred to as *partial CSI* where instantaneous CSI of a few wireless network nodes is known, while other nodes provide statistical CSI.

We next discuss various scenarios of full CSI and partial CSI. We start by considering the case when full CSI is available to the transmitter. In this case, the transmitter can adapt the coding scheme to every realization of the fading coefficients. Now, the case is considered when the transmitter perceives CSI of the authentic channel. In this scenario, a wiretap code of length 2^{nR} code words can be designed, where R represents the transmit rate. For the moment, it is assumed that R is identified as the instantaneous main channel capacity C_M. Then, 2^{nR_e} code words are assigned per bin, where R_e denotes the eavesdropper's equivocation rate. Then, it is easy to show the rate of secure communication as $R_S = R - R_e = C_M - R_e$, which is usually set as a constant value, implying that $R_e = C_M - R_S$ varies according to the main channel condition. As long as the rate of secure communication is less than that of instantaneous secrecy capacity, i.e. $R_S < C_S$, the eavesdropper's channel become worse in comparison to the main's estimate, i.e. $C_E < R_e$, then the transmitter ensures the perfect secrecy by using the wiretap codes. Otherwise, if $R_S > C_S$, then $C_E > R_e$, and secrecy over main channel is compromised.

CSI provides flexibility to the designer to decide optimal transmission strategy and opportunity to choose the correct physical layer security performance metric. It is necessary to have the instantaneous CSI of the authentic channel and the illegitimate channel by which the achievable secrecy rate can be maximized. Once the full CSI of both the authorized receiver and the adversary is accessible, the secure communication can be achieved by maximizing the signal-to-noise ratio (SNR) at the legitimate receiver, while minimizing the SNR at the eavesdropper. However, it is difficult to obtain full CSI of the physical systems because of estimation errors or obscurity of the eavesdropper. Next, we will discuss some other secrecy performance measures of physical layer security.

14.5.2 SECRECY RATE

Since it is complicated to obtain secrecy capacity, as the probability distribution of the transmitter signal X creates an optimization problem, we define the secrecy rate in this case. The difference between the achievable rates of the authentic channel and wiretap channel is referred to as the secrecy rate [34]. It can be written mathematically as

$$R_S = \left[R_M - R_E \right]^+ \tag{14.10}$$

where R_M denotes the achievable secrecy rate of the authentic channel and R_E describes the achievable secrecy rate of adversary's channel link. The maximum secrecy rate R_S refers to the secrecy capacity C_S.

14.5.3 ERGODIC SECRECY CAPACITY/RATE

In general, it is noted that secrecy capacity can be defined only for the channels with predetermined properties. So it is challenging to evaluate the secrecy capacity for time-varying fading channels. Therefore, we need ergodic secrecy capacity to characterize the fading channel where the ergodic features of the channel can be captured by the secure message using channel realizations. Ergodic secrecy capacity can be defined as the average ability of secure communication in fading channels, where this ability of secure communication can be ensured by acclimatization power and rate according to the CSI.

Gopala et al. [35] have examined the secrecy of the system and derived the ergodic secrecy capacity for both the cases where CSI of both the legitimate channel and the adversary's channel is known and only the CSI of the authentic channel is accessible. The probability density function (PDF) of both the legitimate channel and illegitimate channel can be obtained when the full CSI is known, while only the PDF of the legitimate channel can be obtained when partial CSI is known. Mathematically, ergodic secrecy capacity can be formulated for full CSI as [35]

$$\overline{C}_S^{Full} = \max_{E\left\{P\left(|h_M|^2,|h_E|^2\right)\right\}\leq \overline{P}} E\left[C_M - C_E\right]^+$$

$$= \max_{P\left(|h_M|^2,|h_E|^2\right)} \int_0^\infty \int_{h_E}^\infty \left[\log\left(1+|h_M|^2 P\left(|h_M|^2,|h_E|^2\right)\right) - \log\left(1+|h_E|^2 P\left(|h_M|^2,|h_E|^2\right)\right)\right]$$

$$\times f\left(|h_M|^2\right) f\left(|h_E|^2\right) d|h_M|^2 d|h_E|^2$$

(14.11)

where h_M and h_E represent the channel coefficients of the legitimate channel link and illegitimate channel link, respectively; $\overline{P}$ describes the average transmitted power; and $f\left(|h_M|^2\right)$ and $f\left(|h_E|^2\right)$ are the PDFs of the legitimate channel distribution and the wiretap channel distribution. Similarly, the ergodic secrecy capacity can be calculated for partial CSI as [35]

$$\overline{C}_S^{Partial} = \max_{E\left\{P\left(|h_M|^2\right)\right\}\leq \overline{P}} E\left[C_M - C_E\right]^+$$

$$= \max_{P\left(|h_M|^2\right)} \int_0^\infty \int_{h_E}^\infty \left[\log\left(1+|h_M|^2 P\left(|h_M|^2\right)\right) - \log\left(1+|h_E|^2 P\left(|h_M|^2\right)\right)\right]^+ \quad (14.12)$$

$$\times f\left(|h_M|^2\right) f\left(|h_E|^2\right) d|h_M|^2 d|h_E|^2$$

Note that the secrecy capacity cannot be smaller than the achievable ergodic secrecy rate. It can also be noted from Equations 14.11 and 14.12 that both equations are similar, but the obtained CSI decides the optimization of power control policies. The optimal transmission scheme can only be employed when $|h_M|^2 > |h_E|^2$.

14.5.4 Secure Outage Probability (SOP)

One of the critical secrecy measures of physical layer security is known as secure outage probability (SOP), which is widely used to characterize the secure communication where perfect CSI of the licit user (Bob) and eavesdropper (Eve) are unavailable to the transmitter (Alice). The SOP refers to the probability which can be calculated for the event when the targeted secrecy rate $R_S > 0$ is greater than the current secrecy rate C_S. So SOP can be defined mathematically as [36]

$$P_{out}(\gamma_{th}) = P(C_S \leq R_S) \quad (14.13)$$

In Equation 14.13, R_S can be calculated by using the relation $R_S = \log_2(1+\gamma_{th})$, where γ_{th} denotes the threshold SNR. Equation 14.13 can also be expressed as

$$P_{out}(\gamma_{th}) = P\left[\gamma_M \leq (1+\gamma_E)(1+\gamma_{th})-1\right] \tag{14.14}$$

Equations 14.13 and 14.14 explain unsafe transmission between licit users over the main channel and the licit receiver is unable to decode the received signal.

In some cases, it is very difficult to obtain SOP due to convoluted mathematical integrals. Consequently, the lower bound of SOP is defined as

$$P_{out}(\gamma_{th}) = P\left[\gamma_M \leq (1+\gamma_E)(1+\gamma_{th})-1\right]$$
$$\geq SOP^L(\gamma_{th}) \equiv P\left[\gamma_M \leq (1+\gamma_{th})\gamma_E\right] \tag{14.15}$$

14.5.5 STRICTLY POSITIVE SECRECY CAPACITY (SPSC)

Another important secrecy measure is strictly positive secrecy capacity (SPSC), which can be calculated in terms of a probability for the event when positive secrecy capacity, i.e. $C_S > 0$, is achieved. The SPSC can be expressed in terms of SOP as a particular case when $R_S = 0$. As one of the benchmark secrecy metrics, the SPSC explains the scenario where $C_M > C_E$. As long as the same information is not shared with the eavesdropper (Eve) and the legitimate receiver (Bob), the non-zero secrecy capacity is distilled from the randomness. It is evident that the transmitter (Alice) is unable to perceive perfect CSI about eavesdropper (Eve) in this case. SPSC can be defined mathematically as [36]

$$SPSC = P(C_S > 0) = P(\gamma_M > \gamma_E) \tag{14.16}$$

It is interesting to express SPSC in terms of SOP by substituting $R_S = 0$ into Equation 14.13 as

$$SPSC = 1 - P_{out}(0) \tag{14.17}$$

14.5.6 SECRECY OUTAGE CAPACITY (SOC)

The maximum achievable secrecy rate R_S at a certain value of outage probability ε_0 defines the secrecy outage capacity (SOC). The maximum achievable secrecy rate R_S can be expressed mathematically in terms of a certain value of outage probability ε_0 as [37]

$$P_{out}(R_S) = \varepsilon_0 \tag{14.18}$$

So the maximum achievable secrecy rate R_S can be achieved from Equation 14.18 as

$$R_S = P_{out}^{-1}(\varepsilon_0) \tag{14.19}$$

where $P_{out}^{-1}(\cdot)$ represents the inverse function of Equation 14.18.

14.5.7 SECURE REGION (SR)/SECURE OUTAGE REGION (SOR)

The secure region (SR) is utilized to characterize the security performance of long-term fading scenarios such as composite multipath/shadowing fading conditions. SR can be defined as the geometrical region in which the secrecy capacity can never be negative. Assuming the eavesdropper (Eve) is located at a point with coordinates (x_E, y_E), then the secure region can be expressed as [38]

$$SR = \left\{ (x_E, y_E), C_E(x_E, y_E) < C_M \right\}$$ (14.20)

where C_M represents the capacity of the legitimate channel link and $C_E(x_E, y_E)$ describes the capacity of the eavesdropper's channel with the eavesdropper's location at (x_E, y_E). It is clear that Equation 14.20 can be satisfied only when Eve disappears from the secure region. Chang et al. [39] propounded a notion of *secure zone* in which they considered that Eve cannot be present in the vicinity of the transmitter up to a certain radius.

The secure outage region (SOR) was first proposed by Li et al. [40]. It is also defined as the geometric region in which the secure outage probability cannot be higher than the predefined outage probability ε for the given secrecy rate. Thus, SOR is formulated mathematically as

$$SOR = \left\{ \chi_E \middle| P_{out}(C_S \leq R_S) \leq \varepsilon \right\}$$ (14.21)

where χ_E is the position vector that apprises the location of the eavesdropper with respect to the legitimate transmitter.

14.5.8 SECRECY DEGREES OF FREEDOM (SDOF)

The secure region (SR) cannot be defined for the scenario where multiple eavesdroppers exist. In such cases, He et al. [41] and Koyluoglu et al. [42] proposed a new secrecy performance metric called the secrecy degrees of freedom (SDoF), which is defined as the ratio between the asymptotic secrecy rate R_S^∞ and asymptotic SNR λ. Mathematically, SDoF can be expressed as [43]

$$SDoF = \lim_{\lambda \to \infty} \frac{R_S(\lambda)}{\log_2(\lambda)}$$ (14.22)

14.5.9 OTHER SECRECY PERFORMANCE METRICS

Besides the aforementioned secrecy performance metrics, there are some conventional secrecy performance metrics such as average SNR, mean square error (MSE), secrecy energy efficiency (SEE), average bit error probability (ABEP), and signal-to-interference-plus-noise ratio (SINR) which can be used to characterize secrecy performance of physical layer security.

14.5.9.1 Average Signal-to-Noise Ratio (SNR)

The most common and widely used performance measure characteristic of wireless communication is SNR. In general, SNR is associated with data detection as it is measured at the output of the receiver. Due to fading impairments, the SNR of the wireless link is heavily affected by thermal noise. Then the most suitable performance metric is average SNR, which can be calculated by taking the average density PDF of the fading channel. Mathematically, average SNR can be formulated as [44]

$$\bar{\gamma} \triangleq \int_0^\infty \gamma f(\gamma) d\gamma \tag{14.23}$$

where γ is the SNR and $f(\gamma)$ is the PDF of the fading channel.

14.5.9.2 Mean Square Error (MSE)

In wireless communication, the mean square error (MSE) is defined as the average of the squares of the errors. It can be measured at the output of the receiver and calculated by averaging the squared difference between the actual transmitted signal and the estimated signal. If the $\hat{X} = g(\theta)$ is the estimator of the received signal X, then MSE can be calculated as [45]

$$MSE \equiv E\left[\left(X - \hat{X}\right)^2\right] = \int_0^\infty \left(X - \hat{X}\right)^2 f(x) dx \tag{14.24}$$

where $f(x)$ is the PDF of X and $E(\circ)$ represents the expectation function.

14.5.9.3 Signal-to-Interference-Plus-Noise Ratio (SINR)

Signal-to-interference-plus-noise ratio (SINR) [46], or signal-to-noise-plus-interference ratio (SNIR), [47] is a performance metric which provides upper bounds on channel capacity. The ratio between the transmitted power and the sum of interference power and noise power defines the SINR. SINR reduces to SIR for zero noise power, while SINR reduces to SNR for zero interference. Mathematically, SINR can be formulated as [46]

$$SINR = \frac{P}{I + N} \tag{14.25}$$

where P, I, and N represent transmitted power, interference power, and noise power, respectively.

14.5.9.4 Average Bit Error Probability (ABEP)

Average bit error probability (ABEP) is the most powerful performance metric which characterizes the nature of the wireless system behaviour. ABEP is defined as the probability of the bit errors that arise during bit transmissions. ABEP is one of the most challenging performance metrics to compute since conditional BEP is the non-linear function of the instantaneous SNR. Mathematically, ABEP can be formulated as [48]

$$\overline{P_b}(E) \triangleq \int_0^\infty P_b(E/\gamma) f(\gamma) d\gamma \qquad (14.26)$$

Where $P_b(E/\gamma)$ is the conditional BEP over the AWGN channel and $f(\gamma)$ is the PDF of the fading channel.

14.5.9.5 Secrecy Energy Efficiency (SEE)

Recent studies propose a novel performance metric called the secrecy energy efficiency (SEE) [49]. This proposed metric takes into account the consumed energy and power allocation schemes. The ratio between the secrecy capacity and energy consumed by the wireless system defines the SEE. Mathematically, the SEE can be formulated as [49]

$$SEE = \frac{C_S}{\xi^T} \text{ bits/joule} \qquad (14.27)$$

where ξ^T represents the total consumed energy by the wireless system and C_S is the secrecy capacity.

14.6 WIRELESS FADING CHANNELS

Investigation of physical layer secrecy performance over a wireless channel is a complicated study as wireless channels are heavily impacted by various time-varying effects such as multipath fading and shadowing. Due to scattering, diffraction, and reflections from obstacles, the transmitted signal cannot reach the receiver through the direct path. Thus, the transmitted signal takes multiple paths to reach the receiver, and the signal received through these paths is added in-phase. If the amplitude and phase of the received signal are treated as random variables, then the power of the received signal will also be a random variable. These random fluctuations arise because wireless communication transmission is of a broadcast nature and has a significant impact on secrecy performance. Therefore, various diversity techniques such as maximal ratio combining (MRC), selection combining (SC), switch and stay combining (SSC), and equal gain combining (EGC) are proposed to mitigate the fading effects. For instance, Figure 14.8 shows the wiretap channel model with MRC diversity at both the authentic receiver and at the adversary [36]. Despite having a negative impact on system performance, fading also has constructive implications for physical layer security. Fading alone can be used as a tool to guarantee secure communication, even when the adversary's channel is more efficacious than the authentic channel [50, 51]. Fading effects such as scattering, diffraction, and reflections can be characterized by using different distribution function such as Rayleigh, Nakagami-n, Nakagami-m, Weibull, κ-μ, α-μ, and α-η-κ-μ. However, none of these channels have the capability to capture inhomogeneous diffuse scattering environments. Thus, composite (long-term) fading distributions are introduced, which are capable of characterizing inhomogeneous random fluctuations. In general, composite (long-term) fading channels

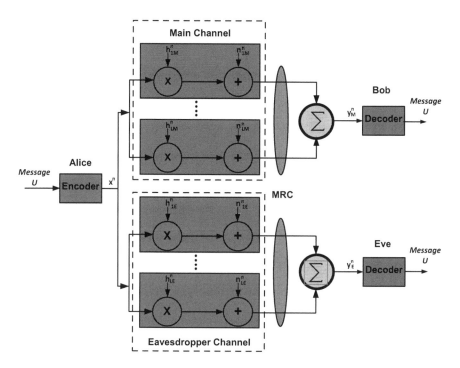

FIGURE 14.8   MRC wiretap model for physical layer security [36].

refer to composite multipath/shadowing fading channels. The PDF of the composite multipath/shadowing fading channel can be evaluated by averaging the conditional PDF of multipath fading over the PDF of shadowing as

$$p(\gamma) = \int_0^\infty p(\gamma/\theta) \cdot p(\theta)d\theta \qquad \gamma \geq 0 \qquad (14.28)$$

where $p(\gamma/\theta)$ is the conditional PDF of multipath fading and $p(\theta)$ is the PDF of shadowing. For composite Weibull/lognormal fading channel, $p(\gamma/\theta)$ and $p(\theta)$ can be written as [36]

$$p(\gamma/\theta) = \frac{\beta\Lambda}{2}\left(\frac{1}{\theta}\right)^{\beta/2} \gamma^{\beta/2-1} \exp\left(-\Lambda\left(\frac{\gamma}{\theta}\right)^{\beta/2}\right) \quad \gamma \geq 0 \qquad (14.29)$$

where β represents the shape parameter and measures the fading severity, $\Lambda = \Gamma\left(1 + 2/\beta\right)^{\beta/2}$, and $\Gamma(\cdot)$ is the gamma function [52, 53].

$$p(\theta) = \frac{1}{\sigma\theta\sqrt{2\pi}} \exp\left(-\frac{(\ln\theta - \mu)^2}{2\sigma^2}\right) \qquad \theta \geq 0 \qquad (14.30)$$

where μ and σ are the parameters which determine mean and standard deviation of the associated $\ln\theta$, respectively.

Recently, new fading distributions have been proposed to characterize the practical scenarios such as mobile-to-mobile (M2M), body area network (BAN), Internet of Things (IoT), and vehicle-to-vehicle (V2V) communication. Next we will discuss a few of them.

14.6.1 THE A-H-K-M FADING CHANNEL

The α-η-κ-μ distribution is introduced recently by Yacoub [54]. The probability density function (PDF) of the current SNR, $f_\gamma(\gamma)$, can be written as [24]

$$f(\gamma) = \frac{\alpha \gamma^{\frac{\alpha\mu}{2}-1}}{2^{\mu+1}\Gamma(\mu)\bar{\gamma}^{-\frac{\alpha\mu}{2}}} \exp\left(-\frac{\gamma^{\frac{\alpha}{2}}}{2\bar{\gamma}^{\frac{\alpha}{2}}}\right) \sum_{r=0}^{\infty} \frac{r! c_r}{(\mu)_r} L_r^{\mu-1}\left(2\left(\frac{\gamma}{\bar{\gamma}}\right)^{\frac{\alpha}{2}}\right) \tag{14.31}$$

where α is the parameter which describes channel non-linearity, η denotes the parameter which can be calculated by taking the ratio of in-phase total power and quadrature scattered waves of the multipath clusters, κ represents the parameter which determines the ratio between the dominant components power and total power of scattered waves, μ is the parameter which signifies number of multipath clusters, $L_n^m(\circ)$ denotes the Laguerre polynomial [53], $\Gamma(\circ)$ denotes the gamma function [53], $(a)_n$ is the Pochhammer symbol [52], and c_r is evaluated using Yacoub [54, equations (15), (30), and (31)] with the aid of p and q which are defined in Mathur et al. [25].

Also, from Equation 14.30 the cumulative distribution function (CDF) of γ is calculated as

$$F(\gamma) = \frac{\left(\frac{\gamma}{\bar{\gamma}}\right)^{\frac{\alpha\mu}{2}} \exp\left(-\frac{\gamma^{\frac{\alpha}{2}}}{2\bar{\gamma}^{\frac{\alpha}{2}}}\right)}{2^{\mu+1}\Gamma(\mu+1)} \sum_{r=0}^{\infty} \frac{r! m_r}{(\mu+1)_r} L_r^\mu\left(2\frac{(\mu+1)}{\mu}\left(\frac{\gamma}{\bar{\gamma}}\right)^{\frac{\alpha}{2}}\right) \tag{14.32}$$

where m_r is computed using Yacoub [54, equations (16), (33), and (34)].

14.6.2 DOUBLE SHADOWED K-M FADING CHANNELS

The double shadowed κ-μ distribution was recently introduced by Bhargav et al. [55]. It encompasses other well-known fading conditions, e.g. double shadowed Rician, shadowed Rician, shadowed Rayleigh, Nakagami-q, Rician, and Rayleigh distributions. The PDF of the instantaneous SNR, $f(\gamma)$, can be given as

$$f(\gamma) = \frac{\gamma^{-m_t}(m_t-1)^{m_t}\xi^\mu \gamma^{\mu-1}}{B(m_t,\mu)\left(\gamma\xi+(m_t-1)\gamma\right)^{-m_t+\mu}}\left(\frac{m_d}{m_d+\mu\kappa}\right)^{m_d} \tag{14.33}$$

$$\times {}_2F_1\left(m_d, m_t+\mu; \mu; \frac{\xi\mu\kappa\gamma}{(m_d+\mu\kappa)\left(\gamma\xi+(m_t-1)\gamma\right)}\right)$$

where $\xi = \mu(\kappa+1)$, $_2F_1(\cdot;\cdot;\cdot;\cdot)$ denotes the Gauss hypergeometric function [52], m_d denotes the Nakagami-m parameter, and m_t is the inverse Nakagami-m parameter.

Also, from Equation 14.33 the CDF of γ can be written as

$$F(\gamma) = \sum_{i=0}^{\infty} \frac{(m_d)_i (m_t+\mu)_i \xi^{\mu+i} \mu^i \kappa^i \gamma^{\mu+i}}{(\mu)_i i! B(m_t,\mu)(m_d+\mu\kappa)^i (\mu+i)(\bar{\gamma}(m_t-1))^{\mu+i}}$$

$$\times \left(\frac{m_d}{m_d+\mu\kappa}\right)^{m_d} {}_2F_1\left(m_t+\mu+i,\mu+i;\mu+1+i;-\frac{\xi\gamma}{\bar{\gamma}(m_t-1)}\right)$$

(14.34)

For the special case of $\mu \to 1$ and $\kappa \to k$, the PDF of the double shadowed Rician (DSR) can be computed from Equation 14.33 as

$$f^{DSR}(\gamma) = \frac{\bar{\gamma}^{-m_t} m_t (m_t-1)^{m_t} (1+k)}{(\gamma(1+k)+(m_t-1)\bar{\gamma})^{m_t+1}} \left(\frac{m_d}{m_d+k}\right)^{m_d}$$

$$\times {}_2F_1\left(m_d, m_t+1; 1; \frac{k(1+k)\gamma}{(m_d+k)(\gamma(1+k)+(m_t-1)\bar{\gamma})}\right)$$

(14.35)

where k is the Rician parameter.

The CDF of γ can be written using Equation 14.35 as

$$F^{DSR}(\gamma) = \sum_{i=0}^{\infty} \frac{(m_d)_i (m_t+1)_i (1+k)^i k^i \gamma^{i+1}}{(1)_i i!(m_d+\kappa)^i (i+1)(\bar{\gamma}(m_t-1))^{i+1}}$$

$$\times \left(\frac{m_d}{m_d+\kappa}\right)^{m_d} {}_2F_1\left(m_t+1+i, 1+i; 2+i; -\frac{(1+k)\gamma}{\bar{\gamma}(m_t-1)}\right)$$

(14.36)

14.7 IMPACT OF FADING ON SECRECY PERFORMANCE

It can be found that channel randomness intuitively renders the security performance of the system deteriorates. The physical layer security metrics highly depend on the fading parameters such as shape parameter, mean, and variance. Most IoT applications such as V2V communications and intelligent transportation systems (ITS) require secure communications between vehicles. This attracts our attention to discuss physical layer security performance in practical fading scenarios.

Here, the three-node system model is considered in which each node is accoutred a single antenna, as shown in Figure 14.9. Out of these three nodes, two nodes are licit users called Alice and Bob, and the third node is mentioned as Eve. Alice is trying to communicate secretly with Bob through the main channel, and Eve is trying to hear the secret information through the wiretap channel. It is assumed that both the channel links, i.e. main channel and wiretap channel, endure identical fading.

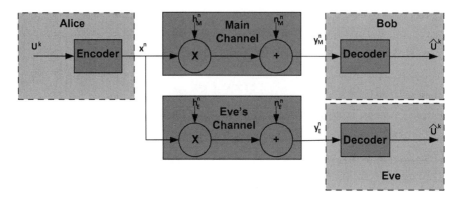

FIGURE 14.9 SISO system model with single eavesdropper [36].

Constantly, Alice is striving to transmit the secret message U^k to Bob. To enhance secrecy over the authentic channel link, the secret message U^k is encrypted into code word X^n. If y_M and y_E denote the signals received by Bob and Eve, respectively, then, y_M and y_E can be written as

$$y_M(i) = h_M(i)x(i) + n_M(i) \qquad (14.37)$$

$$y_E(i) = h_E(i)x(i) + n_E(i) \qquad (14.38)$$

where $h_M(i)$ and $h_E(i)$ are time-varying quasi-static composite Weibull/lognormal shadowing fading coefficients in legitimate and illegitimate channel links, respectively; and $n_M(i)$ and $n_E(i)$ represent the complex Gaussian noise with zero mean at Bob and Eve with variances σ^2_M and σ^2_E, respectively. The channel coefficient $h_M(i)$ is not dependent on the output, which is referred to as CSI. Here, it is also noted that the fading coefficients do not change during transmission of secret message (i.e., $h_M(i) = h_M$, $\forall i$ and $h_E(i) = h_E$, $\forall i$) as quasi-static fading is considered.

14.7.1 SOP AND SPSC

14.7.1.1 If Authentic and Adversary's Channel Experience Composite Weibull/Lognormal Shadowing Fading

As both the authentic and adversary's channels are subject to composite Weibull/lognormal shadowing fading, then the PDF of SNRs at the licit receiver and at the adversary can be expressed as [36]

$$f(\gamma_M) = \frac{\beta\Lambda_\beta}{2\sqrt{\pi}} \sum_{i=1}^{N} \omega_i \psi_{\gamma_M}(\gamma_M;\xi_i) \quad \omega_i > 0 \qquad (14.39)$$

$$f(\gamma_E) = \frac{\beta\Lambda_\beta}{2\sqrt{\pi}} \sum_{i=1}^{N} \omega_i \psi_{\gamma_E}(\gamma_E;\zeta_i) \quad \omega_i > 0 \qquad (14.40)$$

where β is the shape parameter; ω_i are the weights of the corresponding the sample points with $\sum_{i=1}^{N} \omega_i = 1$, $\xi_i = \exp\left(-\frac{\beta}{2}\left(\frac{\sigma_M t_i \sqrt{2}}{+\mu_M}\right)\right)$, $\zeta_i = \exp\left(-\frac{\beta}{2}\left(\frac{\sigma_E t_i \sqrt{2}}{+\mu_E}\right)\right)$; t_i are the roots of the Nth order Hermite polynomial, $\Lambda_n = \Gamma(1+2/n)^{n/2}$; and μ_b and σ_b are shadowing parameters with $b = \{M, E\}$.

Also, the CDFs of γ_M and γ_E can be written as

$$F(\gamma_M) = \frac{1}{\sqrt{\pi}} \sum_{i=1}^{N} \omega_i \left(1 - \frac{\psi_{\gamma_M}(\gamma_M; \xi_i)}{\xi_i \gamma_M^{\frac{\beta}{2}-1}}\right) \omega_i > 0 \tag{14.41}$$

$$F(\gamma_E) = \frac{1}{\sqrt{\pi}} \sum_{i=1}^{N} \omega_i \left(1 - \frac{\psi_{\gamma_E}(\gamma_E; \zeta_i)}{\zeta_i \gamma_E^{\frac{\beta}{2}-1}}\right) \omega_i > 0 \tag{14.42}$$

Using Equations 14.40 and 14.41 in Equation 14.15, the SOP^L can be evaluated for a random set of parameters $\{\mu_M, \mu_E, \sigma_M, \sigma_E\}$ as

$$SOP^L(\gamma_{th}) = \frac{1}{\sqrt{\pi}} \sum_{j=1}^{N} \omega_j - \frac{1}{\pi} \sum_{i=1}^{N} \sum_{j=1}^{N} \frac{\omega_i \omega_j \zeta_i}{\zeta_i + \xi_j(1+\gamma_{th})^2}^{\beta} \tag{14.43}$$

Similarly, SPSC can be obtained using Equations 14.17 and 14.43 as

$$SPSC = 1 - \frac{1}{\sqrt{\pi}} \sum_{j=1}^{N} \omega_j + \frac{1}{\pi} \sum_{i=1}^{N} \sum_{j=1}^{N} \frac{\omega_i \omega_j \zeta_i}{\zeta_i + \xi_j} \tag{14.44}$$

Figure 14.10 demonstrates the behaviour of SOP^L as a function of the main channel parameter μ_M and eavesdropper channel parameter μ_E. It can be observed that the Alice–Bob channel link is gradually vulnerable to eavesdropping when fading severity increases from 1.5 to 4.5. It is also observed that the rise in SOP^L is 6% for the superior eavesdropper's channel $(\mu_M < \mu_E)$ when fading severity increases from 1.5 to 4.5, and reduce by 30% when the Alice–Bob channel link is stronger $(\mu_M > \mu_E)$. This is because Alice can communicate secretly with Bob in a high fading regime with compared to a low fading regime. Figure 14.11 shows the SPSC as a function of the Alice–Bob channel parameter μ_M and eavesdropper channel parameter μ_E. As expected, an improvement in the SPSC can be observed with increasing fading severity for the superior Alice–Bob channel link $(\mu_M > \mu_E)$, whereas deterioration in the SPSC can be observed for the superior eavesdropper's channel $(\mu_M < \mu_E)$.

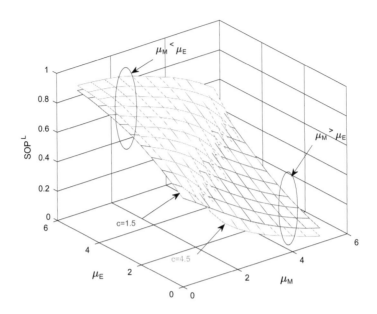

FIGURE 14.10 Behaviour of SOP^L as a function of main channel parameter μ_M and eavesdropper channel parameter [36].

14.7.1.2 If Main Channel and Eavesdropper's Channel Experience Composite κ-μ/Gamma Shadowing Fading

As both the authentic and adversary's channel are subject to composite κ-μ/Gamma shadowing fading, then the PDF of SNRs at the authentic receiver and adversary can be written as [56]

$$f(\gamma_M) = \sum_{p=0}^{\infty} \frac{\xi_{Mp}}{p!} \gamma_M^{\left(\frac{\mu_M+b+p}{2}-1\right)} H_{0,2}^{2,0} \left[\frac{\mu_M(1+\kappa_M)}{\Omega_M} \gamma_M \left| \begin{matrix} - \\ \left(\frac{b-\mu_M-p}{2},1\right) \left(\frac{-b+\mu_M+p}{2},1\right) \end{matrix} \right. \right]$$

(14.45)

$$f(\gamma_E) = \sum_{p=0}^{\infty} \frac{\xi_{Ep}}{p!} \gamma_E^{\left(\frac{\mu_E+b+p}{2}-1\right)} H_{0,2}^{2,0} \left[\frac{\mu_E(1+\kappa_E)}{\Omega_E} \gamma_E \left| \begin{matrix} - \\ \left(\frac{b-\mu_E-p}{2},1\right) \left(\frac{-b+\mu_E+p}{2},1\right) \end{matrix} \right. \right]$$

(14.46)

where $\xi_p = \dfrac{\mu^{\mu+b+3p}(1+\kappa)^{\left(\frac{\mu+b+p}{2}\right)}\kappa^p}{\Gamma(\mu+p)\Gamma(b)e^{\mu\kappa}\Omega^{\left(\frac{\mu+b+p}{2}\right)}}$, b is the shape parameter, μ_b and κ_b are shadowing parameters with $b = \{M,E\}$, Ω_b is the average SNR of shadowing, and $H_{p,q}^{m,n}(\circ)$ is Fox's H function, as defined by Yacoub [57, Equation (1.1)].

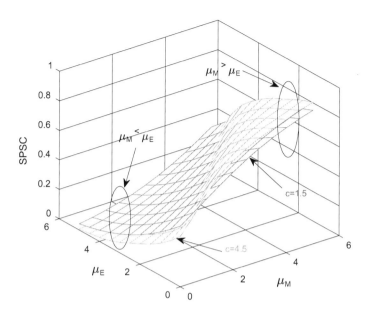

FIGURE 14.11 Behaviour of SPSC as a function of main channel parameter and eavesdropper channel parameter [36].

Also, the CDFs of γ_M and γ_E can be written as

$$F(\gamma_M) = \sum_{p=0}^{\infty} \frac{\xi_{Mp}}{p!} \left(\frac{\Omega_M}{\mu_M (1+\kappa_M)} \right)^{\left(\frac{\mu_M + b + p}{2} \right)} H_{1,3}^{2,1} \left[\frac{\mu_M (1+\kappa_M)}{\Omega_M} \gamma_M \Bigg|_{(b,1)}^{(1,1)} \quad \substack{(1,1) \\ (\mu_M + p,1)} \quad (0,1) \right]$$

(14.47)

$$F(\gamma_E) = \sum_{p=0}^{\infty} \frac{\xi_{Ep}}{p!} \left(\frac{\Omega_E}{\mu_E (1+\kappa_E)} \right)^{\left(\frac{\mu_E + b + p}{2} \right)} H_{1,3}^{2,1} \left[\frac{\mu_E (1+\kappa_E)}{\Omega_E} \gamma_E \Bigg|_{(b,1)}^{(1,1)} \quad \substack{(1,1) \\ (\mu_E + p,1)} \quad (0,1) \right]$$

(14.48)

On substituting Equations 14.45 and 14.47 into Equation 14.15, SOP^L can be evaluated as

$$SOP^L = \sum_{p,t=0}^{\infty} \frac{\xi_{Ep}\xi_{Mp}}{p!\,t!} \left(\frac{\Omega_M}{\mu_M (1+\kappa_M)} \right)^{\left(\frac{\mu_M + b + t}{2} \right)} \left(\frac{\Omega_E}{\mu_E (1+\kappa_E)} \right)^{\left(\frac{\mu_E + b + p}{2} \right)}$$

(14.49)

$$\times H_{3,3}^{2,3} \left[\theta \Bigg|_{(b,1)}^{(1,1)} \quad \substack{(1,1) \quad (1-b,1) \quad (1-\mu_E - p,1) \\ (b,1) \quad (\mu_M + t,1) \quad (0,1)} \right]$$

where $\theta = \dfrac{\mu_M \Omega_E \left(1+\kappa_M\right)}{\mu_E \Omega_M \left(1+\kappa_E\right)}\left(1+\gamma_{th}\right).$

Substituting $\gamma_{th} = 0$ into Equation 14.49, we have obtained SPSC as

$$SPSC = 1 - \left[\begin{array}{l} \displaystyle\sum_{p,t=0}^{\infty} \dfrac{\xi_{Ep}\xi_{Mp}}{p!t!}\left(\dfrac{\Omega_M}{\mu_M\left(1+\kappa_M\right)}\right)^{\left(\frac{\mu_M+b+t}{2}\right)}\left(\dfrac{\Omega_E}{\mu_E\left(1+\kappa_E\right)}\right)^{\left(\frac{\mu_E+b+p}{2}\right)} \\ \times H_{3,3}^{2,3}\left[\theta\left|\begin{array}{ccc}(1,1) & (1-b,1) & (1-\mu_E-p,1) \\ (b,1) & (\mu_M+t,1) & (0,1)\end{array}\right.\right] \end{array}\right] \quad (14.50)$$

In Figures 14.12 and 14.13, the behaviours of SOP^L and SPSC are compared as a function of the main channel parameters κ_M and μ_M, in the case of $\Omega_M > \Omega_E$ with that in case of $\Omega_M < \Omega_E$ for fixed $\{\kappa_E = 2, \mu_E = 1, \gamma_{th} = 5dB, b = 1\}$, respectively. In Figure 14.12, it can be observed that SOP^L decreases with increasing values of $\{\kappa_M, \mu_M\}$ when $\Omega_M > \Omega_E$ and does not vary so much when $\Omega_M < \Omega_E$. Similarly, one can observe that SPSC improves with increasing $\{\kappa_M, \mu_M, \Omega_M\}$. Furthermore, one can also find that a large value of Ω_E yields higher SOP^L and lower SPSC. These results confirmed that the PLS performance ameliorates for the superior Alice–Bob channel link ($\Omega_M > \Omega_E$), whereas the superior Alice–Eve channel link ($\Omega_M < \Omega_E$) worsens the PLS performance.

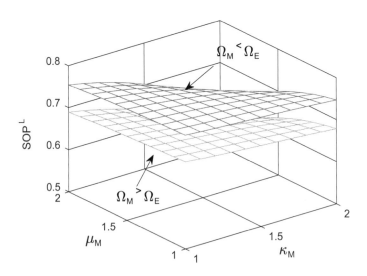

FIGURE 14.12 SOP^L versus with single eavesdropper.

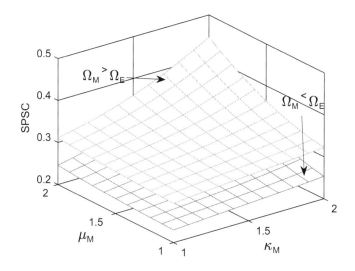

FIGURE 14.13 SPSC versus with single eavesdropper.

14.8 CONCLUSION

Secure communication has become necessary for upcoming wireless technologies such as 5G, IoT, and V2V communication. Traditional cryptographic techniques have failed to provide secure communication for these technologies. In contrast, physical layer security offers many advantages over conventional cryptographic approaches. Physical layer security provides an information-theoretic-based secrecy, which helps researchers to design practical wireless communication systems. In this chapter, we discussed various secure system models proposed by different researchers. We also provided a detailed discussion on different secrecy performance metrics of the physical layer security. We not only provided the theoretical aspects but also provided the mathematical formulation for these secrecy performance metrics. We also discussed different scenarios for eavesdropping, such as active eavesdropping and passive eavesdropping. In particular, the influence of fading on the physical layer secrecy was demonstrated. Various fading scenarios were discussed in this chapter as well.

REFERENCES

1. Zhang J, Duong TQ, Woods R, Marshall A (2017), Securing wireless communications of the Internet of things from the physical layer, an overview, *Entropy*, 19(8), pp. 1–16.
2. Heng S, So-In S, Nguyen TG (2017), Distributed image compression architecture over wireless multimedia sensor networks, *Wireless Communications and Mobile Computing*, pp. 1–21.
3. Nguyen TG, So-In C, Nguyen NG, Phoemphon S (2017), A novel energy-efficient clustering protocol with area coverage awareness for wireless sensor networks, *Peer-to-Peer Networking and Applications*, 10(3), pp. 519–536.

4. Mukherjee A (2015), Physical-layer security in the Internet of things: Sensing and communication confidentiality under resource constraints, *Proceedings of the IEEE*, 103(10), pp. 1747–1761.

5. Naira K, Asmib S, Gopakumar A (2016), Analysis of physical layer security via co-operative communication in Internet of things, *Procedia Technology*, 24, pp. 896–903.

6. Abomhara M, Køien GM (2014), Security and privacy in the Internet of Things: Current status and open issues, *International Conference on Privacy and Security in Mobile Systems (PRISMS)*, Aalborg, pp. 1–8.

7. Granjal J, Monteiro E, Silva JS (2015), Security for the Internet of things: A survey of existing protocols and open research issues, *IEEE Communications Surveys and Tutorials*, 17(3), pp. 1294–1312.

8. Zhou L, Chao H (2011), Multimedia traffic security architecture for the Internet of things, *IEEE Network*, 25(3), pp. 35–40.

9. Jing Q, Vasilakos A, Wan J, Lu J, Qiu D (2014), Security of the Internet of things: Perspectives and challenges, *Wireless Networks*, 20(8), pp. 2481–2501.

10. Zhang K, Liang X, Lu R, Shen X (2014), Sybil attacks and their defenses in the Internet of things, *IEEE Internet of Things Journal*, 1(5), pp. 372–383.

11. Skarmeta AF, Hernández-Ramos JL, Moreno MV (2014), A decentralized approach for security and privacy challenges in the Internet of things, *IEEE World Forum on Internet of Things (WF-IoT)*, Seoul, pp. 67–72.

12. Roman R, Najera P, Lopez J (2011), Securing the Internet of things, *Computer*, 44(9), pp. 51–58.

13. Suo H, Wan J, Zou C, Liu J (2012), Security in the Internet of things: A review, *International Conference on Computer Science and Electronics Engineering*, Hangzhou, pp. 648–651.

14. Mahajan P, Sachdeva A (2013), A study of encryption algorithms AES, DES and RSA for security, *Global Journal of Computer Science and Technology*, 13(15), pp. 15–22.

15. Shannon E (1949), Communication theory of secrecy systems, *The Bell System Technical Journal*, 28(4), pp. 656–715.

16. Wyner AD (1975), The wire-tap channel, *The Bell System Technical Journal*, 54(8), pp. 1355–1387.

17. Csiszár Körner J (1978), Broadcast channels with confidential messages, *IEEE Transactions on Information Theory*, 24(3), pp. 339–348.

18. Cheong SLY, Hellman M (1978), The Gaussian wire-tap channel, *IEEE Transactions on Information Theory*, 24(4), pp. 451–456.

19. Hero A (2003), Secure space-time communication, *IEEE Transactions on Information Theory*, 49(12), pp. 3235–3249.

20. W. of Science Thomson Reuters, Web of science [v.5.21]-all databases. https://webofknowledge.com. [Online; accessed 04 July 2019].

21. Bhargav N, Cotton SL, Simmons DE (2016), Secrecy capacity analysis over κ-μ fading channels: Theory and applications, *IEEE Transactions on Communications*, 64(7), pp. 3011–3024.

22. Romero-Jerez M, Lopez-Martinez FJ (2017), A new framework for the performance analysis of wireless communications under Hoyt (Nakagami-q) fading, *IEEE Transactions on Information Theory*, 63(3), pp. 1693–1702.

23. Jameel F, Wyne S, Krikidis I (2017), Secrecy outage for wireless sensor networks, *IEEE Communication Letters*, 21(7), pp. 1565–1568.

24. Lei H, Jhang H, Ansari IS, Gao C, Guo Y, Pan G, Qaraqe KA (2016), Performance analysis of physical layer security over generalized-K fading channels using a mixture gamma distribution, *IEEE Communications Letters*, 20(2), pp. 408–411.

25. Mathur A, Ai Y, Bhatnagar MR, Cheffena M, Ohtsuki T (2018), On physical layer security of α-η-κ-μ fading channels, *IEEE Communications Letters*, 22(10), pp. 2168–2171.

26. Lei H, Ansari IS, Gao C, Guo Y, Pan G, Qaraqe KA (2016), Physical-layer security over generalised-*K* fading channels, *IET Communications*, 10(16), pp. 2233–2237.

27. Lei H, Ansari IS, Zhang H, Qaraqe KA, Pan G (2016), Security performance analysis of SIMO generalized-K fading channels using a mixture gamma distribution, *IEEE 84th Vehicular Technology Conference (VTC-Fall)*, Montreal, QC, pp. 1–6.

28. Alexandropoulos GC, Peppas KP (2018), Secrecy outage analysis over correlated composite Nakagami-*m* /gamma fading channels, *IEEE Communications Letters*, 22(1), pp. 77–80.

29. Sun J, Li X, Huang M, Ding Y, Jin J, Pan G (2018), Performance analysis of physical layer security over κ-μ shadowed fading channels, *IET Communications*, 12(8), pp. 970–975.

30. Srinivasan M, Kalyani S (2018), Secrecy capacity of κ-μ shadowed fading channels, *IEEE Communications Letters*, 22(8), pp. 1728–1731.

31. Bassily R, Ekrem E, He X, Tekin E (2013), Cooperative security at the physical layer: A summary of recent advances, *IEEE Signal Processing Magazine*, 30(5), pp. 16–28.

32. Bloch M, Barros J (2011), *Physical-Layer Security: From Information Theory to Security Engineering*, 1st edn., Cambridge University Press, Cambridge.

33. Shiu YS, Chang SY, Wu HC, Huang SCH, Chen HH (2011), Physical layer security in wireless networks: A tutorial, *IEEE Wireless Communications*, 18(2), pp. 66–74.

34. Wang HM, Zheng TX (2016), *Physical Layer Security in Random Cellular Networks*, 1st edn., Springer.

35. Gopala PK, Lai L, Gamal HE (2008), On the secrecy capacity of fading channels, *IEEE Transactions on Information Theory*, 54(10), pp. 4687–4698.

36. Singh R, Rawat M (2019), Performance analysis of physical layer security over Weibull/lognormal composite fading channel with MRC reception, *AEU – International Journal of Electronics and Communications*, doi:10.1016/j.aeue.2019.152849.

37. Prabhu VU, Rodrigues MRD (2011), On wireless channels with M-antenna eavesdroppers: Characterization of the outage probability and outage secrecy capacity, *IEEE Transactions on Information Forensics and Security*, 6(3), pp. 853–860.

38. Marina N, Hjorungnes A (2010), Characterization of the secrecy region of a single relay cooperative system, *IEEE Wireless Communication and Networking Conference*, Sydney, NSW, Australia, pp. 1–6.

39. Chang N, Chae C, Ha J, Kang J (2012), Secrecy rate for MISO Rayleigh fading channels with relative distance of eavesdropper, *IEEE Communications Letters*, 16(9), pp. 1408–1411.

40. Li W, Ghogho M, Chen B, Xiong C (2012), Secure communication via sending artificial noise by the receiver: Outage secrecy capacity/region analysis, *IEEE Communications Letters*, 16(10), pp. 1628–1631.

41. He X, Yener A (2011), Gaussian two-way wiretap channel with an arbitrarily varying eavesdropper, *IEEE Global Communications Conference Workshops (GLOBECOM Workshops)*, Houston, TX, USA, pp.845–858.

42. Koyluoglu OO, Gamal HE, Lai L, Poor HV (2008), On the secure degrees of freedom in the *K*-user Gaussian interference channel, *IEEE International Symposium on Information Theory*, Toronto, ON, Canada, pp. 384–388.

43. He B, Zhou X, Abhayapala TD (2013), Wireless physical layer security with imperfect channel state information: A survey [Online]. http://arxiv.org/abs/1307.4146.

44. Simon MK, Alouini MS (2005), *Digital Communication over Fading Channels*, 2nd edn., Wiley, New York.

45. Roberts C, Vandenplas C (2017), Estimating components of mean squared error to evaluate the benefits of mixing data collection modes, *Journal of Official Statistics*, 33(2), pp. 303–334.

46. Haenggi M, Andrews JG, Baccelli F, Dousse O, Franceschetti M (2009), Stochastic geometry and random graphs for the analysis and design of wireless networks, *IEEE Journal on Selected Areas in Communications*, 27(7), pp. 1029–1046.

47. Franceschetti M, Meester R (2007), *Random Networks for Communication: From Statistical Physics to Information Systems*, 1st edn., Cambridge University Press.

48. Peppas KP (2012), A new formula for the average bit error probability of dual-hop amplify-and-forward relaying systems over generalized shadowed fading channels, *IEEE Wireless Communications Letters*, 1(2), pp. 85–88.

49. Ng WK, Lo ES, Schober R (2012), Energy-efficient resource allocation for secure OFDMA systems, *IEEE Transactions on Vehicular Technology*, 61(6), pp. 2572–2585.

50. Chen X, Lei L (2013), Energy-efficient optimization for physical layer security in multiantenna downlink networks with QoS guarantee, *IEEE Communications Letters*, 17(4), pp. 637–640.

51. Barros J, Rodrigues MRD (2006), Secrecy capacity of wireless channels, *IEEE International Symposium Information Theory*, Seattle, WA, pp. 356–360.

52. Wolfram Research, http://functions.wolfram.com [accessed 20 June 2019].

53. Gradshteyn S, Ryzhik IM (2007), *Table of Integrals, Series and Products*, 7th edn., Academic Press, Burlington, VT.

54. Yacoub MD (2016), The α-η-κ-μ fading model, *IEEE Transactions on Antennas and Propagation*, 64(8), pp. 3597–3610.

55. Bhargav N, Silva CRND, Cotton SL, Sofotasios PC, Yoo SK, Yacoub MD (2018), On shadowing the κ-μ fading model, submitted to, *IEEE Transactions on Wireless Communications*.

56. Lingwen Z, Jiayi Z, Liu L (2013), Average channel capacity of composite κ-μ/gamma fading channels, *China Communications*, 10(6), pp. 28–34.

57. Kilbas A (2004), *H-Transforms: Theory and Applications*, 1st edn., CRC Press.

15 Tenable Irrigation System with Internet of Things

Upendra Kumar, Smita Pallavi, and Pranjal Pandey

CONTENTS

15.1 INTRODUCTION

India is the second largest country by population and 70% of its population lives in rural areas. So agriculture plays an important role for survival as well as development of country. In India, even in this 21st century, irrigation depends on the monsoon, which is an insufficient source of water. There is no any scientific method for irrigation. In agriculture, depending upon the soil type and crop type, water is required. In agriculture, irrigation is one of the most important issues faced by farmers especially in the areas where the groundwater level is very low or the field is dry. There are few automated systems for irrigation with pre-defined sets of instructions.

Automated systems sometimes start the water pump even in rain. This leads to water wastage. This chapter focuses on reducing the water wastage and minimizing the manual labour on agricultural land. A soil sensor will collect the moisture content at regular time intervals and the water pump will be started accordingly. When the soil absorbs the required amount of water, the water pump will automatically switch off using the Internet of Things (IoT). The data regarding the moisture content at a particular time and pump status will be sent to the cloud for the future purpose of data analysis. The data of soil status can also be monitored remotely using android apps. This ecosystem would enable the cultivators to constantly monitor the wetness level in the soil, while controlling the water supply tenuously over the web. When the soil moisture reaches below a certain level, the controlled sprinklers would turn on automatically, thus ensuring the agricultural justified irrigation using the Internet of Things. Figure 15.1 renders the network of services in different genres provided by IoT connectivity.

FIGURE 15.1   The network of services in different genres provided by IoT connectivity.

15.1.1 Motivation

The continuous demand for food leads to the discussion of the problem of agriculture. This is one of the important factors in the Indian economy. The population of India is increasing day by day and food is a basic need for everyone. After going through several articles regarding issues of Indian agricultural, it has been found that the water supply is the one that needs to be solved. Due to the inefficient use of groundwater for irrigation purpose, the water level has decreased exponentially which is a matter of concern. So, being engineers it's our duty to try to solve the problems in our surroundings using science and technology. So, in this chapter irrigation is redefined as drop only required amount of water to the soil or to the crop accordingly. The main objectives of this chapter are to help the farmers and reduce their efforts by reducing water wastage and minimizing the manual labour. This chapter is written in view of perennial plant irrigation land and gardening land.

15.2 BRIEF LITERATURE REVIEW PERTAINING TO IOT

Indeed, it was in 1982 that a network of sensible devices was conceptualized with replacement of the Coke machine at Carnegie Altruist University with a web-connected device to report the inventory and chillness of the beverages. Papers in omnipresent computing in 1991, describing the enhancement of nodes from home appliances have continuously led to the Internet of Things becoming the backbone for modern precision agriculture monitoring. IoT is the collective network of non-standard devices that keeps an area unit interconnected via the web facilitating the exchange of data between devices, thereby paving the way for opportunities for the integration between the physical world with computer-based systems. It results in economical work, high productivity, reduced human effort, and reduces issues of human error to nearly zero. The Internet of Things has unleashed numerous productive methods for agricultural tasks like cultivation of soil and maintenance of farm stocks with the use of cheap, easy-to-install sensors and an abundance of insightful data that they offer. It is with the presence of the Internet of Things in agriculture that have built up this prolific security to deliver 24/7 visibility of the soil and crop health with water level maintenance and energy consumption levels from remote locations too.

There are many issues directly associated with agriculture like grid, crop, soil monitoring, irrigation, pesticide, fertilization applications, and cattle farming [1]. Information and communication technology can be a practical tool to show the rural farming community how to replace some of the traditional techniques with easy-to-use technical applications. IoT has been harnessed for this purpose to assist in pre-weather detection, moisture levels of the soil, air flow and pressure, temperature monitoring and to provide solutions to maintain the equilibrium. The long geographical distances between the open market and the actual farm also leads farmers put in extra efforts to raise production and the crop managers must be rewarded for their toil. Jaguey et al. [2] proposed an agriculture based on the CPS design technology that includes three layers: the physical layer, the network layer, and the decision layer. Further, Nisha and Megala proposed [3] the concept of agricultural automation using an embedded system to devise a low-cost system with the aid of WEB and GSM technologies. Soil moisture sensors and temperature sensors are embedded together to detect the amount of water present in the agricultural land. All the parameters of temperature, water level, and the soil moisture content can be monitored on the user's web page through microcontrollers and the recorded information is sent by SMS using General Packet Radio Service (GPRS) for remote locations. Moreover, TonKe [4] presented smart agriculture based on cloud computing and IoT in which an agricultural information cloud is constructed on the basis of cloud computing and subsequently implemented using radio-frequency identification (RFID) and IoT. Over and above, after reviewing substantial state-of-the-art research work, we proceed to build our agro-justice model with the following possible outcomes:

- To provide an autonomous irrigation management system with real-time monitoring of the soil conditions.
- Continuous monitoring of the farms without actual presence in adverse weather conditions.
- Proper management of resource allocation such as water for irrigation is ensured to aid in the high production at farms.
- Real-time analysis at the right place and at the right time gives agricultural justice in a highly engaged collaborative environment.

Chaparro et al. [5] proposed prediction of the extent of wildfires using remotely sensed soil moisture and temperature trends. Further, in Gutiérrez et al. [6], an automated irrigation system using a wireless sensor network and GPRS module is given. Furthermore, Junjin et al. [7] propose the design and research on intelligent irrigation system. Udaykumar and Kumar [8] developed a wireless sensor network (WSN)-based system for precision agriculture. Moreover, Sales and Arsenio [9] proposed a wireless sensor and actuator system for smart irrigation on the cloud. A precise irrigation system based on a wireless sensor network is also given in Khriji et al. [10]. In Anil et al. [11] an automated irrigation system is proposed for the home garden.

15.3 DEVICES IMPLEMENTING IOT

IoT devices can be categorized as wearable ones and embedded ones controlled by the microcontrollers/microprocessors. The Tizen SDK, the most popular in wearable devices, comes ported with a wearable emulator, thus developing wearable solutions. Alternatively, an embedded system performs pre-defined tasks of specific requirements on tools like Arduino, Raspberry Pi, Intel Edision, and Intel Galileo to name a few.

15.3.1 CLOUD PLATFORM

Services like the online payment gateway can now be easily integrated with the hardware platform for an embedded board inside a vending machine. They can be further utilized to detect the location and payment services, all conglomerated together using the Internet of Things. A real-world scenario could be to make a device discoverable over the web, then assign a fixed IP address and maintain the router.

Yaler, Axeda, and Open IoT are the trending communication tools and provide infrastructural support for web services. Google integrates location services with its cloud and hence upgradation on social platforms and personalized searches are much easier to put in action. Thus, we can mark that cloud APIs have a great potential in IoT in all levels of architecture starting from firmware to hardware to more top-level architecture.

15.3.2 IMPLEMENTATION USING IoT

A system implemented using IoT can be justified in the sense of things that are connected with a wireless network. Here, the soil moisture sensor is responsible for the actual moisture content of the soil. The soil sensor works on the principle of a voltage divider. This moisture content value is sent to the microcontroller, i.e. the brain of the system, which is responsible for turning off the motor according to the soil situation. So, here the different components are connected through a wired or wireless network. This is the Internet of Things.

15.4 IOT SECURITY ISSUES

Security concerns in IoT are of paramount importance. Web APIs or particularly APIs are considered as the vital components to connect these devices to the Internet. IoT devices can be tackled by handheld devices as well as with modern websites. However, IoT devices must consider the following security issues:

- Privacy
- Issues pertinent to hardware
- Issues related to data encryption
- Issues pertaining to web interfacing
- Issues pertaining to lack of network awareness
- Issues of insecure software
- Issues related to side channel attacks
- Use of poorly functioning IoT devices
- Issues regarding a corporation's protection

It is pertinent to mention that important personal information is collected by devices and some devices also transfer this information across the network. Therefore, some encryption approach must be implemented for the purpose of security. Further, IoT systems must be equipped to tackle security issues related to hardware. Moreover, use of unencrypted network services should be avoided. Furthermore, simple default passwords and weak session management are also of the prime concern while considering the security of the IoT system because these factors can render high vulnerabilities. Besides this, there are number of organizations which are not properly taking care of the proper configuration of IoT devices. Insecure software and side channel attacks are also responsible to render the IoT services vulnerable.

15.5 HARDWARE SUPPORT TO THE AGRI-IOT MODEL

There are many hardware components required in an IoT-based agricultural (Agri-IoT) model. Some components of paramount importance are briefly described next.

15.5.1 ARDUINO

Arduino is the latest open source physical computing platform with an integrated development environment (IDE). Arduino was incepted at the Ivrea Interaction Design Institute, and since then the Arduino board has evolved in manifold to adapt to ever-changing needs and challenges. A worldwide community of Arduino users has exploited this open source platform and has made contributions in fields of artistry, medical, security, accessibility, robotics, and other areas. Arduino combines the Atmega microcontroller family with an inbuilt boot loader for plug and play embedded programming. Arduino software comes with an IDE and a serial communication window to help burning the program and get the serial data onto Arduino. Figure 15.2 shows the Arduino board pin connectivity. Table 15.1 shows the Arduino Uno technical specifications.

15.5.2 ARDUINO UNO

The Uno microcontroller board is based on the ATmega328P, which operates at 5 V. It also has pins to enable interrupts, receive and transmit data, two-wire interface (TWI) communication, and LEDs.

15.5.3 WI-FI NETWORK SOLUTIONS (ESP8266)

The ESP8266 series comprises of boards fabricated to provide a self-contained Wi-Fi network solution. The ESP8266 highly integrated chip is also a solution to

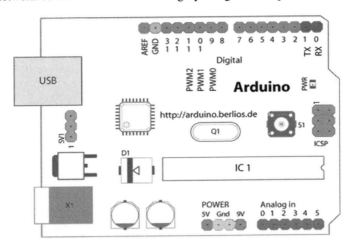

FIGURE 15.2 The Arduino board pin connectivity.

TABLE 15.1
Arduino Uno Technical Specifications

Microcontroller	AT mega 328 P
Operating voltage	7–12 V
Recommended input voltage	6–20 V
Limit of input voltage	14 (of which 6 provide PWM Output)
Digital I/O pins	6
PWM digital I/O pins	6
Analogue input pins	
DC current per I/O pin	20 mA
DC current for 3.3 V pin	50 mA
Flash memory	32 KB (AT mega 328 P) of which 0.5 KB used for boot loader
SRAM	2 KB (AT mega 328 P)
EEPROM	1 KB (AT mega 328 P)
Clock speed	16 MHz
Length	68.6 mm
Width	53.4 mm
Weight	25 gm

minimize space occupancy, save energy, and eliminate cellular/Bluetooth and other interference. The various pin connections to embed ESP8266 to Arduino Uno are as follows. Vcc and CH_PD pin of ESP8266-01 should be connected to 3.3 V of Arduino. The ground pin of ESP8266 should be connected to the ground pin of Arduino Uno. Transfer and receiver pins are applicable for serial communication between Arduino and ESP8266-01. So, the TX pin of ESP8266 is connected to the RX pin of Arduino and vice versa.

15.5.3.1 Characteristics of ESP8266
- Embedded with 32-bit microcontroller
- Working power range 3.0 to 3.6 volts
- System on a Chip (SoC) with capabilities for 2.4 GHz Wi-Fi (802.11b/g/n IEEE standards)
- Built in TCP/IP stack
- SPI and UART communication enabled
- Leading platform for Internet of Things
- Low cost

Dual functionality
a. Better microcontrollers with standalone IDE
c. Self-contained to host entire application
d. Wi-Fi adapter for other microcontrollers

   e. GPIOs to interface with sensors in advanced versions
   f. Multiple versions

ESP8266 finds its varied application domains as smart power plugs, in home automation, industrial wireless control, monitoring of houses and infants, for network
cameras, improvising wireless location-aware devices and positioning system signals
and similar realms of sensor networks and wearable electronic devices. Figure 15.3
shows the pin and block diagram of ESP 8266 and Figure 15.4 displays the pin connectivity and correlation of Arduino Uno with ESP8266. Further, Figure 15.5 illustrates the multiple version of ESP8266.

15.5.3.2   The AT Commands

Communication in the default mode of ESP8266 with serial configuration is done by
AT commands. These instructions are based on the Hayes command set as indexed
in Table 15.2.

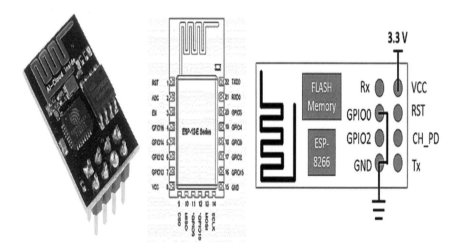

FIGURE 15.3   The pin and block diagram of ESP8266.

ESP8266	Arduino Uno
Vcc	3.3 V
CH_PD	
GND	GND
RX	TX
TX	RX

FIGURE 15.4   The pin connectivity and correlation of Arduino Uno with ESP8266.

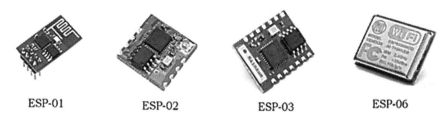

ESP-01   ESP-02   ESP-03   ESP-06

FIGURE 15.5   Multiple versions of ESP8266.

15.5.4 Hardware Support of Soil Moisture Sensor

We have implemented the soil moisture sensor as we observe that reflected micro-wave radiation is used for remote sensing in hydrology and agriculture. The moisture sensor is a low-tech sensor which records the amount of moisture present in the soil surrounding it. The YL-69 sensor has a middleware circuit which needs to be wired to be the two pins on the YL-38 bridge and allows to record outputs as an analogue readout of the resistance between the sensor's probes and the second is a digital output. The next hardware to be implemented is a submersible water pump. The soil moisture sensor is used to measure the loss of moisture over time due to evaporation. The submersible pump has a sealed motor and the whole assembly is submerged in the water tank from where fluid is to be pumped. Figure 15.6 shows the soil moisture sensor YL-69 YL-38 and Figure 15.7 displays the submersible water pump.

15.5.5 Software Used in Agri-IoT Implementation

This section will describe the pertinent software for Agri-IoT implementation. Some important software for this purpose are given next.

15.5.5.1 Arduino IDE

The Arduino Software or Arduino IDE comprises of a text console, a text editor for writing code, a toolbar with buttons for common functions, a message area, and a series of menus. The communication of Arduino and Genuino is established to facilitate the uploading of programmes and their interactions. In this application, a programme is written using Arduino Software (IDE). Further, .ino extension files are produced from Arduino Software and they are edited using text editors. Cutting, pasting, searching, and replacing are the operations that the editor features. Next, feedbacks are displayed in the message area while saving and exporting. This area also displays errors. In addition, complete error messages and other information along with text output are displayed in the console. Configured board and serial ports are displayed in the bottom right corner of the window. Verifying and uploading programmes, save, open, create sketches, and open the serial monitor functions are provided in the toolbar button.

- *Verify:* Checks code and verifies errors for compilation.
- *Open:* Displays a menu of all the sketches in your sketchbook. *Save*: Saves your sketch.
- *Serial Monitor*: Opens the serial monitor.
- *New*: Creates a new sketch.

TABLE 15.2
Index of All Known AT Commands

Function	AT Command	Response
Working	AT	OK
Restart	AT+RST	OK [System Ready, Vendor:www.ai-thinker.com]
Firmware version	AT+GMR	AT+GMR 0018000902 OK
List Access Points	AT+CWLAP	AT+CWLAP +CWLAP:(4,"RochefortSurLac",-38,"70:62:b8:6f:6d:58",1) +CWLAP:(4,"LiliPad2.4",-83,"f8:7b:8c:1e:7c:6d",1) OK
Join Access Point	AT+CWJAP? AT+CWJAP="SSID","Password"	Query AT+CWJAP? +CWJAP:"RochefortSurLac" OK
Quit Access Point	AT+CWQAP=? AT+CWQAP	Query OK
Get IP Address	AT+CIFSR	AT+CIFSR 192.168.0.105 OK
Set Parameters of Access Point	AT+ CWSAP? AT+ CWSAP= <ssid>,<pwd>,<chl>, <ecn>	Query ssid, pwd chl = channel, ecn = encryption
WiFi Mode	AT+CWMODE? AT+CWMODE=1 AT+CWMODE=2 AT+CWMODE=3	Query STA AP BOTH
Set up TCP or UDP connection	AT+CIPSTART=? (CIPMUX=0) AT+CIPSTART = <type>,<addr>,<port> (CIPMUX=1) AT+CIPSTART= <id><type>,<addr>, <port>	Query id = 0-4, type = TCP/UDP, addr = IP address, port= port
TCP/UDP Connections	AT+ CIPMUX? AT+ CIPMUX=0 AT+ CIPMUX=1	Query Single Multiple
Check join devices' IP	AT+CWLIF	
TCP/IP Connection Status	AT+CIPSTATUS	AT+CIPSTATUS? no this fun
Send TCP/IP data	(CIPMUX=0) AT+CIPSEND=<length>; (CIPMUX=1) AT+CIPSEND= <id>,<length>	
Close TCP / UDP connection	AT+CIPCLOSE=<id> or AT+CIPCLOSE	
Set as server	AT+ CIPSERVER= <mode>[,<port>]	mode 0 to close server mode; mode 1 to open; port = port
Set the server timeout	AT+CIPSTO? AT+CIPSTO=<time>	Query <time>0~28800 in seconds
Baud Rate*	AT+CIOBAUD? Supported: 9600, 19200, 38400, 74880, 115200, 230400, 460800, 921600	Query AT+CIOBAUD? +CIOBAUD:9600 OK
Check IP address	AT+CIFSR	AT+CIFSR 192.168.0.106 OK
Firmware Upgrade (from Cloud)	AT+CIUPDATE	1. +CIPUPDATE:1 found server 2. +CIPUPDATE:2 connect server

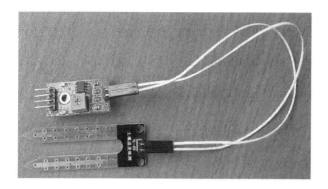

FIGURE 15.6　Soil Moisture Sensor YL-69 YL-38.

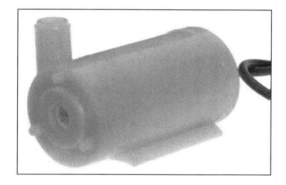

FIGURE 15.7　Submersible water pump.

There are five menus which display additional commands such as file, edit, sketch, tools, and help. The menus are context sensitive.

15.5.5.2　Virtuino Android App

Virtuino is a robust project visualization application. It can control more than one Arduino board at a time over Bluetooth, Internet, Wi-Fi, and SMS. It can create visual interfaces for LEDs, charts, switches, counters, and analogue instruments. The Virtuino App is responsible to show all the data on the mobile phone. It rings an alarm when required and sends an SMS to the registered mobile number when triggered. Further, Figure 15.8 shows the Virtuino App, and the code definition is given in Figure 15.9. Furthermore, the code window of Virtuino App is shown in Figure 15.10. In addition, the Boolean connectivity App Code window is given in Figure 15.11, and Figure 15.12 renders the Setup App Code window.

15.6　WORKING PRINCIPLE OF SMART AGRI-IOT

The working principle of smart Agri-IoT is described in following sections.

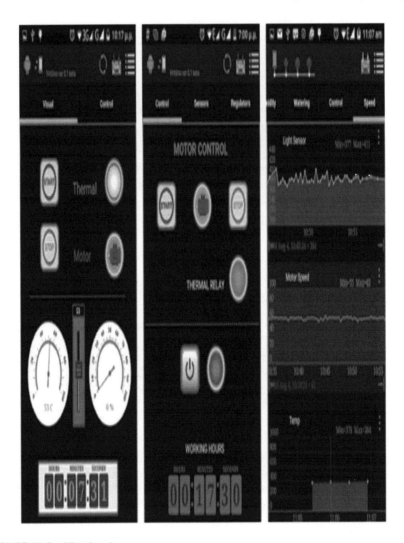

FIGURE 15.8 Virtuino App.

15.6.1 INITIAL SETUPS IN ARDUINO IDE SOFTWARE

Arduino IDE is used for Arduino programming, and code will be uploaded to the Arduino after successful compilation. Different steps of this phenomenon are in the following:

Step 1:   Install the Arduino IDE.
Step 2:   File->Preferences->Additional Boards Manager URLs:
          http://arduino.esp8266.com/stable/package_esp8266com_index.json
Step 3:   Tools->Boards->Boards Manager->
          Download the "esp8266 version 2.2.0"
Step 4:   Tools->Upload Speed->115200
          Port->choose preferred COM ports.

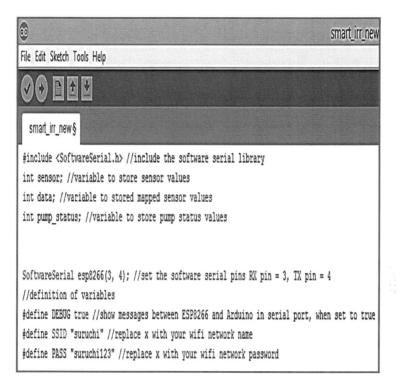

FIGURE 15.9 Code Definition.

15.6.2 INTERFACING ESP8266-01 WITH ARDUINO

Step 1: First upload BareMinimum code to Arduino Uno board.

Step 2: Connect the pins of Arduino Uno to ESP8266-12 pins. Here, the Programme code is directly uploaded into ESP8266 module, so Arduino board is used as a Flash Burner.

Step 3: Reset the ESP8266 and upload the programme code, while connecting to GND.

Step 4: Disconnect the GPIO 0 from GND after upload is done.

Step 5: Debugging Techniques.

Step 6: Connect TX pin of ESP8266 with TX pin of Arduino and Connect RX pin of ESP8266 with RX pin of Arduino.

Step 7: Select the correct COM port.

Step 8: Open serial monitor and set baud rate 115200 and both NL &CR. Type "AT" command, if response = "OK". It is working fine.

15.6.3 WORKING CODE

To achieve the goal, five functions have been written.

1. *sendAT()*
2. *connectwifi()*

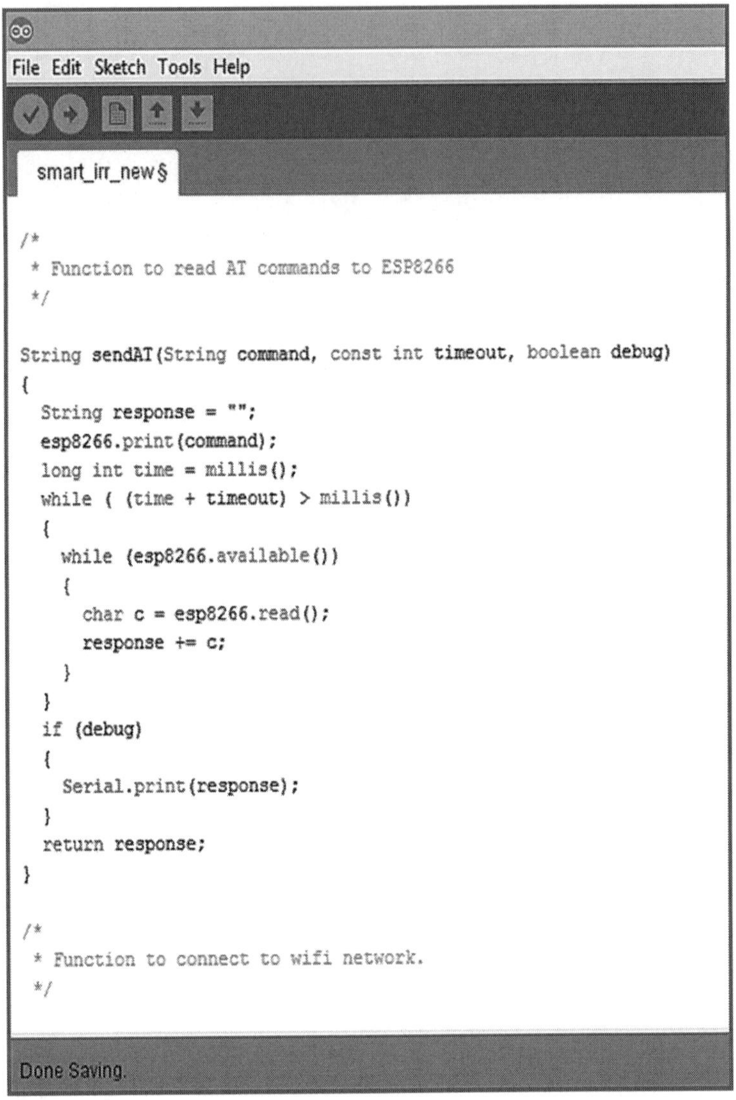

```
File  Edit  Sketch  Tools  Help

smart_irr_new §

/*
 * Function to read AT commands to ESP8266
 */

String sendAT(String command, const int timeout, boolean debug)
{
  String response = "";
  esp8266.print(command);
  long int time = millis();
  while ( (time + timeout) > millis())
  {
    while (esp8266.available())
    {
      char c = esp8266.read();
      response += c;
    }
  }
  if (debug)
  {
    Serial.print(response);
  }
  return response;
}

/*
 * Function to connect to wifi network.
 */
```

Done Saving.

FIGURE 15.10   Virtuino App Code window.

3. *setup()*
4. *loop()*
5. *updateTS()*

15.6.3.1   sendAT

This function is written for the purpose of sending the AT commands to the ESP8266. It has three arguments: String, integer, and Boolean. AT commands are written in String, and that is sent to the ESP module using the command esp8266.print (command), as esp8266 is the instance of the serial software and the command contains

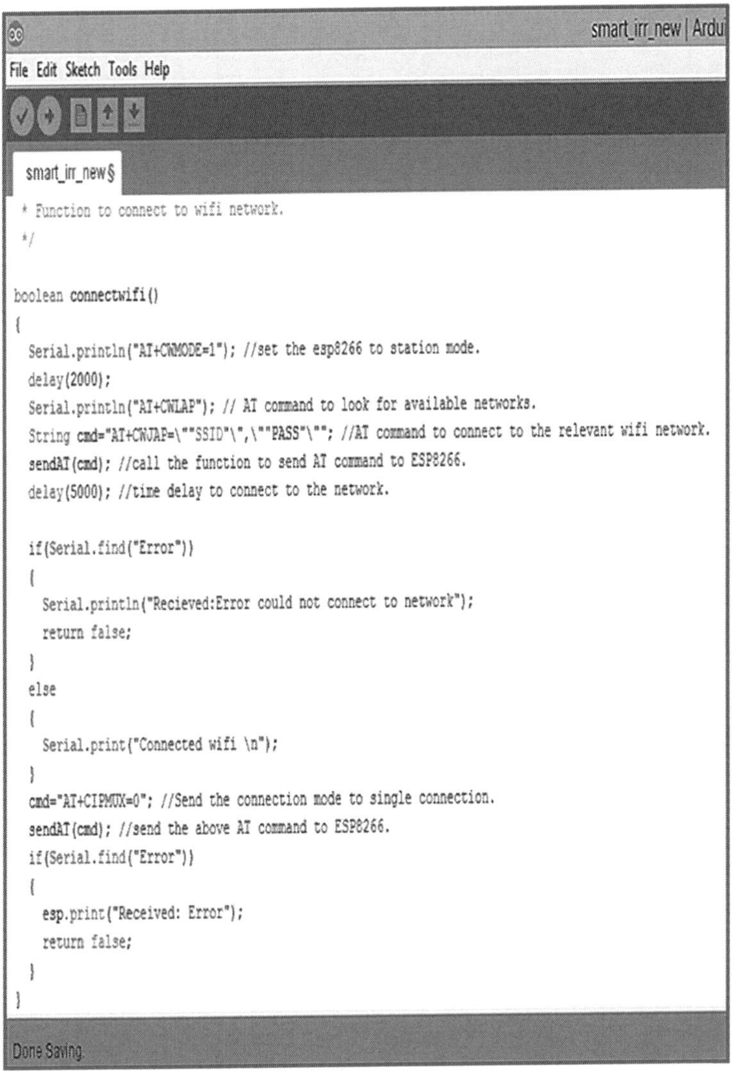

```
                                                              smart_irr_new | Ardu
File Edit Sketch Tools Help

  smart_irr_new§

 * Function to connect to wifi network.
 */

boolean connectwifi()
{
  Serial.println("AT+CWMODE=1"); //set the esp8266 to station mode.
  delay(2000);
  Serial.println("AT+CWLAP"); // AT command to look for available networks.
  String cmd="AT+CWJAP=\""SSID"\",\""PASS"\""; //AT command to connect to the relevant wifi network.
  sendAT(cmd); //call the function to send AT command to ESP8266.
  delay(5000); //time delay to connect to the network.

  if(Serial.find("Error"))
  {
    Serial.println("Recieved:Error could not connect to network");
    return false;
  }
  else
  {
    Serial.print("Connected wifi \n");
  }
  cmd="AT+CIPMUX=0"; //Send the connection mode to single connection.
  sendAT(cmd); //send the above AT command to ESP8266.
  if(Serial.find("Error"))
  {
    esp.print("Received: Error");
    return false;
  }
}

Done Saving
```

FIGURE 15.11   Boolean connectivity App Code window.

the AT commands in String form. Integer denotes the timeout time, that is after this amount of time, if not responded debug=false and try again. Table 15.2 renders index of all known AT commands.

15.6.3.2 connectwifi()

This function is written for connecting the esp8266 to the hotspot available. Here, the SSID= "suruchi" and PASS ="suruchi123" is already given to the program. So, AT+WJAP="SSID","PASS" is the command to connect to the Internet. Thus, using send AT() function(above) this command has been sent to the esp8266. If any error occurred, it will be handled in the described manner as written in the following snapshot.

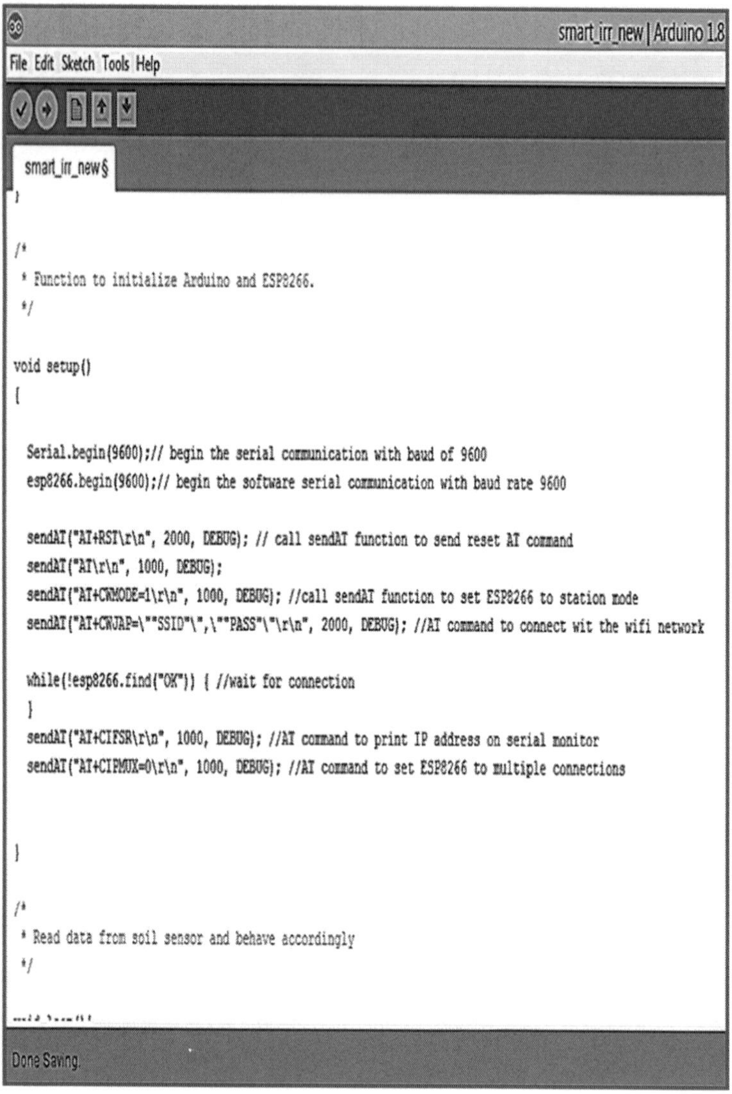

```
                                                    smart_irr_new | Arduino 1.8
File  Edit  Sketch  Tools  Help

  smart_irr_new §
}

/*
 * Function to initialize Arduino and ESP8266.
 */

void setup()
{

  Serial.begin(9600);// begin the serial communication with baud of 9600
  esp8266.begin(9600);// begin the software serial communication with baud rate 9600

  sendAT("AT+RST\r\n", 2000, DEBUG); // call sendAT function to send reset AT command
  sendAT("AT\r\n", 1000, DEBUG);
  sendAT("AT+CWMODE=1\r\n", 1000, DEBUG); //call sendAT function to set ESP8266 to station mode
  sendAT("AT+CWJAP=\""SSID"\",\""PASS"\"\r\n", 2000, DEBUG); //AT command to connect wit the wifi network

  while(!esp8266.find("OK")) { //wait for connection
  }
  sendAT("AT+CIFSR\r\n", 1000, DEBUG); //AT command to print IP address on serial monitor
  sendAT("AT+CIPMUX=0\r\n", 1000, DEBUG); //AT command to set ESP8266 to multiple connections



}

/*
 * Read data from soil sensor and behave accordingly
 */

void loop(){

Done Saving.
```

FIGURE 15.12   Setup App Code window.

15.6.3.3   setup()

This is the inbuilt function of the Arduino programming language. This is the initialization of the whole programme. First, this setup() section starts executing in this code. Comments are well written to understand the setup section.

15.6.3.4   loop()

This function is the soul of the whole code which runs repeatedly. And it behaves in the way we want our code. Here, analogRead() is used to read the analogue value of the soil sensor module. It is connected with the A0 pin. So, analogRead(A0) is stored in the sensor variable which is the actual analogue value from soil sensor module. Now, the soil sensor module works on the principle of voltage divider, i.e. the lower the voltage, the lower the resistance and the higher the concentration of ion, that is more moisture content. So, data=map(sensor,0,1023,100,0); is responsible for mapping the data to its actual moisture content. Serial.print ("Soil Moisture: "); will print the moisture content to the serial monitor .if(data< "value") here value is the required or ideal moisture content for a particular type of gardening or for a particular soil type. It is determined by the agricultural scientist. For example, for the wet type of soil 30 is the required moisture content. So, the pump will be directed to start once the moisture content is lower than the required moisture content. Further, the motor enable pin is connected with pin 8 of Arduino. The code module of this purpose is as follows:

```
if(data<value) //check if sensor value is less than the
ideal "value " for the particular soil
{
digitalWrite(8,HIGH); //switch on the water pump
    pump_status=100; //update pump status variable value to
    100
}
```

This programme module will enable the water pump on. Since, the pin is made HIGH when the moisture content is below the required limit, and pump status is another variable used to store the status of the pump whether the pump is on or off. The pump status =100 implies that the pump is on, otherwise it is off. In addition, the value of moisture and the pump status is stored for every interval of 10 seconds on the cloud using updateTS() function which will be discussed further. The programme module for this task is

```
else
{
    digitalWrite(8,LOW); //switch off the water pump
    pump_status=0; //update pump status variable value to 0
}
```

Since the loop() function is executing continuously so the if-else statement is also checked continuously. Once the moisture content is above the required point, the pump will be turned off using digitalWrite(8,LOW), and the pump status is again 0.

All the data on the cloud can be further used for data mining.

15.6.3.5   updateTS()

This function is responsible for updating the ThingSpeak cloud with the sensor data and pump status value. The following command is used to start a TCP connection to the ThingSpeak free cloud using its API (api.thingspeak.com) through port number 80 to a channel named srchsmn on ThingSpeak.

sendAT("AT+CIPSTART=\"TCP\",\"api.thingspeak.com\",80\r\n", 1000, DEBUG);

15.7   EXPERIMENTAL WORK

Different components required for the sake of experimentation are listed in Table 15.3 and a description of components is given in Table 15.4. The Wi-Fi module is connected with the Internet by an Internet service provider like mobile hotspot or a Wi-Fi router. The soil sensor reads moisture content and sends the data to the Arduino. Further, Arduino changes the sensor reading into the moisture percentage.

TABLE 15.3
List of Components

Components	Quantity
Arduino Uno	1
Arduino Uno Cable	1
Arduino Uno Barrel Connector	1
ESP8266	1
Water Pump	1
YL-38 Soil Moisture Sensor Module	1
YL-69 Soil Moisture Probe	1
Motor Driver PCB	1
Male to Female Connector	8
Female to Female Connector	6
Male to Male Connector	6

TABLE 15.4
Components Description

Component 1	Pin	Pin Description		Pin	Pin Description	Component 2
Arduino Uno	A0	Analog Read	→	A0	Analog Data	YL – 38 Soil Moisture Sensor Module
	5V	Vcc		Vcc	·	
	GND	Ground		GND	Ground	
Arduino Uno	8	Digital I/O	→	A2	Input 2	L293D Motor Driver connected to Water Pump
	GND	Ground		A1	Input 1	
	5V	Vcc		ENA	Enable	
	GND	Ground		GND	Ground	
Arduino Uno	0 (Rx)	Receiver	→	Tx	Transmitter	ESP8266
	1 (Tx)	Transmitter		Rx	Receiver	
	3V3	3.3 V		Vcc	·	
	3V3	3.3V		CH_PD	Chip Enable	

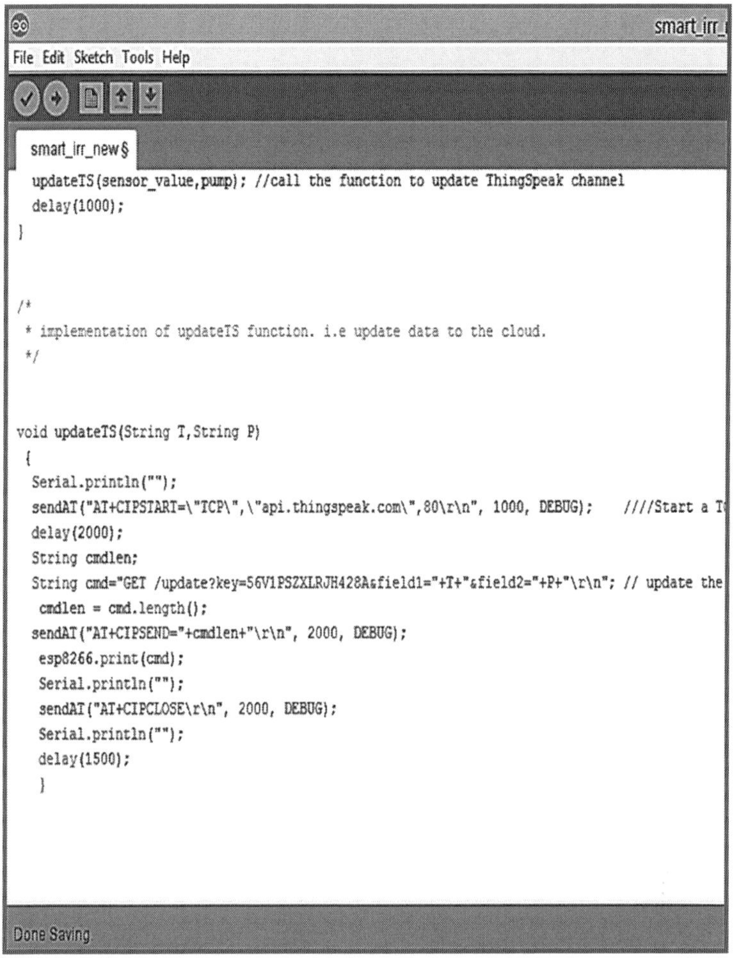

```
smart_irr_new§

  updateTS(sensor_value,pump); //call the function to update ThingSpeak channel
  delay(1000);
}


/*
 * implementation of updateTS function. i.e update data to the cloud.
 */


void updateTS(String T,String P)
{
  Serial.println("");
  sendAT("AT+CIPSTART=\"TCP\",\"api.thingspeak.com\",80\r\n", 1000, DEBUG);    ////Start a T
  delay(2000);
  String cmdlen;
  String cmd="GET /update?key=56V1PSZXLRJH428A&field1="+T+"&field2="+P+"\r\n"; // update the
  cmdlen = cmd.length();
  sendAT("AT+CIPSEND="+cmdlen+"\r\n", 2000, DEBUG);
  esp8266.print(cmd);
  Serial.println("");
  sendAT("AT+CIPCLOSE\r\n", 2000, DEBUG);
  Serial.println("");
  delay(1500);
  }

Done Saving.
```

FIGURE 15.13   Updates to ThingSpeak App Code window.

The system works according to the loop() function (described earlier), and Arduino keeps running and changes the state of the motor from on to off and vice versa continuously as per the requirements of the soil. The Wi-Fi module is keep sending the pump status and the moisture content to the cloud. Moreover, the data has both private and public view. In public view anyone can access the data and make conclusion (https://thingspeak.com/channels/476222). In addition, one can check when the pump is started and how frequently the soil needs water. We can also monitor the whole situation from Virtuino App which is connected to the ThingSpeak channel through channel ID. It is also possible to set the alarm in Virtuino App according to the data so that we can get the SMS as well as alarm. The limit is flexible and can be set by us according to the need. The system will be always in ON condition until the power supply is ON and is working continuously to fulfil the need of water to the soil. The system will turn off when the need is fulfilled.

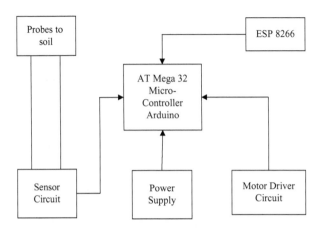

FIGURE 15.14 Connectivity of smart irrigation.

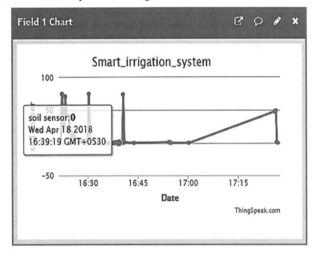

FIGURE 15.15 Soil sensor status.

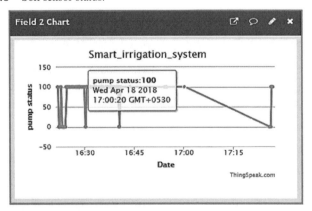

FIGURE 15.16 Pump status.

The status of this work and all past data at ThingSpeak cloud is accessible on the link https://thingspeak.com/channels/476222 (public view). The ThingSpeak App Code window is shown in Figure 15.13, and connectivity of smart irrigation system is given in Figure 15.14. Further, Figure 15.15 renders the soil sensor status, and Figure 15.16 shows the pump status. In addition, Figure 15.17 shows a snapshot of the ThingSpeak channel.

FIGURE 15.17   Snapshot of ThingSpeak channel at on April 18, 2018.

15.8   CONCLUSIONS AND FUTURE ENHANCEMENT

This chapter preludes the development of an automated irrigation system. The chapter concludes that an irrigation system becomes fully automated and this automation proves comfortable and efficient for farmers as they can operate as well as monitor it from remote locations, e.g. home or market. Farmers can also keep track of the moisture level every second and behave accordingly. They will also receive a message if there is any problem in the system. The ESP8266 is the best device for IoT projects as it is small, compact, lightweight, easily programmable, and easily installable. This will save time as well as labour. Not only this, it will also minimize the water wastage and would help in increasing the groundwater level and productivity. However, there is lots of room to improve the performance of the smart irrigation system in the area of multiple sensors, action time, and combined system effect.

REFERENCES

1. Joaquin Gutierrez Jaguey et al., Smartphone irrigation sensor. *Sensors Journal*, 15(9), 2015.
2. G. Nisha and J. Megala, Wireless sensor network based automated irrigation and crop field monitoring system. *Sixth International Conference on Advanced Computing (ICOAC)*, 2014.
3. Vidadala Srija and P. Bala Murali Krishna, Implementation of agricultural automation system using web and GSM technologies. *International Journal of Engineering & Science Research*, 5(9), 1201, 2015.
4. Fan TongKe, Smart agriculture based on cloud computing and IOT. *Journal of Convergence Information Technology (JCIT)*, 8(2).
5. David Chaparro, Merce Vall-llossera, Maria Piles, Adriano Camps, Christoph Rüdiger and Ramon Riera-Tatch, Predicting the extent of wildfires using remotely sensed soil moisture and temperature trends. *IEEE Journal of Selected Topics in Applied Earth Observations and Remote Sensing*, 9(6), 2016.
6. Joaquín Gutiérrez et al., Automated irrigation system using a wireless sensor network and gprs module. *IEEE Transactions on Instrumentation and Measurement*, 63(1), 2014.
7. Ruan Junjin, Peng Liao and Chen Dong, The design and research on intelligent irrigation system. *7th International Conference on Intelligent Human-Machine Systems and Cybernetics*, 2015.
8. Santosh Kumar and R. Y. Udaykumar, Development of WSN system for Precision agriculture. *IEEE Sponsored 2nd International Conference on Innovations in Information Embedded and Communication SystemsICIIECS'15*.
9. Nelson Sales and Artur Arsenio, Wireless sensor and actuator system for smart irrigation on the cloud. Agricultural Communications Documentation Center, 2015.
10. S. Khriji, D. El Houssaini, M. W. Jmal, C. Viehweger, M. Abid and O. Kanoun, Precision irrigation based on wireless sensor network. *IET Science, Measurement and Technology*, IJS. 2014.
11. Aravind Anil et al., *Project HARITHA - An Automated Irrigation System for Home Gardens*, 2012.

16 Privacy and Security Challenges Based on IoT Architecture

Umang Shukla

CONTENTS

16.1   FUNDAMENTALS OF IOT

Communication plays an important role in day-to-day life. Transferring message between persons via the Internet either in the government, personal, or business domain is one of the important task in daily activities if we can consider examples of email communication, applications that send report and administrative data, and Facebook and WhatsApp applications. Human-to-machine (H2M) communication is possible when sensors are collecting and tracking data from environments and sending them to humans. In machine-to-machine (M2M) communication, point-to-point commutation between machines use wired or wireless communication that is does not always need to rely on the Internet and also have limited integrated options based on communication standards. On the other hand, smart nodes can collect data and apply intelligence-based decision system.

Ashton introduced the Internet of Things (IoT) at Auto-ID in the late 1990s. He and his team discovered how to connect smart nodes to the Internet using a radio-frequency identification (RFID) tag [1]. IoT would become popular based on two laws: The first law is the well-known Moore's law, which is based on the number of transistors on a chip doubling in 2 years. This statement enhanced industry to produce more powerful CPUs and other processing units of the same size. In early 1971 Intel placed 300 transistors on a processor. In 2012, processors contained 1.4 billion transistors. In 2018 IBM computer chips could fit 30 billion transistors. The second law, known as Koomey's law, states that the number of computations considered as kilowatt-hour-based double every one-and-a-half years. Based on these two laws, a more powerful device in small size is possible and the energy consumption needed to perform a computation regularly reduces.

16.2   BASIC ELEMENTS OF IOT

IoT is not only about connecting sensors and visualizing data on smartphones or other devices. There is a much larger impact than that. For example, a smart transport system is useful to manage traffic flow, emergency services, and reduce fuel consumption. At home, a smart refrigerator can send reminders when milk is low. IoT is not considered for connecting devices to Internet. It collects data and gives advice to users based on data analysis. There are unique identity and standard protocols for receiving notifications on your mobile devices based on alerts if temperature changes or fitness tracking. Connectivity is between physical devices to digital media. It has the characteristics of any time, anywhere, and any content [2].

Functional blocks performing tasks such as detection of objects, actuation based on sensing data, conducts device-to-server and server-to-server communication, and provide different services with semantic architecture (Figure 16.1) [3].

- **Device:** There are different task that can be handled by devices such as sensing, actuation, control, and monitoring activities in a system. The two types of connectivity possible are wired and wireless.
  - Device to other device communication or send data to server
  - Collecting input and output sensors
  - Internet-based connectivity
  - Based on data storage
  - Audio- and video-based
- **Communication:** Web services are provided on different devices. There are many business applications available which have different web services in such a communication model. The main categories are as per the following:
  - D2D is known as device-to-device communication.
  - D2S is known as device-to-server for collection of data and transfer.
  - S2S is known as server-to-server for internal communication and analysis.
- **Services:** Different aspects include modelling, controlling, publishing, and analysis of data.
- **Security:** The major parts of include authentication, privacy, device security, and the access control mechanism (Figure 16.2). A traditional security policy cannot be directly applied to smart devices because of many issues such as scalability and their dynamic nature.

16.3 CHARACTERISTICS

There are specific key characteristics in the Internet of Things such as dynamic and self-adapting, self-configuration, interoperable communication, unique identification, context awareness, and smart decision-making [2].

- **Dynamic and self-adapting:** Smart devices are made in such a way as to adapt to changes and make decisions based on the surrounding environment. Let's take a college management system for example, which has a

FIGURE 16.1   Basic elements of IoT.

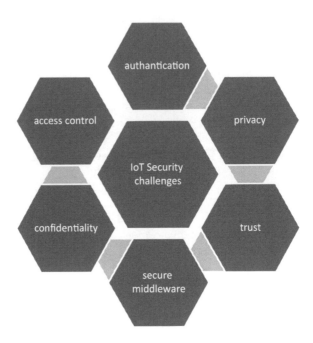

FIGURE 16.2   Security in IoT.

number of motion sensors for temperature detection in each classroom. As per collection of data, the system will provide changes in classroom. Many data samples can be taken as per the dynamic and self-adapting nature, and energy consumption reduced.

- **Self-configuring:** In many situations, devices are required to configure themselves because of placement in remote areas. Providing large access to node for proper collaboration is critical.
- **Interoperable communication protocols** are required to upgrade existing communication protocols for suitable IoT communication with the infrastructure.
- **Unique identity:** Every IoT device requires a unique label such as an ID, IP address, or URI. IoT systems having smart nodes enable communicating with users and environmental contexts. Collecting and updating data of particular devices becomes possible via unique ID, which allows control and verification of information as well.
- **Integrated information network (IIN):** An IIN is an information network which enables data exchange between different devices and from device to system. Let's take the example of a temperature sensor node which is able to share data with applications and different nodes in the network. At this moment, connected devices sharing data enhance intelligence in the IoT domain.
- **Context awareness:** Based on the context of the surrounding environment, nodes collect information and extract knowledge from the collected data. This knowledge is utilizes for decision-making in the IoT domain.

- **Smart decision-making system:** IoT is multi-hop in nature. To increase the lifetime of a node energy efficient for communication in an application is required. All nodes communicate and transfer data themselves and make proper decisions.

16.4 TAXONOMY FOR OBJECTS

There are different categories considered for the taxonomy of objects such as power management, communication, local user interface, functional attributes, and hardware and software resources [16] (Figure 16.3).

16.4.1 POWER MANAGEMENT

Sensor node power management is divided into two main categories. The first is energy provision and the second is energy consumption. For a fixed-size battery that does not allow replacement, the node will be discarded from network. For example, during war or some other harsh environmental area where replacement cannot be possible. In contrast, with a replaceable fixed-size battery replacement is possible. Energy from a environmental source is known as a solar power [64]. There are Bluetooth low-energy, solar-powered devices available in the market. Vibration-based power shake flashlights have been developed. For power consumption, a duty cycle can be based on sleep and wakeup modes. Data driven is another approach where a data-reduction algorithm is required to manage unnecessary data avoidance [65].

16.4.2 COMMUNICATION

There are mainly two types of interfaces: wired and wireless. There are two different scenarios possible between the object and pickup point for data transfer. Either the object initiate or the pickup point can initiate communications. In communication, security is achieved based on authentication of object or pickup point. For transferring sensitive data over a network proper encryption for the data stream is required.

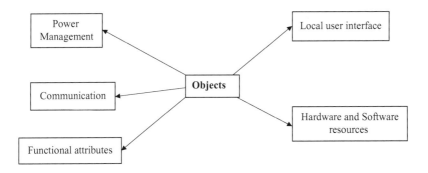

FIGURE 16.3 Taxonomy for objects.

One of the major physical specifications divided as rate of data exchange and range as PAN, LAN, or type of the network.

16.4.3 FUNCTIONAL ATTRIBUTES

Sensors interact with a system based on collected data, for example temperature sensor and position data sensor for a specific application. At the sensor level there are two categories based on sensors: with memory and without memory. Actuators react in environment to emit light, produce cold, and move objects. Sensors and actuators as a hybrid form one of the categories. They collect data and react based on data.

16.4.4 LOCAL USER INTERFACE

Interaction with the user directly with use of a button on the sensor is known as active interaction. In passive interaction, the user communicates through a display screen, or light or sound or vibration. Basic functionality provides for direct use for sensors. Some of the objects that do not have any interface, then directly communicate through a local point.

16.4.5 HARDWARE AND SOFTWARE RESOURCES

Smart sensors come with computing, input/output, and communication in a single integrated system. Microcontrollers are found in microwaves, washing machines, and refrigerators. Based on the hardware point of view (Table 16.1), devices are categorized into two sections: high-end devices and low-end devices. Low-end devices require a small memory footprint and a trade-off between performance and API; support for heterogeneous hardware; network connectivity such as IEEE 802.15.4, Bluetooth, or DASH7; security; real-time operation; and a energy efficient solution [67]. System updates to devices are known as 'patches'. Over-the-air delivery of software updates to multiple nodes is complicated. Outdated devices or devices without updates can become targets of malware such as the Mirai Botnet [66].

16.5 ISSUES IN TRADITIONAL TCP/IP LAYER APPROACH

The TCP/IP protocol was originally developed for traditional computer networks, and the Internet of Things has heterogeneity of devices with low-end and resource-constrained devices. These devices have less computational power and required long-time operational on a given power backup. At the IP network layer a minimum MTU around 1500 bytes or higher is sent in comparison to the sensor network support of around 127 bytes or lower. Even IPv6 supports a minimum of 1280 bytes. To notify multiple nodes in a group and sending query without knowing exact node are both features supported in a TCP/IP multicast. Many wireless MAC protocols do not support ACK for multicast and do not having lost packet recovery management. They also having different data rates or link layers. In the sleeping mode, a node may lose multicast packets. A multicast needs to send packets over the network via multiple hops which overload at the node and forwarding packets requires more

TABLE 16.1
Hardware Platforms in IoT

Hardware Platform	Processor	Flash Memory	Communication	Environments	Programming
Arduino Uno	ATmega328	32 kb	802.11, 802.15.4, BLE4, serial	Own IDE	Wiring with predefined examples
Intel Edison	Quark	4 gb	802.11, 802.15.4, BLE4, serial	Intel XDX, Eclipse	Wiring, C, C++, and Node js
Beagle Bone Black	Sitara	4 gb	802.11, 802.15.4, BLE4, serial	Android, Cloud9 IDE	C, C++, Python, java, Node js
Raspberry Pi 3	Broadcom BCM2835 SoC	4 gb	802.11, 802.15.4, BLE4, serial	Different operating systems available such as windows IoT, Raspbian	Ruby, Python, C, C++, Java
mbed - ARM Cortex M3 Core- LPC1768 v 5.1	ARM Cortex m3	512 kb	802.11, 802.15.4, BLE4, serial	C, C++ SDK	C, C++

power consumption. Congestion control and reliable data transfer are both important tasks at the transport layer, which transfers efficient bulk data over point-to-point communication, providing reliability for in sequence delivery of every byte in the stream. The sensor transmits a small amount of data at this stage; establishing a connection mechanism for small interaction is unacceptable. The node acts as a actuator sending data with low latency requirement. The traditional handshaking process causes delay. Most proposed IoT applications are developed based on the resource-oriented request–response communication model, for example, sending temperature data of the living room to a user. The TCP/IP communication model requires the client and server to be available at same time. The sensor might be in sleep mode for power savings, and required to dynamic or intermittent communication model. At this stage the sensor communication model needs to find a solution based on caching for efficient data transfer or selected proxy-based communication which can transmit or request/response on behalf of the sensor. In the http protocol, the client is required to identify which data should forward or perform reverse proxy node. Heterogeneity of the node is demanded for different mechanisms. In case of topology, changes are needed to reconfigure proxies to restart the cache mechanism [55].

A flexible, layered architecture is needed because the number of connected smart devices in the future could be in the trillions. Many different architectures have still not converted to a proper reference model [4–6, 9]. IoT-A [10] is an architecture regarding an analysis collaborative from industry. Many basic models propose a three-layer architecture [11]: perception, network, and application layers. More layers are adopted to the IoT architecture [12]. As per Figure 16.4, architectures among them is the five-layer model approach, which interchanges the application or business layer [3, 13].

The objects layer at the bottom has different devices and actuators, which is also considered as the perception layer in some research papers. At this layer different

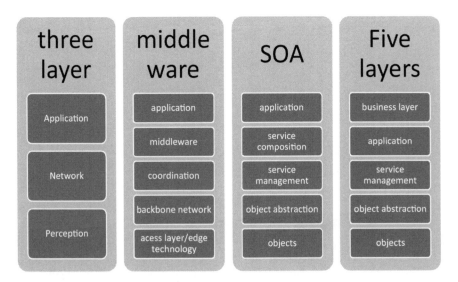

FIGURE 16.4 IoT architecture.

sensors collect data and they can also transfer data to another sensor node. After collecting data from the sensor, the nodes main task is to trigger the above layer with the collected information which is analyzed at the application level. In the smart home or classroom, humidity, or motion sensors are examples of the perception layer [14].

The object abstract layer provides a connectivity mechanism which provides secure communication with a range of technologies such as GSM, IEEE 802.15.4, BLE 4.0, and infrared.

The service management layer provides different services for transferring data from server to device or device-to-device with reference to unique addresses and names. Service provider independent from device implementation details and hardware specification. Applicability of collected data for providing specific data for prediction and action performance [8, 14].

The application layer uses the CoAP protocol for the web which utilizes resource-constrained sensors for the connected node. [15]. At the same time the application layer deals with the user interface which provides all solutions at one place and real-time data analysis [7].

At top layer major concern about business application domain, IoT is collecting data from sensor and prediction some fundamental and compressed data or selected data for give decision to user.

In the home automation system, energy consumption analysis-based prediction can trigger other sensors to automatically make decisions [11].

16.6   STANDARDS AND NETWORKING PROTOCOLS

Rapid growth of the IoT is due to availability of the Internet and smart objects. IoT devices have limited power and minimal storage resources. There are many challenges in communication as per the following [17, 68]:

1. Unique addressing and identification of objects
2. Limited power communication
3. Low memory communication in routing protocol
4. Faster communication
5. Mobility in smart objects

See Table 16.2.

TABLE 16.2
Protocol in Each Layer

Application		MQTT, SMQTT, CoAP, AMQP
Network	Encapsulation	6LowPAN 6TiSCH 6Lo Thread
	Routing	RPL CORPL CARP
Data link		BLE, ZigBee, NFC, Weightless, Homeplug GP, Z-Wave

TABLE 16.3
Data Link Layer Protocols

Standards	ZigBee	BLE	Z-Wave	NFC	HomePlay GP
IEEE spec	802.15.4	802.15.1	ITU-T	ISO 13157	IEEE1901- 2010
Max. signal rate	250 kb/s	305 kb/s	40 kb/s or 100 kb/s	424 kb/s	10 Mbps
Nominal range	10 m	50 m	30 m	5 cm	100 m
Network type	WPAN	WPAN	WPAN	P2P	WPAN
Power consumption	40 mA	12.5 mA	2.5 mA	50 mA	0.5 w

16.6.1 DATA LINK LAYER

At this layer the BLE, Z-wave, ZigBee, NFC, HomePlug GP, and Wi-Fi protocols are available (Table 16.3). Each protocol works under individual alliances of companies for covering network and application domain can be identified for a potential standard [69].

16.6.1.1 ZigBee

ZigBee is a short range, less complex, and less power consuming wireless communication. IEEE 802.15.4 standard is applied for the protocol and it can send data at a lower rate for many devices. Application developments include intelligent control of lighting, gas sensing, and notification. ZigBee has star, cluster tree, and mesh topologies. There are three types of ZigBee devices: (1) PAN coordinator, which can start the network and be responsible to route data; (2) full function device can operate in three modes as the PAN coordinator, coordinator, and device; (3) reduced function device can communicate with a full function device.

16.6.1.2 Bluetooth LE (BLE)

BLE is designed for low-power application. At the physical layer BLE still uses adaptive FHSS (frequency hopping spread spectrum) and the number of channels reduced 79 to 40. An SIG (special interest group) officially announced the support of mesh model in 2017, which can provide many-to-many device communications for home applications such as healthcare, beacon, and entertainment industry.

16.6.1.3 Z-Wave

Z-Wave was developed by Zensys for home applications. It has two types of devices: controlling and slave. There is a single controlling device for many slaves devices. Delay is relaxed up to 200 milliseconds. In the application domain, they support smart lighting, locks, energy management at home, and door opening and closing. It also supported by more than 50 brands, including Amazon, Nest, and GoControl.

16.6.1.4 Near Field Communication (NFC)

NFC was initially developed and standardized by the end of the 20th century for transportation. The main concept behind deployment is ticketing based on secure microcontrollers. There are many NFC based payment operations available, for example, Google Pay (Android Pay) and Apple Pay. NFC is also used for communication between mobile phones.

16.6.1.5 HomePlug GP

HomePlug GP provides power line communications for broadband application such as IPTV, gaming, and Internet content distribution. It manages in-house communication between electric systems and applications. IEEE 1901-2010 is the standard for power line network, and Homeplug 1.0, initially introduced in 2001 has peak PHY-rate of 14Mbit/s. HomePlug AV provides sufficient bandwidth for HDTV and VoIP also having peak data rate 200 Mbit/s at the physical layer. AV2 was introduced earlier in 2012 with repeating functionalities and power saving. Green PHY, used for smart grid, has a peak of 10Mbit/s. It is also available for home network for sharing data with power utility.

16.6.2 NETWORK LAYER ROUTING PROTOCOLS

16.6.2.1 RPL

It is a challenging task to develop a routing algorithm for low power and lossy radio connections for a battery-dependent node. Another challenge is frequent topology changes due to mobility. RPL assumes that the sensor needs to identify the path to the controller. Second, some or all sensors are energy constrained. Another assumption is that some of the sensors may not identify the path to the controller which can transfer to another sensor and which will transfer to the controller.

RPL builds upon DODAG, the Destination Oriented Directed Acyclic Graph [34]. It constructs a single path from the individual leaf node to the root.

- Each node sends the DODAG Information Object (DIO) advertising to the root node.
- All nodes transfer this message after the Destination Advertisement Object (DAO) for advertisement to their own parent node.
- The DAO transfers to the root node which makes a decision on destination.
- In such a case, adding a new node sends a DODAG Information Solicitation request which gets DAO-ACK for confirmation.

16.6.2.2 CORPL

In cognitive networks which use DODAG topology for generation for cognitive are required to modify RPL and required to build opportunistic forwarding to pass the packet by choosing a multiple forwarder set which coordinates between devices. Based on comparison, select hop to send. It also sends updates regarding changes in forward sets via sending DIO messages. As per update message, the node adapted changes dynamically [35].

16.6.2.3 CARP

CARP, the Channel-Aware Routing Protocol, is a distributed routing protocol.

- It is actually designed for underwater communication which uses light-weight packets.
- It performs computation of link quality based on collected historical data from neighbouring sensors.
- At the start of the network, the sink node sends a HELLO packet to all other nodes in the network.
- In other approach, data forwarding via hop by hop at each node calculation of next hop is identified independently.
- CARP does not supporting reusability of any data that was previously collected.
- Enhancement for allowing a sink node to recollect old data is known as E-CARP [36].

16.6.3 ENCAPSULATION PROTOCOLS AT NETWORK LAYER

16.6.3.1 6LoWPAN

6LoWPAN is known as IPv6 over Low-Power Wireless Personal Area Network. 6LoWPAN aims to be an Internet protocol that works for smallest devices that have limited processing and less power. It defines encapsulation and header compression for transferring data IPv6 over IEEE 802.15.4 oriented network. The address and the packet size in IPv6 requires a maximum transmission unit of at least 1280 octets and an IEEE 802.15.4 standard packet size of 127 octets. As per address, IPv6 assigned 128 bits, and IEEE 802.15.4-based devices mostly use either of IEEE 64 bits extension or 16-bit addresses that are unique within PAN. As per routing, mesh routing prefers for PAN and IPv6 domain to PAN domain. There are several protocols that have been proposed as LOAD and LOADng. As per application in smart grid or smart billing system. Thread is largest group of more than 50 companies for home automation [37].

16.6.3.2 6TiSCH

A working group in the Internet Engineering Task Force (IETF) known as 6TiSCH is aiming for IEEE 802.15.4e Time-Slotted Channel Hopping. The working group proposed improvement on delay, scalability, reliability, and low energy consumption performance tuning for IPv6 over IEEE 802.15.4e TSCH [38].

16.6.3.3 6Lo

An active working group based on IPv6 over Networks of Resource-Constrained Nodes is aiming to develop a set of standards for IPv6. Currently in development is a data link specification about IPv6 over BLE, over NFC, and over 802.11ah [39, 40].

16.6.3.4 IPv6 over BLE

BLE uses low energy consumption and is popular in smartphones, wearables, and laptops, and there are applications in health care, home automation, fitness tracking, and smart shopping malls. Bluetooth wireless technology is available with two variants: basic rate and enhanced data rate. Up to 2021, Bluetooth-based shipments will reach more than 5 billion. BLE transfers a small amount of data periodically. The latest version of Bluetooth 5 provides a higher data rate, extended range, and higher advertisement capacity. Also the support mesh enables a many-to-many BLE networking solution. The mesh enable update is available as a software update to BLE 4.0, 4.1, 4.2, and 5.0. 6LoWPAN supports reduced overhead and stateless IPv6 address autoconfiguration. In BLE, the central node is considered as the 6LoWPAN border router (6LBR) and the peripheral node as 6LoWPAN node (6LN) [41].

16.6.4 APPLICATION LAYER PROTOCOLS

16.6.4.1 MQTT

MQTT was developed by IBM and is a publish/subscribe protocol that runs over TCP [42]. MQTT is a lightweight protocol because of that many IoT applications have adopted it. MQTT uses text for topic names and it will tend to increase overhead. Clients can provide topic names, which are also available for subscription. The role of broker is to validate the client for specific subscriptions. There is much ongoing research on security-based MQTT (SMQTT) [43].

16.6.4.2 AMQP

Advanced Message Queuing Protocol (AMQP) runs over TCP. It has the same mechanism as the subscriber and publisher architecture as per MQTT, but the main difference is that with AMQP the broker categories are exchange and queues. Exchange provides publisher messages and transfers them into multiple queues which have different roles and conditions. A subscriber can get data as per available in the queue [44].

16.6.4.3 CoAP

CoAP is used in most IoT applications as an alternate of HTTP [11]. Unlike HTTP, it uses efficient XML for transfer data which is suitable for smart sensors and also faster than existing solutions of HTML. Reducing the size of the header identifies correct resources, transfers data, and manages data failure or correction. The different types of CoAP messages are confirmable, reset, acknowledgement, and non-confirmable. Piggybacked is used in response in the acknowledgement itself. It also supports a security feature with implementation of Datagram Transport Layer Security [45].

16.6.4.4 XMPP

The Extensible Messaging and Presence Protocol (XMPP) was mainly designed for chatting and message exchange. It has IETF standards and is also highly efficient for transferring data. It follows any architecture based on application that demands publisher/subscriber or request/response architecture. There are many extension parts like efficient xml interchange format, sensor data, and discovery for publish/subscribe authentication and security. [46].

16.7 IOT APPLICATIONS

Here we will explore smart applications that make human life simpler [47]. All of these applications are not complete or available; some are only at the have conceptual level. The quality of life in our society in the home, medical, and government sectors will benefit if they utilize IoT.

The most popular application has been the development of home automation, which has more than 300 industries and startups working on home-based applications. There are many examples such as Nest and AlertMe. Next is wearables such as the Sony Smart B Trainer, and Fitbit is a smart health care system. Other potential applications are city traffic, water distribution, waste management, pollution control, and electricity distribution. Connected car is under the development by popular companies like Google and Apple.

16.8 CATEGORIES OF TECHNOLOGICAL CHALLENGES

In industry there are many different parameters for smooth execution and reliability of systems. There are many challenges [4, 8, 12, 15, 47] as shown in Figure 16.5, such as security, connectivity between different devices, compatibility between different technologies, and standards must be required for common approach for application development. There must also be different analysis based on real-time data analysis. For example, home automation is not only able to turn things on and off, but on or off it can also give proper analysis to the user on how they can change day-to-day activities. It will also give basic statements as per energy consumption and activity analysis based on collected data from different sensors.

16.8.1 SECURITY

IoT has major issues regarding security, which is required to maintain in application development standards from all over the world. There are many examples of failed security such as getting access to smart cars, smart home devices, and tracking personal information of user.

FIGURE 16.5 Technology challenges.

16.8.2   CONNECTIVITY

Another challenging phase of the IoT environment is having different nodes that must be connected and transfer data using communication protocols. Currently communication connectivity is based on identifying low connectivity and lost connection in such a network, as many devices are connected which also demand maintenance of a cloud server.

16.8.3   COMPATIBILITY AND LONGEVITY

As the connected device concept is getting popular, many people started making different applications on their own. But from a system point of view, there are big challenges to identifying a common approach.

16.8.4   STANDARDS

In a connected device network, there are many network communication protocols, and data aggregation standards must be required:

*   Handle unstructured data
*   Communication protocol and architecture
*   Device-to-device data transfer and store mechanism
*   Technical skills to leverage newer aggregation tools

16.8.5   INTELLIGENT ANALYSIS AND ACTIONS

After collection of the data which tends to decision-making system. AI requires more data for analysis and prediction, many cloud-based services are available as open source, and real-time data mining is playing an important role for identification of event-based analysis. There are many challenges in the data model without having knowledge about outliers, and many software are based on traditional data analysis.

16.9   TESTBEDS AND SIMULATIONS

In this section, we will review different facilities and support available for setting up experiments for research on the Internet of Things, which includes different simulators, live projects, and testbeds being used for different IoT development. The entire review is centred on availability of testbeds for the public, as shown in Table 16.4 [48, 70].

For heterogeneous elements, support scalability, providing energy, efficacy, and customizing protocol testing, there are main three categories: the full stack simulator, which provides end-to-end support of all IoT elements; big data processing aspects for applications; and network simulators (Table 16.5) [49, 50–52, 70].

16.10   PRIVACY AND SECURITY

Nowadays connected smart devices have the ability to sense data of the surrounding environment and they also connect to cloud storage. In regular life users not only

TABLE 16.4
Testbeds in IoT and WSN

Testbed	Type of Network	Support/Feature	Remarks
FIT IoT-LAB	IoT	Medium, nearly 2,700 sensors, lab environments	It is large scale and multi-site with multiple user support. Mobility provided by programmable robots.
Smart Santander	IoT	Larger node support as 20,000 Real city environments	Implemented in city of Spain, having sensors with RFID and QR code.
JOSE	IoT	Larger sensor nodes, Providing infrastructure as services Outdoor environments	Located in JAPAN; provides excremental at application layer; transmits data using IEEE 802.11, 3G, and LTE.
Motelab	Wireless sensor network, mesh network, RFID	Three floors of laboratories engaged for deployment of 190 TelosB motes deployed	One of the first publicly accessible indoor facilities. It is used as framework to build other testbeds like CCNY, CWSNET, INDRIYA.
NetEye		Total 130 TelosB motes deployed in one room on 15 benches per 1 foot indoor	Supports multi-hop networks by FCFS scheduling technique, and works on different power levels.
TutorNet		91 TelosB, 13 MicaZ motes deployed in one room	Reservation of node is possible in this testbed to set an experiment.

For heterogeneous elements, support scalability, providing energy, efficacy, and customize protocol testing. There are main three categories: full stack simulator which provides end-to-end support of all IoT elements; big data processing aspects for applications; and last network simulators [49][50][51][52][70].

communicate with a single desktop or office laptop, but variety of devices. Easy connectivity to the Internet is possible now, so we can think that many devices in the world take the place of major real-life applications. IoT devices can collect information regarding each user's personal data, analyze it, and based on it appropriate action will be taken [2]. At this point IoT applications can transform daily life, but users need to understand that they have to compromise privacy and security at a certain point. For example, users do not frequently change passwords because they have so many devices to manage and they often use the default password of the device. Let's take some more real-life examples to understand importance of privacy and security in IoT devices, because with more and more devices it is difficult to manage security policy at each service. In simple desktop is really hard to understand standard of security for unknown person form technology. What if there is a ransomware attack on connected things in a connected home of IoT appliances?

TABLE 16.5
Simulators in IoT and WSN

Simulator	Characteristic	Remarks
Cooja	Available as part of Contiki OS; researcher can reproduce real scenario on simulator	Emulates both WSN and IoT scenario; popular for application layer protocol implantation such as MQTT and CoAP
NS-3	Having support for 6LoWPAN over 802.15.4	Lack of support for application layer
GlomoSIM	Wired and wireless network simulator	Provides different lib for larger scale simulation
QualNet	Commercial version of GlomoSIM	Suitable for Scalable Network Technologies (SNT)
SensorSIM	A framework to simulate for senor network	Fast, can build power usage in sensor nodes
TOSSIM	Support to TinyOS	Power TOSSIM is an extension; can simulate issues of the energy consumption in WSN
Emstart	Has Linux microkernel provision of fault mitigation amongst nodes	Execution of modules separately possible
OmNet++	Discrete event model build in C++ provides module programming model	Both commercial and academic licences available

Suppose a kitchen is connected with different devices. Whether it's a Crockpot, refrigerator, coffee maker, or a light bulb, you can simply control all of them from anywhere at any time from your smartphone. Yes, it sounds convenient, but it could also unlock a world of hurt for an unaware user. For example, take the IoT Crockpot Android application. There were many vulnerabilities found in Belkin WeMo home automation devices and solutions were provided to the Android application. It will create a need for safety guidelines for all IoT devices and application.

Security challenges of individual devices include [53]:

- Smart refrigerator also having privacy issue regarding compromising user's personal details.
- Video monitoring and movement tracking data also compromised.
- Mirai malware affected many devices with botnet for denial of service.
- Major cyberattacks on many IoT devices as a domain name server (DNS) infrastructure. There is evidence that they affected many reputed services like Twitter, *The New York Times*, Netflix, Spotify, SoundCloud, and many more.

16.11   SECURITY ON IOT ARCHITECTURE

There is a need to implement security mechanisms in such a manner that will enable Internet protocols of IoT applications [54]. In the following sections, we will explore the security issue based on IoT layer and existing solutions.

16.11.1 PERCEPTION LAYER

Perception layer is categories as sensor or controller and establishes communication with the upper layer. There are data sensors for location, identification, distance, audio/video, and many more for observation of different activities. But there are many issues in the perception layer that affect the node or abnormal node identification in a heterogeneous network. There is much research ongoing in the same direction with solutions such as a heterogeneous algorithm with a security mechanism. Establish a public key-based authentication in cryptography creates issue for power. There are many probabilities that affect a node or trust establishment between nodes. Identification of sensitive data before sending to other node algorithms also provides proper privacy to user personal data.

16.11.2 NETWORK LAYER

Every device in IoT requires unique identification, so as per the growth rate of devices and sensor-based applications they cannot use IPv4 because they require large address space. IPv6 with low power consumption known as 6LoWPAN must be used. Also, there are proposed security solutions without any modification in standard of security and gateways [56]. Header authentication and encapsulated security payload are required for secure and authentic communication endpoints. Small MTU for low energy nodes [55] can be managed with header compression.

Some parameters are required to manage multicast and mesh network routing in IoT networks.

16.11.3 TRANSPORT LAYER

A major aspect of the transport layer is to provide reliable data transfer and manage congestion in communication, but in IoT applications it is hard to identify different activities based on low power nodes, low latency, and less amount of data in communication. Datagram Transport Layer Security provides two-way authentication and works with X.509 certification using RSA key and reduced header size for energy saving while maintaining standards on traditional Internet. There are many works for additional security bits as a combination of DTLS and CoAP generalized as Lithe [57].

16.11.4 APPLICATION LAYER

IoT applications require different types of data from nodes and also sending back this data to nodes, so the communication module is based on request–response. Nodes based on Linux are affected by Mirai, which leads to DDoS attack. After that MalwareMustDie identified the IRCTelnet. There are many attacks of malicious code that try to observe and get users' personal information. There are also many phishing attacks and sniffing to identify network behaviour. Security measurements, data authentication, risk assessment, and identify intrusion are required to measure the security level of any application [58].

16.12 PROBABILITY-BASED TECHNIQUES FOR TRUST BUILDING IN IOT APPLICATIONS

The Bayesian network was considered as a random variable probability-based directed graph. IoT is a combination of different devices that can be considered as a node. This node might get affected by attacks and stolen private data of users. With the same identity they are sending data to another node. So, the identification node trust level is very important for security reasons [59].

First, establish a structure level of trust of new node or any node trust level based on four steps, which shown in Figure 16.4. Second, with determination of parameters, define three levels of node trust, basic trust and distrust, and also integration of prior probability as well as allocation of conditional probability. At the next level we have to find inference of posterior probability based on Bayes' theorem using prior and conditional probability. At the end is evaluation of trust node based on threshold probability.

$$P(A \mid B) = \frac{P(B \mid A)P(A)}{P(B)}$$

Another research [60] proposed a method for identifying authentication of a node in communication based on joint probability. Here there is concern about information sharing between nodes in the IoT infrastructure. It is based on conditional probability data authentication between the server and nodes. This method provides better efficiency by 5.2%, and a 7.8% reduction in process time compared with conventional schemes. Communication overhead was also reduced by 3.5%.

Fitness belt and wearable devices are used in day-to-day activities like health analysis, reminders, getting email, and other measurements. We share data without concern about personal data discloser. Many companies that utilize this sensitive data are required to manage privacy. Researchers have found that singular data and non-sensitive data collected from different sources and aggregate data can lead to other personal information as well. Using the Bayesian network, they have calculated posterior probabilities based on prior and conditional probability for risk measurement of sharing data with a third party [61].

Water controlling systems have different sensors and these sensors are required to be secure while transferring data to each other. Networks are required to identify intrusion within a short time. Bayesian inference can successfully identify normal or intrusive activity [62].

Here they have computed intrusion detection and trusted node detection using the Bayesian network as per figure above perception, transport, and application having set of nodes. From the perception layer, failure leads to the next layer (transport), and transport layer failure leads to the application layer. Applying a Markov model based on the current situation, they are not considering the all over security mechanism. Simulation results shown in research provide confidence, authentication, and integrity for decision-making processes based on the Markov model and also they have introduced aspect-oriented programming. As per the result shown in Wang et al. [63], applied security more running node and less energy consumption.

16.13   SUMMARY

Security and privacy issues are not only limited to a single layer of IoT architecture. Each application requires different solutions regarding privacy and security. Different cryptographic and header authentication require more processing power at nodes which means less power and minimum computation. The probability-based model provides less power consumption and reduces server overhead. User privacy is more important without compromising data utility. A probability-based approach prevents unauthorized access by a third party.

ABBREVIATIONS

6LoWPAN	IPv6 over Low-Power Wireless Personal Area Networks
ACK	acknowledgement
AMQP	Advanced Message Queuing Protocol
BLE	Bluetooth Low Energy
CoAP	Constrained Application Protocol
CSMA/CD	Carrier Sense Multiple Access with Collision Detection
CVP	Cut Vertex Portioning Routing Protocol
DAD	duplicate address detection
DSDV	Destination-Sequenced Distance-Vector Routing
DSL	Digital Subscriber Line
DSR	Dynamic Source Routing
DTLS	Datagram Transport Layer Security
ETSI	European Telecommunications Standards Institute
GHC	generic header compression
GPRS	General Packet Radio Service
H2H	human to human
H2M	human-to-machine
HD	high-definition
HTML	Hypertext Markup Language
HTTP	Hypertext Transfer Protocol
ICMP	Internet Control Message Protocol
ICT	Information and Communications Technology
IEEE	Institute of Electrical and Electronics Engineers
IETF	Internet Engineering Task Force
IGMP	Internet Group Management Protocol
IoT	Internet of Things
IP	Internet Protocol
IPV6	Internet Protocol Version 6
LAN	local area network
M2M	machine-to-machine
MANET	mobile ad hoc network
MQTT	Message Queue Telemetry Transport
ND	Neighbour Discovery
NFC	near field communication

OSPF	Open Shortest Path First
P2P	peer to peer
QoS	quality of service
QR	quick response
RARP	Reverse Address Resolution Protocol
RFID	radio-frequency identification
RIP	Routing Information Protocol
RoLL	Routing Protocol for Low Power and Lossy Networks
RREQ	route request
SDOs	standards developing organizations
TCP	Transmission Control Protocol
UDP	User Datagram Protocol
UMTS	Universal Mobile Telecommunications System
URL	Uniform Resource Locator
UTRAN	Universal Terrestrial Radio Access Network
WAN	wide area network
WPAN	wireless personal area network
WSN	wireless sensor network
XL	approximate link state routing protocol

REFERENCES

1. Mattern, F., & Floerkemeier, C. (2010). From the Internet of computers to the Internet of things. In: From Active Data Management to Event-Based Systems and More (pp. 242259). Springer, Berlin, Heidelberg.
2. Ray, P. P. (2016). A survey on Internet of things architectures. Journal of King Saud University-Computer and Information Sciences.
3. Al-Fuqaha, A., Guizani, M., Mohammadi, M., Aledhari, M., & Ayyash, M. (2015). Internet of things: A survey on enabling technologies, protocols, and applications. IEEE Communications Surveys and Tutorials, 17(4), 2347–2376.
4. Singh, D., Tripathi, G., & Jara, A. J. (2014, March). A survey of Internet-of-things: Future vision, architecture, challenges and services. In: Internet of Things (WF-IoT), 2014 IEEE World Forum on (pp. 287–292). IEEE.
5. Lin, J., Yu, W., Zhang, N., Yang, X., Zhang, H., & Zhao, W. (2017). A survey on Internet of things: Architecture, enabling technologies, security and privacy, and applications. IEEE Internet of Things Journal.
6. Sethi, P., & Sarangi, S. R. (2017). Internet of things: Architectures, protocols, and applications. Journal of Electrical and Computer Engineering, 2017.
7. Beevi, M. J. (2016, February). A fair survey on Internet of Things (IoT). In: Emerging Trends in Engineering, Technology and Science (ICETETS), International Conference on (pp. 1–6). IEEE.
8. Chaqfeh, M. A., & Mohamed, N. (2012, May). Challenges in middleware solutions for the Internet of things. In: Collaboration Technologies and Systems (CTS), 2012 International Conference on (pp. 21–26). IEEE.
9. Krco, S., Pokric, B., & Carrez, F. (2014, March). Designing IoT architecture (s): A European perspective. In: Internet of Things (WF-IoT), 2014 IEEE World Forum on (pp. 79–84). IEEE.
10. EU. (2014 September 18). FP7 Internet of things architecture project [Online]. http://www.iot-a.eu/public.

11. Atzori, L., Iera, A., & Morabito, G. (2010). The Internet of things: A survey. Computer Networks, 54(15), 2787–2805.

12. Khan, R., Khan, S. U., Zaheer, R., & Khan, S. (2012, December). Future Internet: The Internet of things architecture, possible applications and key challenges. In: Frontiers of Information Technology (FIT), 2012 10th International Conference on (pp. 257–260). IEEE.

13. Yang, Z., Yue, Y., Yang, Y., Peng, Y., Wang, X., & Liu, W. (2011, July). Study and application on the architecture and key technologies for IOT. In: Multimedia Technology (ICMT), 2011 International Conference on (pp. 747–751). IEEE.

14. Wu, M., Lu, T. J., Ling, F. Y., Sun, J., & Du, H. Y. (2010, August). Research on the architecture of Internet of things. In: Advanced Computer Theory and Engineering (ICACTE), 2010 3rd International Conference on (Vol. 5, pp. V5-484). IEEE.

15. Sheng, Z., Yang, S., Yu, Y., Vasilakos, A., Mccann, J., & Leung, K. (2013). A survey on the ietf protocol suite for the Internet of things: Standards, challenges, and opportunities. IEEE Wireless Communications, 20(6), 91–98.

16. Dorsemaine, B., Gaulier, J. P., Wary, J. P., Kheir, N., & Urien, P. (2015, September). Internet of things: A definition & taxonomy. In: Next Generation Mobile Applications, Services and Technologies, 2015 9th International Conference on (pp. 72–77). IEEE.

17. Zeng, D., Guo, S., & Cheng, Z. (2011). The web of things: A survey. Journal of Communications, 6(6), 424–438.

18. Salman, T., & Jain, R. (2015). Networking protocols and standards for Internet of things. In: Internet of Things and Data Analytics Handbook (2015), 215–238.

19. Mirzoev, D. (2014). Low rate wireless personal area networks (lr-wpan 802.15. 4 standard). arXiv preprint arXiv:1404.2345.

20. Ahmed, N., Rahman, H., & Hussain, M. I. (2016). A comparison of 802.11 ah and 802.15. 4 for IoT. ICT Express, 2(3), 100–102.

21. Nobre, M., Silva, I., & Guedes, L. A. (2015). Routing and scheduling algorithms for WirelessHARTNetworks: A survey. Sensors, 15(5), 9703–9740.

22. Yassein, M. B., Mardini, W., & Khalil, A. (2016, September). Smart homes automation using Z-wave protocol. In: Engineering & MIS (ICEMIS), International Conference on (pp. 1–6). IEEE.

23. Mackensen, E., Lai, M., & Wendt, T. M. (2012, October). Bluetooth Low Energy (BLE) based wireless sensors. In: Sensors, 2012 IEEE (pp. 1–4). IEEE.

24. Narendra, P., Duquennoy, S., & Voigt, T. (2015, October). BLE and IEEE 802.15. 4 in the IoT: Evaluation and interoperability considerations. In: International Internet of Things Summit (pp. 427–438). Springer, Cham.

25. Aburukba, R., Al-Ali, A. R., Kandil, N., & AbuDamis, D. (2016, March). Configurable ZigBee-based control system for people with multiple disabilities in smart homes. In: Industrial Informatics and Computer Systems (CIICS), 2016 International Conference on (pp. 1–5). IEEE.

26. Weyn, M., Ergeerts, G., Berkvens, R., Wojciechowski, B., & Tabakov, Y. (2015, October). DASH7 alliance protocol 1.0: Low-power, mid-range sensor and actuator communication. In: Standards for Communications and Networking (CSCN), 2015 IEEE Conference on (pp. 54–59). IEEE.

27. Pinomaa, A., Ahola, J., Kosonen, A., & Nuutinen, P. (2015, March). HomePlug green PHY for the LVDC PLC concept: Applicability study. In: Power Line Communications and its Applications (ISPLC), 2015 International Symposium on (pp. 205–210). IEEE.

28. Brandt, A., & Buron, J. (2015). Transmission of IPv6 packets over ITU-T G.9959 networks. IETF RFC 7428, February 2015, https://www.ietf.org/rfc/rfc7428.txt

29. Elsaadany, M., Ali, A., & Hamouda, W. (2017). Cellular LTE-A technologies for the future Internet-of-things: Physical layer features and challenges. IEEE Communications Surveys and Tutorials, 19(4), 2544–2572.

30. Vangelista, L., Zanella, A., & Zorzi, M. (2015, September). Long-range IoT technologies: The dawn of LoRa™. In Future Access Enablers of Ubiquitous and Intelligent Infrastructures (pp. 51–58). Springer, Cham.

31. Lin, J., Shen, Z., & Miao, C. (2017, July). Using blockchain technology to build trust in sharing LoRaWAN IoT. In: Proceedings of the 2nd International Conference on Crowd Science and Engineering (pp. 38–43). ACM.

32. Poole, I. (2014). Weightless wireless — m2m white space communications-tutorial. http://www.radioelectronics.com/info/wireless/weightless-m2m-white-space-wireless communications/basics-overview.php.

33. Bush, S. (2015 September). Dect/ule connects homes for iot. http://www.electronicswee kly.com/news/design/communications/dect-ule-connectshomes-iot-2015-09/.

34. Lahbib, A., Toumi, K., Elleuch, S., Laouiti, A., & Martin, S. (2017, October). Link reliable and trust aware RPL routing protocol for Internet of things. In: Network Computing and Applications (NCA), 2017 IEEE 16th International Symposium on (pp. 1–5). IEEE.

35. Aijaz, A., & Aghvami, A. H. (2015). Cognitive machine-to-machine communications for Internet-of-things: A protocol stack perspective. IEEE Internet of Things Journal, 2(2), 103–112.

36. Basagni, S., Petrioli, C., Petroccia, R., & Spaccini, D. (2015). CARP: A channel-aware routing protocol for underwater acoustic wireless networks. Ad Hoc Networks, 34, 92–104.

37. Babu, H. R., & Dey, U. (2014). Routing protocols in IPv6 enabled LoWPAN: A survey. International Journal of Scientific and Research Publications, 4(2).

38. Dujovne, D., Watteyne, T., Vilajosana, X., & Thubert, P. (2014). 6TiSCH: Deterministic IP-enabled industrial Internet (of things). IEEE Communications Magazine, 52(12), 36–41.

39. Rghioui, A., Khannous, A., Bouchkaren, S., & Bouhorma, M. (2014). 6lo technology for smart cities development: Security case study. International Journal of Computer and Applications, 92(15).

40. Gomez, C., Paradells, J., Bormann, C., & Crowcroft, J. (2017). From 6LoWPAN to 6Lo: Expanding the universe of IPv6-supported technologies for the Internet of things. IEEE Communications Magazine, 55(12), 148–155.

41. Spörk, M. (2016). IPv6 over Bluetooth Low Energy Using Contiki (Doctoral dissertation. Master's thesis, Graz University of Technology, Graz, Austria).

42. Locke, D. (2010). Mq telemetry transport (mqtt) v3. 1 Protocol specification. IBM Developerworks Technical Library.

43. Singh, M., Rajan, M. A., Shivraj, V. L., & Balamuralidhar, P. (2015, April). Secure mqtt for Internet of Things (IoT). In: Communication Systems and Network Technologies (CSNT), 2015 Fifth International Conference on (pp. 746–751). IEEE.

44. OASIS. (2012). Oasis advanced message queuing protocol (amqp) version 1.0. http:// docs.oasisopen.org/amqp/core/v1.0/os/amqp-core-complete-v1.0-os.pdf.

45. Bormann, C., Castellani, A. P., & Shelby, Z. (2012). CoAP: An application protocol for billions of tiny Internet nodes. IEEE Internet Computing, 16(2), 62–67.

46. Saint-Andre, P. (2011). Extensible messaging and presence protocol (XMPP): Core.

47. Díaz, M., Martín, C., & Rubio, B. (2016). State-of-the-art, challenges, and open issues in the integration of Internet of things and cloud computing. Journal of Network and Computer Applications, 67, 99–117.

48. Gluhak, A., Krco, S., Nati, M., Pfisterer, D., Mitton, N., & Razafindralambo, T. (2011). A survey on facilities for experimental Internet of things research. IEEE Communications Magazine, 49(11).

49. Al-Fuqaha, A., Guizani, M., Mohammadi, M., Aledhari, M., & Ayyash, M. (2015). Internet of things: A survey on enabling technologies, protocols, and applications. IEEE Communications Surveys and Tutorials, 17(4), 2347–2376.

50. Chandrasekaran, V., Anitha, S., & Shanmugam, A. (2013). A research survey on experimental tools for simulating wireless sensor networks. International Journal of Computer and Applications, 79(16).

51. Gluhak, A., Krco, S., Nati, M., Pfisterer, D., Mitton, N., & Razafindralambo, T. (2011). A survey on facilities for experimental Internet of things research. IEEE Communications Magazine, 49(11).

52. Park, S., Savvides, A., & Srivastava, M. B. (2000, August). SensorSim: A simulation framework for sensor networks. In: Proceedings of the 3rd ACM International Workshop on Modeling, Analysis and Simulation of Wireless and Mobile Systems (pp. 104–111). ACM.

53. Fu, K., Kohno, T., Lopresti, D., Mynatt, E., Nahrstedt, K., Patel, S., & Zorn, B. (2017). Safety, security, and privacy threats posed by accelerating trends in the Internet of things. Technical Report. Computing Community Consortium.

54. Yang, Y., Wu, L., Yin, G., Li, L., & Zhao, H. (2017). A survey on security and privacy issues in Internet-of-things. IEEE Internet of Things Journal.

55. Shang, W., Yu, Y., Droms, R., & Zhang, L. (2016). Challenges in IoT networking via TCP/IP architecture. NDN Project, Tech. Rep. NDN-0038.

56. Raza, S., Voigt, T., & Roedig, U. (2011). 6LoWPAN extension for IPsec.

57. Raza, S., Shafagh, H., Hewage, K., Hummen, R., & Voigt, T. (2013). Lithe: Lightweight secure CoAP for the Internet of things. IEEE Sensors Journal, 13(10), 3711–3720.

58. Swamy, S. N., Jadhav, D., & Kulkarni, N. (2017, February). Security threats in the application layer in IOT applications. In: I-SMAC (IoT in Social, Mobile, Analytics and Cloud)(I-SMAC), 2017 International Conference on (pp. 477–480). IEEE.

59. Lin, Q., & Ren, D. (2016, October). Quantitative trust assessment method based on Bayesian network. In: Advanced Information Management, Communicates, Electronic and Automation Control Conference (IMCEC), 2016 IEEE (pp. 1861–1864). IEEE.

60. Lee, S. H., & Jeong, Y. S. (2016). Information authentication selection scheme of IoT devices using conditional probability. Indian Journal of Science and Technology, 9(24).

61. Torre, I., Koceva, F., Sanchez, O. R., & Adorni, G. (2016, December). Fitness trackers and wearable devices: How to prevent inference risks? In: Proceedings of the 11th EAI International Conference on Body Area Networks (pp. 125–131). ICST (Institute for Computer Sciences, Social-Informatics and Telecommunications Engineering).

62. Sun, F., Wu, C., & Sheng, D. (2017). Bayesian networks for intrusion dependency analysis in water controlling systems. Journal of Information Science and Engineering, 33(4).

63. Wang, E. K., Wu, T. Y., Chen, C. M., Ye, Y., Zhang, Z., & Zou, F. (2015). Mdpas: Markov decision process based adaptive security for sensors in Internet of things. In Genetic and Evolutionary Computing (pp. 389–397). Springer, Cham.

64. Spachos, P., & Mackey, A. (2017). Energy efficiency and accuracy of solar powered BLE beacons. Computer Communications.

65. Khan, J. A., Qureshi, H. K., & Iqbal, A. (2015). Energy management in wireless sensor networks: A survey. Computers and Electrical Engineering, 41, 159–176.

66. Lee, J. (2018). Patch transporter: Incentivized, decentralized software patch system for WSN and IoT environments. Sensors, 18(2), 574.

67. Hahm, O., Baccelli, E., Petersen, H., & Tsiftes, N. (2016). Operating systems for low-end devices in the Internet of things: A survey. IEEE Internet of Things Journal, 3(5), 720–734.

68. Bhat, A. V., & Geetha (2017). Survey on routing protocols for Internet of things. In: Embedded Computing and System Design (ISED), IEEE, Durgapur, India.

69. Al-Sarawi, S., Anbar, M., Alieyan, K., & Alzubaidi, M. (2017, May). Internet of Things (IoT) communication protocols. In: Information Technology (ICIT), 2017 8th International Conference on (pp. 685–690). IEEE.

70. Chernyshev, M., Baig, Z., Bello, O., & Zeadally, S. (2017). Internet of Things (IoT): Research, simulators, and testbeds. IEEE Internet of Things Journal.

Index

Printed and bound by CPI Group (UK) Ltd, Croydon, CR0 4YY

24/10/2024

01778493-0005